# Integration of IoT with C. Computing for Smart Applications

**Integration of IoT with Cloud Computing for Smart Applications** provides an integrative over-view of the Internet of Things (IoT) and cloud computing to be used for the various futuristic and intelligent applications. The aim of this book is to integrate IoT and cloud computing to translate ordinary resources into smart things. Discussions in this book include a broad and integrated per-spective on the collaboration, security, growth of cloud infrastructure, and real-time data monitoring.

**Features**:

- Presents an integrated approach to solve the problems related to security, reliability, and energy consumption.
- Explains a unique approach to discuss the research challenges and opportunities in the field of IoT and cloud computing.
- Discusses a novel approach for smart agriculture, smart healthcare systems, smart cities and many other modern systems based on machine learning, artificial intelligence, and big data, etc.
- Information presented in a simplified way for students, researchers, academicians and sci-entists, business innovators and entrepreneurs, management professionals and practitioners.

This book can be great reference for graduate and postgraduate students, researchers, and academi-cians working in the field of computer science, cloud computing, artificial intelligence, etc.

## Chapman & Hall/CRC Cloud Computing for Society 5.0

*Series Editor: Vishal Bhatnagar and Vikram Bali*

*Digitalization of Higher Education using Cloud Computing*
Edited by: S.L. Gupta, Nawal Kishor, Niraj Mishra, Sonali Mathur, Utkarsh Gupta

*Cloud Computing Technologies for Smart Agriculture and Healthcare*
Edited by: Urmila Shrawankar, Latesh Malik, Sandhya Arora

*Cloud and Fog Computing Platforms for Internet of Things*
Edited by: Pankaj Bhambri, Sita Rani, Gaurav Gupta, Alex Khang

*Cloud based Intelligent Informative Engineering for Society 5.0*
Edited by: Kaushal Kishor, Neetesh Saxena, Dilkeshwar Pandey

*Cloud based Multi-Modal Information Analytics: A Hands-on Approach*
Authors: Srinidhi Hiriyannaiah, Siddesh G M, Srinivasa K G

*Integration of IoT with Cloud Computing for Smart Applications*
Edited by: Rohit Anand, Sapna Juneja, Abhinav Juneja, Vishal Jain, Ramani Kannan

For more information about this series please visit: https://www.routledge.com/Chapman--HallCRC-Cloud-Computing-for-Society-50/book-series/CRCCCS

# Integration of IoT with Cloud Computing for Smart Applications

Edited by
Rohit Anand
Sapna Juneja
Abhinav Juneja
Vishal Jain
Ramani Kannan

CRC Press

CRC Press
Taylor & Francis Group
Boca Raton London New York

CRC Press is an imprint of the
Taylor & Francis Group, an **informa** business

A CHAPMAN & HALL BOOK

Designed cover image: © ShutterStock

First edition published 2023
by CRC Press
6000 Broken Sound Parkway NW, Suite 300, Boca Raton, FL 33487-2742

and by CRC Press
4 Park Square, Milton Park, Abingdon, Oxon, OX14 4RN

*CRC Press is an imprint of Taylor & Francis Group, LLC*

*Library of Congress Cataloging-in-Publication Data*
Names: Anand, Rohit, editor. | Juneja, Sapna, editor. | Juneja, Abhinav,
editor. | Jain, Vishal, 1983- editor. | Kannan, Ramani, editor.
Title: Integration of IoT with cloud computing for smart applications /
edited by Rohit Anand, Sapna Juneja, Abhinav Juneja, Vishal Jain, Ramani Kannan.
Description: Boca Raton : Chapman & Hall/CRC Press, 2023. | Series: Cloud
computing for Society 5.0 | Includes bibliographic references. |
Summary: "Integration of IoT with Cloud Computing for Smart Applications
provides an integrative overview of Internet-of-Things (IoT) and Cloud
Computing to be used for the various futuristic and intelligent
applications. The aim of this book is to integrate IoT and cloud
computing to translate the ordinary resources into smart things.
Discussions in this book include a broad and integrated perspective on
the collaboration, security, growth of cloud infrastructure and
real-time data monitoring"-- Provided by publisher.
Identifiers: LCCN 2023000907 (print) | LCCN 2023000908 (ebook) |
ISBN 9781032328676 (hardback) | ISBN 9781032333434 (paperback) |
ISBN 9781003319238 (ebook)
Subjects: LCSH: Internet of things | Cloud computing. | Application software.
Classification: LCC TK5105.8857 .I5275 2023 (print) | LCC TK5105.8857
(ebook) | DDC 005.3--dc23/eng/20230301
LC record available at https://lccn.loc.gov/2023000907
LC ebook record available at https://lccn.loc.gov/2023000908

ISBN: 978-1-032-32867-6 (hbk)
ISBN: 978-1-032-33343-4 (pbk)
ISBN: 978-1-003-31923-8 (ebk)

DOI: 10.1201/9781003319238

Typeset in Times
by SPi Technologies India Pvt Ltd (Straive)

# Contents

Preface.................................................................................................................................vii

Notes on the Editors...........................................................................................................ix

Contributors .......................................................................................................................xi

**Chapter 1**    Novel Techniques Using IoT and Cloud Computing in Agriculture .........................1

             *T. Ratha Jeyalakshmi, Kunal Dey, and Gokul Thanigaivasan*

**Chapter 2**    A Comparative Analysis on Performance Factors for Communicating
Devices in a MIPv6 Environment under Smart City Model....................................21

             *Shweta Singh, Manish Bhardwaj, Arun Kumar Tripathi, and Shivali Tyagi*

**Chapter 3**    Futuristic Trends in Vehicle Communication Based on IoT and
Cloud Computing.................................................................................................33

             *Chhote Lal Prasad Gupta and Shashank Gaur*

**Chapter 4**    Towards Resolving Privacy and Security Issues in IoT-Based Cloud
Computing Platforms for Smart City Applications ...............................................53

             *Soma Saha and Bodhisatwa Mazumdar*

**Chapter 5**    Challenges and Opportunities Toward Integration of IoT with Cloud Computing....81

             *Manoj Kumar Patra, Anisha Kumari, Bibhudatta Sahoo, and
Ashok Kumar Turuk*

**Chapter 6**    Multi-variant Processing Model in IIoT .................................................................99

             *N. Ambika*

**Chapter 7**    Mobile Health: Roles of Sensors and IoT in Healthcare Technology ...................111

             *Amit Kumar Singh, Anshul Gaur, Maya Datt Joshi, and M. Marieswaran*

**Chapter 8**    IoT and Cloud Computing: Two Promising Pillars for Smart Agriculture
and Smart Healthcare.........................................................................................131

             *A. Sherly Alphonse, S. Abinaya, Ani Brown Mary, and D. Jeyabharathi*

**Chapter 9**    Splitter with Cryptographic Model for Cloud Data Transmission Security ...........147

             *Ashima Arya, Mitu Sehgal, Aarzoo, Sangeeta, and Junaid Rashid*

**Chapter 10**   IoT in Connected Electric Vehicles for Smart Cities..............................................161

             *Manish Bhardwaj, Shweta Singh, and Yu-Chen Hu*

**Chapter 11**  Artificial Intelligence and Machine Learning for Smart Farming Using
Cloud Computing ................................................................................................. 181

*Sapna Juneja, Abhinav Juneja, Arti Sharma, Vishal Jain, and Amena Mahmoud*

**Chapter 12**  Parivem–Parivahan Emulator .............................................................................. 191

*Dimple Chawla and Yukta Malhotra*

**Chapter 13**  Building Integrated Systems for Healthcare Considering Mobile Computing
and IoT ............................................................................................................... 203

*Rohit Anand, Ashy V. Daniel, A. Lenin Fred, Tarun Jaiswal, Sapna Juneja,
Abhinav Juneja, and Ankur Gupta*

**Chapter 14**  Clustering of Big Data in Cloud Environments for Smart Applications ............... 227

*Rohit Anand, Vipin Jain, Anushi Singh, Disha Rahal, Prachi Rastogi,
Avinash Rajkumar, and Ankur Gupta*

**Index** ....................................................................................................................................... 249

# Preface

This book has focused on the emerging and futuristic need of modern-day computational requirements which can be resolved through the amalgamation of cloud computing and Internet of Things (IoT). There is an increasing trend in the technological world to make better and smarter applications based on human and societal needs. These smart applications are nowadays based on data being captured by smart sensors of varied types, which is being further used for smart decision making and to control our processes. We are using IoT-driven systems to manage personal, societal, and environmental processes. The data we generate through sensors is enormous and it needs to be systematically stored and available as and when needed by the user. Cloud computing makes this possible by being omnipresent over the Internet across the globe. So, with these two technologies working cohesively, a lot of new dimensions in modern-day processing are evolving.

The book comprises 14 chapters and each chapter will surely add a lot of value to your journey of exploring IoT and cloud computing for smart applications.

Chapter 1 focuses on how IoT and cloud computing strategies work together to support farmers in their daily tasks. The chapter showcases the primary needs for implementing IoT and cloud-based agriculture. It also discusses some novel techniques that can be used to turn traditional agriculture into smart agriculture.

Chapter 2 provides an insight into data transmission and the various factors to take into account for the security and reliability of data in the context of smart cities. In this chapter to achieve the reliability and security, a proxy is applied in the IoT-based environment with MIPv6 protocol.

Chapter 3 discusses an important privilege of cloud computing, i.e., fog computing, which improves the energy consumption of vehicular communication networks and thus the latency of the communicating packet between the vehicles improves.

Chapter 4 showcases the various security aspects of components of secured cloud computing-based smart systems, including smart meters, smart healthcare management systems, and smart transportation management systems which find their applications in our day-to-day life.

Chapter 5 brings out an introduction to fundamentals of cloud computing and the IoT. The chapter provides an architecture model for integrating the IoT with cloud computing and outlines the different challenges and opportunities this presents and also refers to some research challenges in this field.

Chapter 6 is focused on multi-variant detection in the Industrial Internet of Things (IIoT) which is one of the prominent developments for cloud applications.

Chapter 7 explores the various smart applications of IoT technology in medical applications through the use of smart sensors. These sensors can capture vital data and provide useful decision making with the help of machine learning methods for better monitoring and control of medical-related issues.

Chapter 8 explores the use of IoT and cloud computing for modern-day control and monitoring of our health infrastructure and applications. The chapter opens various issues related to the topic and also offers a vision of exploring the usability of the technology for human convenience in healthcare operations.

Chapter 9 provides a novel method of implementing a splitter with cryptography to ensure the secure transmission of data over cloud-based applications. To ensure security, data packets are split and then a cryptographic algorithm is applied to the portions in the proposed work.

Chapter 10 focuses on the use of IoT for the connectivity of electric vehicles in upcoming smart cities. The IoT diminishes the need for human intervention in monitoring and control of smart systems. The chapter discusses the use of smarter technologies in the implementation of smart solutions for electric vehicles and also discusses various challenges and opportunities.

Chapter 11 explores the various prospects and capabilities of wireless IoT sensors with the adoption of artificial intelligence and machine learning for smart agriculture and farming. The chapter makes a detailed overview of opportunities as well as the challenges which the farming industry can face when combining this approach with a conventional farming approach.

Chapter 12 proposes a smart tag-based system that embeds a pollution-level meter in the vehicle dashboard. The meter displays the pollution level measured using MQ-7 and MQ-2 sensors embedded in the vehicle outlet. These sensors collect the relevant data on pollution levels emitted by the vehicle and store it in the cloud for analysis and sharing.

Chapter 13 proposes an integrated mobile computing environment that takes into account the usage of artificial intelligence and machine learning with the vital data being generated through the smart sensing IoT-based applications. The chapter proposes techniques which can provide an overall solution to the user for their medical and healthcare needs.

Chapter 14 discusses the adoption of clustering techniques in the management of big data over the cloud. There are typical challenges in managing the large volume of data we come across in all modern-day applications. Clustering of similar types of data may simplify the cloud operations to large extent.

All the chapters included in the book explore new horizons for adding new dimensions to the understanding of the technologies related to cloud computing and smart IoT-based systems. The editors are thankful to all the authors for their wonderful contribution and effort in writing meaningful chapters for the readers.

Happy Learning …

Editors<br>
**Rohit Anand**<br>
**Abhinav Juneja**<br>
**Sapna Juneja**<br>
**Vishal Jain**<br>
**Ramani Kannan**

# Notes on the Editors

**Rohit Anand** is currently working as an Assistant Professor in the Department of Electronics and Communication Engineering at G. B. Pant DSEU Okhla-1 Campus (formerly G.B.Pant Engineering College), New Delhi, India. He has more than 21 years' teaching experience including UG and PG Courses. He is a Life Member of the Indian Society for Technical Education (ISTE). He has published thirteen book chapters in reputed books and 19 research papers in Scopus/SCI Indexed Journals. He has chaired sessions in fifteen International Conferences. He received the Best Paper Presentation award in an international conference and has awards such as the Indian and Asian Record holder, Integral Humanism award and Best Teacher award. His research areas include electromagnetic field theory, antenna theory and design, wireless communication, image processing, optical fiber communication, IoT, etc.

**Sapna Juneja** is a Professor in the Department of CS at KIET Group of Institutions, Delhi-NCR, Ghaziabad, India. Earlier she worked as a Professor in the Department of CSE at IITM Group of Institutions and BMIET, Sonepat. She has more than 17 years of teaching experience. She completed her masters and doctorate in Computer Science and Engineering from M.D. University, Rohtak in 2010 and 2018, respectively. Her broad area of research is software reliability of embedded systems. Her areas of interest include software engineering, computer networks, operating systems, database management systems, and artificial intelligence, etc. She has guided several research theses of UG and PG students in Computer Science and Engineering. She is editing a book on recent technological developments.

**Abhinav Juneja** is currently working as a Professor in the Department of CSIT at KIET Group of Institutions, Delhi-NCR, Ghaziabad, India. Before this, he worked as Associate Director & Professor in the Department of CSE at BMIET, Sonepat. He has more than 21 years of experience teaching postgraduate and undergraduate engineering students. He completed his Doctorate in Computer Science and Engineering from M.D. University, Rohtak in 2018 and has a masters in Information Technology from GGSIPU, Delhi. He has research interests in the field of software reliability, IoT, machine learning and soft computing. He has published several papers in reputed national and international journals. He has been reviewer of several journals of repute and has been on various committees of international conferences.

**Vishal Jain** is presently working as an Associate Professor at the Department of Computer Science and Engineering, School of Engineering and Technology, Sharda University, Greater Noida, U. P., India. Before that, he worked for several years as an Associate Professor at Bharati Vidyapeeth's Institute of Computer Applications and Management (BVICAM), New Delhi. He has more than 14 years of academic experience and has earned several degrees: PhD (CSE), MTech (CSE), MBA (HR), MCA, MCP, and CCNA. He has authored more than 95 research papers in professional journals and conferences. He has authored and edited more than 30 books with various reputed publishers, including Elsevier, Springer, Apple Academic Press, CRC, Taylor and Francis Group, Scrivener, Wiley, Emerald, River Publishers, and IGI-Global. His research areas include information retrieval, semantic web, ontology engineering, data mining, ad hoc networks, and sensor networks. He received a Young Active Member Award for the year 2012–13 from the Computer Society of India, Best Faculty Award for the year 2017, and Best Researcher Award for the year 2019 from BVICAM, New Delhi.

**Ramani Kannan** is currently working as a Senior Lecturer in the Center for Smart Grid Energy Research, Institute of Autonomous Systems, University Teknologi PETRONAS (UTP), Malaysia.

Dr. Kannan completed a Ph.D. (Power Electronics and Drives) from Anna University, India in 2012, M.E. (Power Electronics and Drives) from Anna University, India in 2006, B.E. (Electronics and Communication) from Bharathiyar University, India in 2004. He has more than 15 years of experience in prestigious educational institutes. Dr. Kanan has published more than 130 papers in various reputed national and international journals and conferences. He is the editor, co-editor, guest editor and reviewer of various books including for Springer Nature, Elsevier, etc. He has received an award for best presenter in CENCON 2019, IEEE Conference on Energy Conversion (CENCON 2019), Indonesia.

# Contributors

**Aarzoo**
PIET
Samalkha, Panipat, India

**S. Abinaya**
Vellore Institute of Technology
Chennai, India

**A. Sherly Alphonse**
Vellore Institute of Technology

**N. Ambika**
St. Francis College

**Rohit Anand**
G.B. Pant DSEU Okhla-1 Campus
New Delhi, India

**Ashima Arya**
CSIT Department
KIET Ghaziabad, India

**Manish Bhardwaj**
KIET Group of Institutions Delhi-NCR
Ghaziabad, India

**Dimple Chawla**
Vivekananda Institute of Professional Studies
New Delhi, India

**Ashy V. Daniel**
Mar Ephraem College of Engineering &
    Technology
Marthandam, India

**Kunal Dey**
Jain Deemed to be University
Bangalore, India

**A. Lenin Fred**
Mar Ephraem College of Engineering &
    Technology
Marthandam, India

**Anshul Gaur**
Uttarakhand Technical University
Uttarakhand, India

**Shashank Gaur**
Bansal Institute of Engineering and Technology
Lucknow, India

**Ankur Gupta**
Vaish College of Engineering
Rohtak, India

**Chhote Lal Prasad Gupta**
Bansal Institute of Engineering and Technology
Lucknow, India

**Yu-Chen Hu**
Providence University
Taiwan

**Vipin Jain**
Teerthanker Mahaveer University
Moradabad, India

**Vishal Jain**
Sharda School of Engineering and Technology,
    Sharda University
Greater Noida, India

**Tarun Jaiswal**
National Institute of Technology
Raipur, India

**D. Jeyabharathi**
Sri Krishna College of Technology
Coimbatore, India

**Maya Datt Joshi**
GLA University
Mathura, India

**Abhinav Juneja**
KIET Group of Institutions
Ghaziabad, India

**Sapna Juneja**
KIET Group of Institutions
Ghaziabad, India

**Anisha Kumari**
National Institute of Technology
Rourkela, India

**Amena Mahmoud**
Kafrelsheikh University
Egypt

**Yukta Malhotra**
Vivekananda Institute of Professional Studies
New Delhi, India

**M. Marieswaran**
National Institute of Technology
Raipur, India

**Ani Brown Mary**
Sarah Tucker College
Tirunelveli, India

**Bodhisatwa Mazumdar**
IIT Indore
Indore, India

**Manoj Kumar Patra**
National Institute of Technology
Rourkela, India

**Disha Rahal**
Teerthanker Mahaveer University
Moradabad, India

**Avinash Rajkumar**
Teerthanker Mahaveer University
Moradabad, India

**Junaid Rashid**
Kongju National University
South Korea

**Prachi Rastogi**
Teerthanker Mahaveer University
Moradabad, India

**Soma Saha**
VIT Bhopal University
India

**Bibhudatta Sahoo**
National Institute of Technology
Rourkela, India

**Sangeeta**
PIET, Samalkha
Panipat, India

**Mitu Sehgal**
PIET, Samalkha
Panipat, India

**Arti Sharma**
KIET Group of Institutions
Ghaziabad, India

**Anushi Singh**
Teerthanker Mahaveer University
Moradabad, India

**Shweta Singh**
KIET Group of Institutions DELHI-NCR
Ghaziabad, India

**Amit Kumar Singh**
Biomedical Engineering Department VSB
    Engineering College
Karur, India

**T. Ratha Jeyalakshmi**
Jain Deemed to be University
Bangalore, India

**Gokul Thanigaivasan**
Jain Deemed to be University
Bangalore, India

**Arun Kumar Tripathi**
KIET Group of Institutions DELHI-NCR
India

**Ashok Kumar Turuk**
National Institute of Technology
Rourkela, India

**Shivali Tyagi**
KIET Group of Institutions DELHI-NCR
India

# 1 Novel Techniques Using IoT and Cloud Computing in Agriculture

*T. Ratha Jeyalakshmi, Kunal Dey, and Gokul Thanigaivasan*
Jain Deemed to be University, Bangalore, India

## CONTENTS

1.1 Introduction ........................................................................................................2
1.2 Role of Sensors in Agriculture ..........................................................................4
1.3 Types of Sensors in Agriculture ........................................................................5
    1.3.1 Location Sensors ......................................................................................5
    1.3.2 Electromagnetic Sensors ..........................................................................5
    1.3.3 Optical Sensors ........................................................................................6
    1.3.4 Mechanical Sensors ................................................................................6
    1.3.5 Dielectric Soil Moisture Sensors ............................................................7
    1.3.6 Airflow Sensor ........................................................................................7
    1.3.7 Acoustic Sensors ....................................................................................7
1.4 Needs of the Cloud in IoT-Based Agriculture ..................................................8
    1.4.1 Use of Infrastructure as a Service in Agriculture ..................................8
    1.4.2 Use of Software as a Service in Agriculture ..........................................8
    1.4.3 Use of Platform as a Service in Agriculture ..........................................8
1.5 Types of Cloud Architecture and Their Uses in Agriculture ............................9
    1.5.1 Features of Public Cloud ........................................................................9
    1.5.2 Features of Private Cloud ........................................................................9
    1.5.3 Features of Community Cloud ................................................................9
    1.5.4 Features of Hybrid Cloud ......................................................................9
1.6 Applications ......................................................................................................10
    1.6.1 Crop Recommendation ..........................................................................10
    1.6.2 Cattle Tracking ......................................................................................11
    1.6.3 Precision Farming ..................................................................................11
    1.6.4 Soil Monitoring ....................................................................................12
    1.6.5 Ripening of Fruits ................................................................................13
    1.6.6 Smart Greenhouses ..............................................................................13
        1.6.6.1 Applications of Smart Greenhouses ......................................13
        1.6.6.2 Advantages of Smart Greenhouses ......................................15
    1.6.7 Choosing Crops Based on Soil Texture and Weather ..........................15
    1.6.8 Crop Protection ....................................................................................15
1.7 Future Scope ....................................................................................................16
    1.7.1 Optimized Usage of Scarecrows ..........................................................16
    1.7.2 SONAR for Creature Detection at Night ..............................................16
    1.7.3 Proper Weather Forecasting ..................................................................16

DOI: 10.1201/9781003319238-1

1.7.4   Drones for Farm Management ............................................................... 16
1.7.5   Combining Solar Panels with IoT Devices to Enhance Power Usage ...... 16
1.7.6   Weed-Killing Using New Techniques .................................................. 16
1.7.7   Soil Heat Maintenance ....................................................................... 17
1.8   Conclusion ..................................................................................................... 17
References ................................................................................................................. 17

## 1.1   INTRODUCTION

The Internet of Things (IoT) consists of many devices linked together as a network. The components of IoT are sensors, software, and other technologies which help in the exchange of data between various devices, software and systems connected together over the Internet. IoT finds its usage in various household as well as industrial applications. IoT provides facilities to make the systems work with autonomy which enables us to collect voluminous data that can be analyzed and processed generating various kinds of reports which needs less interaction.

The data in the IoT is collected by sensors. A sensor is a small device that detects or measures external environmental signals. These can be factors such as speed, velocity, pressure, temperature, or light which will either be converted into analogue or digital representation. Data collected from sensors is used for various applications, such as home appliances, business systems, industrial equipment, agriculture, etc. IoT uses sensors in agriculture for effective farming which can be termed smart agriculture. These sensors are known as agriculture sensors. These agricultural sensors provide various relevant data in agriculture which help the farmers to observe and optimize crop production with the various environmental factors and challenges. These agricultural sensors can be easily installed in relevant places like soil, trees, etc., to capture the data. They can be accessed and used properly by the mobile apps which are developed for this purpose. We need either Wi-Fi or cellular towers to use the mobile apps. With the advancement of technology and the rising popularity of the cloud services, use of IoT became an interesting field for researchers. Big data analysis can lead to better precision farming and cultivation management – the 'iFarm' cultivation management system to enable effective farming management [1]. Smartphone applications, web-based applications, and a cloud server make up the system. Using smartphones, farmers in the field can easily refer to their plans and enter the data into the cloud server. Furthermore, the data collected by the system is projected to provide big data for precision agriculture.

The cloud services can be quite advantageous in the agronomic setting as it includes publicly available modules, even various agricultural enablers; FIWARE is specifically the cloud provider used [2]. Using the FIWARE components, a program was created and tested in a semi-arid region of southern Spain on real crops with the goal of lowering irrigation water usage. FIWARE's benefits, as opposed to those of traditional systems, are thoroughly examined and emphasized here. There is also discussion on the advantages of FIWARE over other well-known cloud service providers. An additional advantage for the food and agriculture industries is the FIWARE cloud, which provides a group of specific enablers that allow stakeholders to transfer agricultural information in a standardized, consistent, and secure manner. To show the suggested architecture's performance advantages, prototype was created [3]. The cloud computing concept was reviewed [4] with respect to the features, deployment methodology, cloud service model, cloud advantages, and challenges in agriculture. Also, demonstration of an Internet of Things architecture adapted for applications in precision agriculture. The three-layer architecture is designed to collect the data needed and transmit it to a cloud-based backend for processing and analysis. Depending on the evaluated data, feedback actions may be sent back to the frontend nodes.

A cloud-based database is used to store the agricultural data that is gathered through the IoT devices. The assessment of data, such as fertilizer requirements, crop analysis, market and stock requirements for the crop, is done using big data analysis in the cloud server. A data mining approach is then used to create the forecast, and the information is subsequently sent to the farmer via a mobile app [5] to maximize crop productivity while limiting agricultural costs using projected information. The suggested smart model for agriculture is made to predict crop yield and select the optimal crop sequence based on information about previous crop sequences on the same farmland and the present state of the soil's nutrients. Real-time soil sampling will allow farmers to determine the current fertilizer needs for their agricultural crops.

A cloud-based software architecture [6] has the goal of facilitating the deployment and validation of a whole crop management system. The design contained modules built with Google App Engine that enable information to be quickly collected and processed, as well as agricultural chores to be correctly specified and scheduled. A condensed method for improving agricultural resources using the cloud-based observation of relevant variables also exists [7]. Experts talk about how cloud computing will be used in the agriculture industry. There is evidence that, in light of the growing effects of climate change, Sustainable Precision Agriculture [8] and Environment – which could make use of previous technologies as well as big data analysis – is the force behind the next revolution in precision agriculture. By evaluating each essential stage, from data collection in crop fields to variable rate applications [9], the present condition of sophisticated farm management systems is assessed to enable producers to make the best decisions possible to save costs while preserving the environment. A broad overview of 5G technology in agricultural production was covered in the necessity of smart precision farming [10] and the function of predictive maintenance.

Agriculture cloud and IT services supply farmers with specialized knowledge on crop cultivation, price, fertilizers, disease details, and treatment methods. Scientists working in agriculture will share their findings, ideas for current agricultural practices, and fertilizer usage, as well as the region's history. An agri-cloud-based tool [11] for agriculture increases agricultural output and the accessibility of data related to failed research projects. Agri-CLOUD was developed [12] to help farmers analyze crop problems, acquire the advice they need, and locate the best fertilizers for their crops at the lowest possible cost. The benefits of IoT may be realized through developing unique applications in healthcare sectors, transportation facilities, agricultural fields, and smart environments (home, office, or plant). A paradigm for maximizing agricultural resources (water, fertilizers, pesticides, and human labor) through the usage of IoT appears in [13]. AgriTech is a framework consisting of smart devices, WSN, and the Internet that will automate agricultural activities. With a cellphone in hand, the farmer may better monitor their crops and farmland without ever being there. The farmer may manage agricultural instruments such as an automatic water sprayer to be utilized in the field using smart mobile phones. There is also the feasibility of using categorized raster maps derived from hyperspectral data to generate a work task for precision fertilizer application [14].

An extensive Decision Support System implemented a combination of GIS and cloud computing technology [15] to govern agricultural data needed for fertility of soil to show the distribution of various nutrients and other elements in the soil and their spatial variability so that recommendations can be made for fertilizers using equations of Soil Test Crop Response (STCR) for higher yields. Monitoring environmental conditions, which includes temperature and humidity in agricultural fields with sensors with a CC3200 single chip [16], is a key role in improving the output of efficient crops. Current issues with traditional agriculture include low irrigation water use and management levels that are too low. To address these issues [17], an intelligent big data system that integrates irrigation of water and fertilizer has been established using the Internet of Things, big data, and other technologies. A unique sensor for Nitrogen-Phosphorus-Potassium (NPK) in combination with a Light Dependent Resistor and Light Emitting Diodes for an IoT-based system is incorporated. To track and evaluate the nutrients in the soil, colorimetric theory is applied [18]. To enable quick data retrieval, the built NPK sensor sends data from the chosen agricultural areas to a Google cloud database.

Monitoring and measuring agricultural factors in real time are critical for agricultural development. There is an Android-based [19] cloud-based solution for measuring and monitoring soil moisture utilizing IoT and a multidisciplinary strategy for smart agriculture [20] employing five essential technologies: IoT, sensors, cloud computing, mobile computing, and big data analysis. Farmers will be able to determine the crop's current fertilizer needs through real-time soil sampling. This is a crucial demand for the Indian agricultural industry in order to boost crop yield while reducing fertilizer costs and maintaining soil health [20]. A method for extracting new insights from precision agricultural data using a big data approach exists [21]. In order to collect massive amounts of data in an agricultural big data context, a scenario for the uses of information and communication technology (ICT) services is needed [21]. There is a system design and implementation of a cloud-based remote environment monitoring system with distributed weather stations [22], and an architecture for cloud-based decision support and automation systems that can take data from many sources, make decisions tailored to certain applications, and control field equipment from the cloud [23]. A weather station was created and set up at an edamame farm [24] which compared the meteorological data to that of the Davis Vantage Pro2 that was placed commercially at the same farm. The system's design encourages low-income farmers to include it into their methods of climate-smart farming. The IoT ecosystem and how DA and IoT work together to enable smart agriculture is described in [25] and in addition, upcoming tendencies and chances that are divided into technological advancements, application scenarios, business, and marketability. Applying a cloud construct application in light of agriculture was necessary for the review [26]. This is dependent on agro-cloud, which improves agricultural generation and the accessibility of information related to research expands in the field [26]; doing so will save costs and save time, making communication easier and quicker.

An IoT system makes use of variables such as soil type, groundwater level, local population, daily and seasonal demands of the locals, labor resources of the farmer, range of the same plantation, and range of agricultural land in the area [27], studied to benefit soil scientists and boost agricultural output. To accelerate and mobilize the user contact with an application frontend, an application for smart mobile devices [28] has been released that provides users with weather information, gives them quick access to check the moisture available in the soil, and offers the capability to input applied irrigation quantities into WISE. Since it allows easy access to the application within any cellular data network, this program will be valuable to potential users like irrigation managers, agricultural producers, and researchers. Creating an IoT platform that gathers plant photos and analyses the abrupt changes that occur during the development of an illness, either on the leaves in the shape of light brown and pale green borders or on the leaves [29], enhanced the application. Brown patches that are left behind by the fungus can kill young plants or break adult stems [29]. The goal of cloud computing is to share computations in a networked environment [30]. Cloud computing networks [30] have access to a shared pool of reconfigurable networks, servers, storage, services, applications, and other crucial computing resources.

## 1.2  ROLE OF SENSORS IN AGRICULTURE

Sensors can be used in agriculture to monitor a variety of parameters, including soil temperature at different depths, rainfall, leaf moisture content, atmospheric pressure, chlorophyll content, the direction of the wind, solar radiation, air temperature, relative humidity, etc. IoT can be used to find out the deficiency of nutrients so that proper fertilizer could be applied when it is required. It can be used to apply pesticides at regular intervals. It can also be used to find the ripened fruits. Toxic substances can be detected by which we are moved to bring the remedies for it. Water can be analyzed to find out factors like pH value to make the water suitable for irrigation. The microbes play a big role in boosting the soil nutrients. So, it is necessary to find out the deficiency of microbes which helps to find a resolution. The data which is collected from the sensors is stored in cloud storage so that all the farmers and researchers at various locations can make use of it and benefit.

Application of smart agriculture in association with IoT and cloud computing is depicted in Figure 1.1

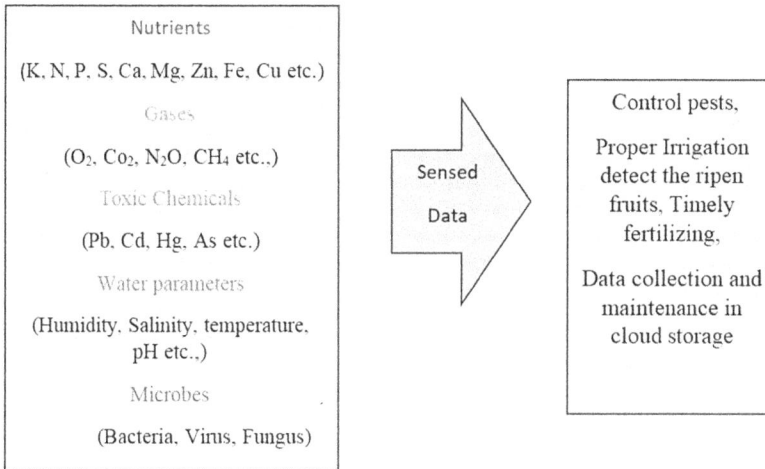

FIGURE 1.1   Smart agriculture.

## 1.3   TYPES OF SENSORS IN AGRICULTURE

Various types of sensors can be used in agriculture to implement the concept of smart agriculture – some of the common ones are as follows.

1. Location Sensors,
2. Electromagnetic Sensors,
3. Optical Sensors,
4. Mechanical Sensors,
5. Dielectric Soil Moisture Sensors,
6. Airflow Sensors,
7. Acoustic Sensors.

### 1.3.1   LOCATION SENSORS

The location sensor can be attached to cattle which can be used for tracking their location which is needed if we find that the animal is missing. It could also be used to find out if it is necessary to give first aid if the animal meets with an accident or it could not move after falling sick unexpectedly or after giving birth. These sensors can also be used for giving the location information of the soil which is deficient in water and other nutrients when it is attached with a soil analyzing sensor.

### 1.3.2   ELECTROMAGNETIC SENSORS

The electrical conductivity, or EC, of a substance, can be used to estimate the concentration of charged molecules in a solution. By running an alternating electrical current between two electrodes and observing the resistance value, the EC can be easily calculated. The ratio of the distance between the electrodes to their area is multiplied by the conductance to arrive at the EC (Table 1.1).

The medium's dielectric permittivity around the sensor is measured by an electromagnetic field. The sensor has got prongs through which a 70-MHz oscillating wave is supplied and this charges the dielectric part. The output value of dielectric permittivity is obtained by a microprocessor available in the sensor which measures the charge. Due to the fact that EC's absolute value depends on a variety of chemical and physical soil characteristics –namely salt content, water content, fineness of the soil etc. – the EC value cannot be used directly as such. In addition to this, there are other aspects

**TABLE 1.1**

**Applications of Electromagnetic Sensors**

| Soil Properties Determined by the Sensor | Applications |
| --- | --- |
| Soil texture, organic content level, drainage conditions | Smart soil sampling |
| Topsoil depth, soil fertility, organic content | Seeding |
| Soil texture, drainage conditions | Nutrient application |
| Soil texture, organic content | Pesticide application |
| Water content of the plant | Estimation of yield |
| Electrolyte in the soil | Soil salinity diagnosis |
| Water retention capacity of the soil | Drainage remedial plan |

of soil qualities that can be assessed by electromagnetic sensors, such as cation exchange capacity (CEC), texture of soil, drainage conditions, salt content, various proportions of organic matter, etc. These electromagnetic sensors can evaluate the ability of the various components of soil to conduct or store electrical charge. Contact and non-contact methods exist to measure this. In contact mode, the electrodes in the sensor are buried in soil so that it can measure the electrical charge. In non-contact mode, the sensor is installed on a device and pulled by a vehicle which is provided with a GPS receiver. It is the preferred sensor because of its accuracy in measuring the electrical charge of the soil. The only demerit is that it is not applicable for the fields when the area is too large or too small. The measured electrical charge of the soil does not have direct effect either on the growth of the crops or its yield. However, it would be useful for the farmers to determine the specific soil properties which may affect the crop yield.

### 1.3.3 Optical Sensors

There are many optical sensors applied in agriculture, such as sensors used to analyze soil properties and sensors used to measure how healthy the plants are by processing the light reflected by the plants. There are two main types of optical sensors, namely Active Optical Sensors (AOSs) and Passive Optical Sensors (POSs). AOSs are fitted with specialized components that transmit light at a certain wavelength to the target and detect light reflections on the photo-detector component of the sensor. The fundamental benefit of an AOS is that measurements may be done at any time, under any background illumination condition. In a POS, the target must be visible under enough ambient light. The sensor's irradiance readings are influenced by the amount of sunshine, azimuth angles, and the sensor's position in relation to the target (view angle, altitude). Clouds and other shadows should not block the distant target while passive satellite sensors are being used. In addition to that the atmospheric conditions also have impact if canopy optical characteristics are measured.

Optical sensors have various applications. They characterize soil by its ability to reflect light. These sensors can detect the reflectance of polarized, mid-infrared, or near-infrared light. The same basic methodology is used by vehicle-based optical sensors as in remote sensing. They can perform a variety of remote sensing tasks that require measuring soil reflectance with a satellite or UAV.

Crop data can be measured and recorded in real time using optical sensors. The amount of organic matter present in the soil determines its nitrogen content. The sensors can measure the nitrogen content of plants and direct the applicators to provide the required amount of nitrogen to the plants.

### 1.3.4 Mechanical Sensors

These sensors are used to give a rough estimation of the soil compaction. When the mechanism found in this type of sensor penetrates the soil, it measures the mechanical resistance with the strain gauges which are found as components of the sensor. This is used to find out the amount of force that is exerted by the roots of the crops to absorb water and this data can be used to determine the optimal tilling and watering method for the crops. When the tractor is used to till the soil in the early stages of the life cycle of the crop, this measured resistance is used to determine the right amount of pressure that is exerted by the tractor to get the maximum yield over a long duration.

### 1.3.5 DIELECTRIC SOIL MOISTURE SENSORS

Soil's dielectric constant is observed by this sensor to measure moisture levels in the soil. This sensor can be either a portable or stationary model, such as a hand-held probe. Portable soil moisture probes can be used to detect soil moisture at various locations while stationary sensors can be installed at the necessary depths and locations in the land. The LC resonance circuit theory is applied in the dielectric property approach. The sensor element in the resonance circuit is connected in series to the external fixed inductance Lf. The capacitance of the sensor element is assumed as LC. The soil dielectric constant changes along with changes in the soil moisture in probe plates. This capacitance value can be achieved by altering the circuit's resonance state, and the soil's volumetric water content will then be determined.

### 1.3.6 AIRFLOW SENSOR

Air permeability is measured using these sensors. They are portable or can be deployed in a fixed place. Airflow sensors are used to measure the soil air permeability on-the-go. Good airflow is needed to control temperature and remove excess humidity from crop growing fields which helps to maintain ideal growing and sustaining conditions essential for producing quality and profitable crops. This is an essential method in indoor farms which are engineered for specific plantings and where climate management systems with airflow and humidity sensing, as well as irrigation and lighting systems, are required. The cultivator must be very careful in controlling all the factors of the growing process in order to achieve consistent, quality crops as well as to protect plants from undesirable effects of mold, mildew, and other fungal diseases. Every soil has different air permeability depending on the structure of the soil, its compaction, water content, and soil type. Airflow sensors can easily measure the air permeability in the soil which is required by the farmers to monitor health of the crops by providing the information necessary to calculate the optimum fertilizer requirements.

### 1.3.7 ACOUSTIC SENSORS

An acoustic sensor use sound pulses to detect the objects as we find in a medical stethoscope. It has the ability to receive and feel vibrations from moving object. It has sufficient intelligence to find the object's movement as well as the distance it moves. Acoustic sensors can be used to monitor flow through the entire seeder to ensure even sowing. An acoustic sensor can also be used to detect pests because it senses the sound produced by them. The sensors are fixed in the field which are connected to a base station. A threshold will be fixed and if the sound intensity of the pest crosses the threshold, the sensor informs the station in the control room by transmitting a signal. From the signal received, the control room will be able to find out the damaged area. It is quite natural that any device installed in the field will become dirty. Some sensors can be covered to protect them from dust, which can prevent them from accurately detecting blockages. Acoustic sensors are not affected by dust and hence they need not be covered by something to protect them from dust. So, they can provide more accurate, consistent information, and since they don't require any thing for protection, they don't need much maintenance. This level of performance also makes acoustic sensors work well with small seeds that flow at a low rate, like mustard, alfalfa, or other cover crops. Reliability, sensitivity, and easy maintenance are the advantages of acoustic sensors when compared to other sensors.

Water is more capable of reflecting sound than soil and hence if water is present on the soil, it will be easily sensed by the sensor. Porosity and resistance for water flow are the most important features that affect acoustic impedance. The strength of the reflected sound wave is directly proportional to the saturation of the soil. The availability of free water in the soil can be detected by varying the reflectance of the sound waves and it is feasible to have robust water sensing for irrigation systems by erecting acoustic sensors.

## 1.4   NEEDS OF THE CLOUD IN IOT-BASED AGRICULTURE

Cloud computing is an on-demand and "pay-per-use" resource like software or hardware over the Internet. One can get to use the needed hardware without even actually buying it and can use IT infrastructures and services like storage space, computational power, software application, and various kinds of database over the Internet without the hassle of maintaining and buying the actual product. From the viewpoint of agriculture, the combination of IoT devices and cloud computing plays a major role. There are various services provided by the cloud, these are:

1. Infrastructure as a Service (IaaS),
2. Software as a Service (SaaS),
3. Platform as a Service (PaaS).

### 1.4.1   USE OF INFRASTRUCTURE AS A SERVICE IN AGRICULTURE

Infrastructure as a Service (IaaS) can be used by big agricultural farms or companies, which can make their own application as well as let that application be used by someone else over the Internet. In IaaS, the customer has the flexibility to choose the infrastructure of the system, including the hardware and software, and change it as and when required. IaaS is helpful for large farms or agricultural organizations, when full flexibility is needed in storing, computing, and analyzing the results, the data of which directly comes as an input from the agricultural farms from the various sensors.

### 1.4.2   USE OF SOFTWARE AS A SERVICE IN AGRICULTURE

With the use of Software as a Service (SaaS), small farm owners can use pre-built software directly in the cloud. One does not need to buy this software but pay for it on a regular basis as per the usage. This software can be used to keep track of all the agricultural work which includes the farm management software, or it can be weather forecasting or pest tracker software. It can be also used to keep track of the sales all around the year, which can be later used to analyze the profits over the year. This software is pre-built and it does not need any configuration in any system, all it needs is an account, so that the data from anywhere can be stored in it. These services can also be used to create communities between people with similar interests and it can be used to connect them all around the globe. The advantage of using SaaS is that, without the need of any installation or buying anything, one can directly start using it without even taking care about the hardware or the necessary software part.

### 1.4.3   USE OF PLATFORM AS A SERVICE IN AGRICULTURE

Platform as a Service (PaaS) can be used in those cases when the customer needs more flexibility over the system they are using. By system, this means that the hardware and the software can now be configured by the customer over the cloud. Virtualization technology is used to share a cloud server between all the customers who requested the service. This service can be used in those cases when the customer needs to get a server with some computation power and storage space to use their custom applications, which can be made specifically to cater to their agricultural needs. PaaS provides much more flexibility when anyone needs to make an application based on specific agricultural needs.

## 1.5 TYPES OF CLOUD ARCHITECTURE AND THEIR USES IN AGRICULTURE

Basically, there are four types of cloud architecture which can find uses in agriculture:

1. Public cloud,
2. Private cloud,
3. Community cloud, and
4. Hybrid cloud.

### 1.5.1 FEATURES OF PUBLIC CLOUD

1. This is a shared platform which is accessible to the general public through the Internet.
2. Cloud service providers administer public clouds as a pay-per-use model.
3. Public cloud facilitates the same storage for multiple users using it at the same time.
4. Use of public clouds can be seen in various businesses, government organizations, and even in the agricultural industry.

### 1.5.2 FEATURES OF PRIVATE CLOUD

1. Private cloud is generally owned to get a private internal network mostly within an organization or for some selected users.
2. It provides high security as compared to the public cloud architecture and data privacy is the key feature of this. It also makes sure that anyone outside the organization cannot get access to the data kept in the private cloud.
3. In the field of agriculture, the sensitive data related to sales of the crops or by-products owned by one agricultural company can be stored.

### 1.5.3 FEATURES OF COMMUNITY CLOUD

Community cloud architecture allows the services and the systems to be available for a particular set of people which may or may not belong to the same organization but will have some common interest. The various organizations and their members who are accessing this cloud architecture own and manage it.

### 1.5.4 FEATURES OF HYBRID CLOUD

Hybrid cloud is an architecture which combines the features of both the public and private cloud. In this architecture, non-sensitive content is stored and computation is done in the public cloud while the sensitive data is stored and computed in the private cloud.

In general, all these cloud architectures and services mentioned above would not be used as a stand-alone system in the field of agriculture. Instead, it is the IoT devices with the sensors that will gather data based on the surrounding environment and interact with the cloud service.

The cloud is not only responsible for storing all this data from the sensors, but it has the capability and computation power for analyzing the data and giving some output based on which some essential decisions can be taken.

## 1.6  APPLICATIONS

### 1.6.1  CROP RECOMMENDATION

Precision agriculture is an improved agricultural strategy that combines data from soil features, soil types, crop production statistics, and meteorological variables to recommend the best crop to produce in a given time period to farmers. This technique can help in reducing the failure of crops and this technique will also help farmers to get insights on the current crop requirements. Farmers may utilize cloud and IoT technologies to choose the best crop to grow. A smart sensor that can detect ripe fruits may be installed to aid in timely harvesting. IoT and cloud computing solutions that may be used together to help farmers with their everyday activities.

The soil dataset that Kulkarni et al. [31] proposed, principally consists of soil physical and chemical parameters, combined with climate information.

Out of 9,000 observations, 2,100 observations have complete dataset and 6,900 observations have missing data. Based on the complete dataset we used EDA and found out that the variables don't have a high correlation among themselves. Table 1.2 shows the summary of the dataset used.

Each observation is trained and tested on the decision tree (DT), logistic regression (LR), and random forest (RF) shown in Table 1.3. The average accuracy of crop prediction by the given dataset and given techniques is approximately 96%

### TABLE 1.2
### Summary of the Dataset Used

|      | N      | P      | K      | Temperature | Humidity | pH   | Rainfall |
|------|--------|--------|--------|-------------|----------|------|----------|
| Mean | 50.55  | 53.36  | 48.15  | 25.62       | 71.48    | 6.47 | 103.46   |
| Std  | 36.92  | 32.99  | 50.65  | 5.06        | 22.26    | 0.77 | 54.96    |
| Max  | 140.00 | 145.00 | 205.00 | 43.68       | 99.98    | 9.94 | 298.56   |
| Min  | 5.00   | 5.00   | 5.00   | 8.83        | 14.26    | 3.50 | 20.21    |

### TABLE 1.3
### DT, LR, RT Accuracy for Crop Recommendation is 95%

**Decision Tree's Accuracy is: 90%**

|              | Precision | Recall | f1-score | Support |
|--------------|-----------|--------|----------|---------|
| Accuracy     | 0.9       | 440    |          |         |
| Macro avg    | 0.84      | 0.88   | 0.85     | 440     |
| Weighted avg | 0.86      | 0.9    | 0.87     | 440     |

**Logistic Regression's Accuracy is: 95%**

|              | Precision | Recall | f1-score | Support |
|--------------|-----------|--------|----------|---------|
| Accuracy     | 0.95      | 440    |          |         |
| Macro avg    | 0.95      | 0.95   | 0.95     |         |
| Weighted avg | 0.95      | 0.95   | 0.95     | 440     |
| Weighted avg | 0.86      | 0.9    | 0.87     | 440     |

**Random Forest's Accuracy is: 99%**

|              | Precision | Recall | f1-score | Support |
|--------------|-----------|--------|----------|---------|
| Accuracy     | 0.99      | 440    |          |         |
| Macro avg    | 0.99      | 0.99   | 0.99     | 440     |
| Weighted avg | 0.99      | 0.99   | 0.99     | 440     |
| Weighted avg | 0.86      | 0.9    | 0.87     | 440     |

### 1.6.2 CATTLE TRACKING

Cattle tracking is a common application of IoT and cloud services in the field of agricultural industry. Cattle generally have tags in their ears to identify and count them, the traditional tags just having the unique ID of that particular cattle, but with the help of IoT devices, tags are now equipped with location and temperature sensors. The ear of the cattle has a specific normal temperature, which can change in the case of any sickness or even in the case of severe injury and blood loss. So, the temperature sensor can keep track of the temperature and sense it constantly while the location sensor will give the current location of the cattle as well as the places it went meaning the tags, containing both the temperature and the location sensor, can be used not only to track down lost cattle but also to alert farmers or concerned persons in case of any injury or sickness of the cattle and can give the exact location of it. Figure 1.2 shows an example of this application.

### 1.6.3 PRECISION FARMING

Precision farming (PF) is the modern method of farm management. It uses recent technologies like GPS, drones, satellite images, IoT, cloud computing, and mobile computing to cultivate crops and nourish the soil adequately to achieve maximum productivity. The main objective of PF is to obtain income, sustainability, and conservation of the environment. In PF, the farmers obtain information about crop status, weather data, environmental conditions, etc., using sensors from the IoT devices. PF uses modern technology to collect real-time data about the status of the crops, soil, air, weather conditions, labor costs, and resources. In this type of farming, data analytics is applied to provide farmers with guidance about the right time for planting, harvesting times, crop rotation, and the management of soil. IoT sensors are used in fields to measure the porosity, moisture content, nutrition, texture, and temperature of the soil and surrounding environment. Satellites and robotic drones

**FIGURE 1.2**   Tag for cattle location and temperature tracking.

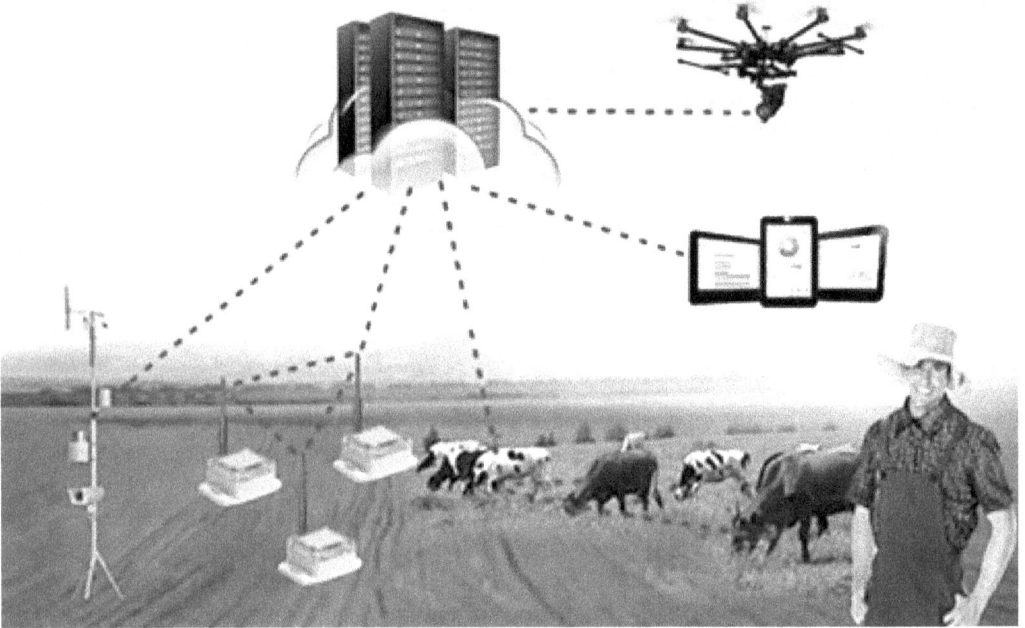

**FIGURE 1.3**   Precision farming as well as cattle tracking can be done using IoT and cloud architecture.

are used to capture images of the crops that can be processed with efficient software and that data can be provided to sensors and mobile apps which can be used to determine the amount of manure, water, pesticide, etc., to be given for a particular crop in a location. This helps the farmer to manage the resources optimally, guaranteeing that the soil contains the correct proportions of materials for optimum health, costs, and conservation of the environment. Thus, PF helps the farmers to improve the quantity and quality of the agricultural products with minimum costs [32]. Figure 1.3 shows an example of precision farming.

### 1.6.4   SOIL MONITORING

With the help of IoT devices soil can be monitored to empower the farmers and other stakeholders to produce the maximum yield from a given land as well as to decrease the disease rate and optimize the resources like irrigation water, fertilizers, etc. [33, 34]. The state-of-art IoT sensors can measure the NPK concentration, temperature of the soil, the water content in the soil, photosynthetic radiation, soil water retention capacity, and soil oxygen content which can be helpful in understanding the porosity of the soil as well. These sensors constantly sense the data from the surrounding soil and this data is transmitted to the cloud for storage and analysis for visualization of the various issues or the trends or also to use deep learning or machine learning to predict something useful which can lead to important decisions related to that land. Here are a few examples where IoT is applied for soil monitoring in agriculture:

1. Soil Temperature: The temperature of the soil is one of the most essential factors which influences root growth, respiration, decomposition, and also the intake of nitrogen from the soil. Various IoT sensors can predict the soil temperature by measuring the air temperature and various other factors, however an accurate way of measurement is to use a device that can be fitted down in the soil. Depending on the structure of the root of the particular crop,

the device can have access to various points and varying depths inside the soil. Even IR technology can be used to get the temperature of the soil.

2. Soil Moisture: For checking the soil moisture content, the dielectric soil moisture sensor can be used. Soil moisture content is needed for many reasons – water content is essential for each plant, with less water affecting the plant as well as excess water, so a balance is needed. The production of the crop is directly related to the water content in the soil. The soil water also plays an important role in supplying water-soluble nutrients to the plants. Soil temperature can also be regulated by the percentage of water present in the soil.

3. Nutrient and pesticide concentration: There are sensors that can measure the content of Nitrogen, Phosphorous, and Potassium (potash) in the soil and also other substances like pesticides. These sensors can be buried in the wet soil, from where they can get the data on the various concentrations of these essential nutrients and pesticides. If any change in concentration is noticed, an alert is sent from the cloud server to the farmers to take necessary actions. To make it more precise, other devices can be used which can precisely measure the amount of fertilizer or other nutrients needed so that it can be added to the soil in the given area.

4. Porosity: The airflow sensors are used to measure the porosity of the soil – based on this data, the water content inside the soil can also be predicted, and the type of crop which can grow in that soil texture can also be determined.

## 1.6.5  RIPENING OF FRUITS

With the help of thermal sensors and optical sensors, trees in big cultivations can be tracked to check for ripened fruits. The optical sensor will sense the data and the thermal sensor will sense the temperature from a distance so that it can sense a larger location at the same time. This data from these IoT devices are transferred to the cloud server, where a deep learning or a machine learning model is used to predict whether a particular fruit has ripened or not. Generally ripening is associated with the change in color of the fruit and sometimes even its internal temperature, both of which can be sensed by these devices.

## 1.6.6  SMART GREENHOUSES

A greenhouse is a bounded space or a building where crops or any other plants are grown in a controlled manner irrespective of the climate and environment outside. Traditional greenhouses require the manual supervision of humans, whereas in a smart greenhouse approach, farmers benefit by automatic monitoring of the crops, agricultural land, and total control of the greenhouse environment. In today's world due to industrialization and urbanization, the availability of land has greatly decreased, and there might be some places where the environment – including the soil and climate – may not be right to grow the crops, thus it is necessary to create greenhouses, which can be reserved mostly for growing the crops all year round without any constraints by the environment outside. With the advancement of technology, one can control, monitor, and analyze the greenhouse using many IoT devices, all of which will be connected via the cloud to the customer's devices. The farmer can control the greenhouse from any part of the world and get accurate data regarding the crops and the greenhouse as a whole through the cloud server to which the IoT devices are connected. For a user-friendly interface, an app can be used which can serve as the user interface between the cloud server and the user for the ease of access to the data and monitoring of the smart greenhouse. Figure 1.4 shows an example of smart greenhouse using cloud technology and IoT devices [32].

### 1.6.6.1  Applications of Smart Greenhouses

A greenhouse is mainly focused on techniques that enhance the yield of crops, fruits, vegetables, etc. In a greenhouse, the environmental parameters are generally controlled in two ways – manual intervention, or a proportional and systematic approach including modern technologies for its automation.

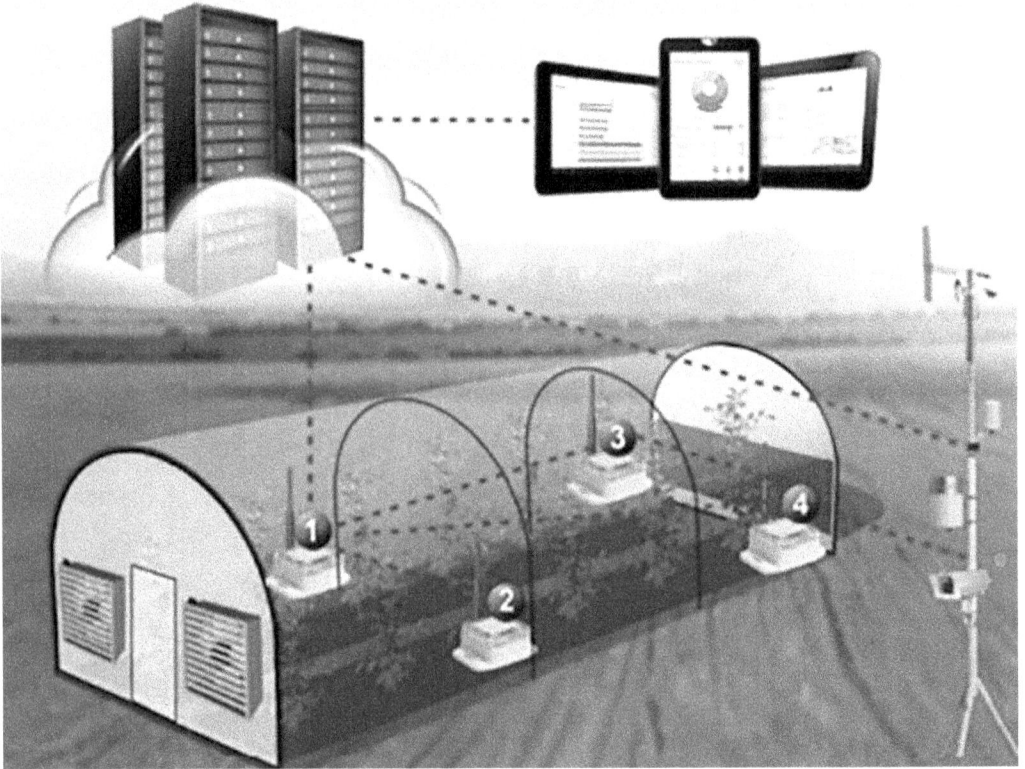

**FIGURE 1.4** Smart greenhouse using IoT devices and cloud technology, which can be also monitored using mobile devices.

Since the manual intervention of humans has disadvantages like energy loss, labor cost, and also production loss, these methods are not as effective as the smart greenhouse approach. A smart greenhouse uses IoT devices to sense the data around it as well as control the environment inside the greenhouse. Thus, it not only monitors the present condition of the greenhouse, but it is also capable of taking control of the environment inside the greenhouse itself. The IoT devices will have sensors attached to them, by which the data is sent to the cloud server and in the cloud server itself, the computation can be based on previous data available so that a smart decision can be taken automatically by the system. Thus, data from the sensors in an IoT device goes to the cloud server, the cloud server makes decisions after the computation of data, and this decision is then carried out by the IoT devices which are directly connected by the cloud server. All of this is without any intervention from humans. In order to control the environment in the smart greenhouse, a variety of sensors are available that can measure environmental parameters based on the needs of plants. A cloud server is used to store the data as well as for computing, with all the IoT devices directly connected to this server.

Various sensors which are generally used to sense the environmental parameters inside a smart greenhouse are given below:

1. Soil Moisture Sensor: Two copper leads serve as the sensor probes. A sample of soil whose moisture content needs to be determined has these leads submerged in it. The amount of moisture or water that is present in the soil affects its conductivity. The conductivity will increase with the increase in the moisture content of the soil, as it forms a path for the electrical conduction between the two copper leads, leading to a closed-circuit that facilitates the flow of electric current.

2. Humidity Sensor: This is used to detect the water vapors in the surroundings. These sensors will send information to the cloud server if there is a change in the relative humidity of the enclosed area.

3. Light Sensor: This sensor is very responsive in the visible light spectrum. An IoT device attached to these light sensors can sense the natural light in the surroundings and record its intensity, which can be sent to the cloud server. If the value is considered to be very low, then necessary decisions can be taken up by the server so as to increase the intensity of light in the smart greenhouse.

These are just a few sensors that are commonly used in smart greenhouses: to make full use of the power of automation, it will require a number of IoT devices with various kinds of sensors.

### 1.6.6.2   Advantages of Smart Greenhouses

Temperature control is very important in any greenhouse since small fluctuations in temperature for even a short period of time can damage plants within a few hours. Smart monitoring of the temperature which can be obtained via a smart greenhouse can help in controlling the temperature to a greater extent, thus protecting the crops and plants from any damage. Automatic soil moisture control helps in keeping track of the water requirement of each plant based on their needs and thus different kinds of crops can be cultivated in the same place with varying water or nutrient requirements, all of which can be taken care of by the IoT devices and the cloud technology.

Remote monitoring of the devices is also possible in a smart greenhouse, so any device or system failure can be quickly detected so that necessary steps can be taken easily.

The controlling and monitoring of all the systems inside a smart greenhouse senses different parameters of the environment which affect the growth of a plant. With the combination of IoT devices and cloud technology, these data can be sensed, a decision can be taken, and an action performed automatically. This system is much more profitable as it will help in optimizing all the available resources in the greenhouse.

### 1.6.7   Choosing Crops Based on Soil Texture and Weather

Soil monitoring can help in gathering a large amount of data. The weather data can also be collected from satellites. With the combination of these two, a cloud server can be used to analyze the data and predict which crop will grow well in which region, based on its soil texture and weather, and all the data can be extracted using IoT devices. This application is very useful when the agricultural industry or the government wants to allocate land for a particular type of crop to get the maximum profit.

### 1.6.8   Crop Protection

Domestic as well as wild animals cause heavy damage to the crops. Farmers apply various methods to protect their crops from these animals. Traditional methods used by farmers are electric fences and scarecrows. The former method is dangerous to human lives whereas the latter is ineffective in the course of time. Using IoT, devices can be built which have sensors to drive the animals which enter into the fields. The device should be able to recognize animals of different sizes and should produce sounds on sensing the animals' entry which will frighten the animals. Pathogens also play a vital role in destroying the crops. Numerous plant pathogens directly affect crop quality and ends in the reduction of food supply. This has a very high impact when the crop is a staple food for people in undeveloped regions which lack resources. Generally, plant diseases are managed by traditional methods like applying chemicals, crop rotation, good seeds, using resistant varieties, and foliar fungicidal treatment. Fungicides are the mostly used measure in disease control, but repeated use of fungicides will lead to insensitivity in plants. Hence, it is the need of the hour to design tools to help to detect pathogens fast with accuracy and sensitivity. This will definitely reduce excess amounts of chemical usage on the crops thereby decreasing the adverse impacts on the environment. Various biosensing techniques for plant pathogen detection are available in practice. Electro-chemical biosensors are used to detect target pathogens that form and spread under different conditions in the environment which includes air, water, and also on seeds. These sensors can be used in different places, such as greenhouses, crop fields, and also in post-harvest storage. Among biosensors, Colorimetric biosensors, fluorescence-based assays, and surface plasmon resonance-based biosensors are the most commonly used optical biosensors for plant pathogen detection.

## 1.7  FUTURE SCOPE

### 1.7.1 OPTIMIZED USAGE OF SCARECROWS

Traditional scarecrows are used to scare birds or animals but with a very low success rate. Modern technology can be used to optimize this. A camera sensor can be attached to the scarecrow model, which will be able to detect motion. In the server, a machine learning model can run and analyze the video footage to identify unwanted birds or animals, which can be driven away by a sudden motorized action and sound from the scarecrow.

### 1.7.2 SONAR FOR CREATURE DETECTION AT NIGHT

SONAR can be used to detect movement in a field at night. A thermal camera won't work as most of the nocturnal creatures are cold-blooded, so they will adapt their body temperature to the surrounding environment. SONAR will easily detect movement in dark and will work better than infrared cameras because in a dense farm it will not function properly. After significant movement is detected at night in the field, necessary actions can be taken against it, like setting up an alarm or some motorized action to drive the creatures away.

### 1.7.3 PROPER WEATHER FORECASTING

Weather forecasting is easy using so many applications present, but there is always an issue because the applications only provide weather information for a larger town or city as a whole and not for a specific area. This causes some uncertainty in the weather prediction for each local area, so a collaboration can be made with these companies so that they can give exact information about the weather for a particular area, and this can help farmers to plan the timings for spraying pesticides or fertilizers in a rainy season.

### 1.7.4 DRONES FOR FARM MANAGEMENT

Drones might be incredibly beneficial in farmland since they can spray pesticides or just transmit footage to allow farmers to monitor what is going on in real time.

### 1.7.5 COMBINING SOLAR PANELS WITH IoT DEVICES TO ENHANCE POWER USAGE

IoT equipment may be situated far from the farmers and will require power. As a result, a technique for supplying power to the equipment is required. Solar panels with strong batteries can therefore be added, depending on the particular requirements.

### 1.7.6 WEED-KILLING USING NEW TECHNIQUES

Modern machine learning models and deep learning models can be trained to differentiate between different plants. One such method can be used to differentiate between weeds and crops. A device with a normal camera sensor can be used to detect weeds: the model will be available in the cloud and the weed detection will also be carried out in the cloud server. Small, powered lasers can be used to target exactly the weeds from the device. These devices need not be attached everywhere, but they can be attached in a given portion of land, where they can occasionally monitor for the weeds, which, when detected, can be removed by targeting a laser on to it.

### 1.7.7 Soil Heat Maintenance

During winter or the rainy season, when the soil temperature around the roots of the plant reduces, it may affect the plant as well as its growth. Temperature sensors can be used to keep track of the soil temperature. A long, enclosed pipe can be built which runs throughout the soil (in the rows where plants are sown), which can be used to flow water at a certain temperature to regulate the temperature of the soil. Note that this will be an enclosed pipe and the water will not directly enter the soil, instead it will flow through the pipe covering all the areas where the temperature drop has been noticed.

## 1.8  CONCLUSION

With the advent of modern technologies, farming can be made smart farming. By this, food production can be optimized, which in turn will increase the economy of the world. This will help bring harmony across the world. Cloud technology can help anyone to access the server from anywhere over the Internet, thus farmers or agricultural industrialists can monitor the agricultural land even remotely by using a mobile app. With the power of machine learning and deep learning, the gathered data in the server can be analyzed to get meaningful insights which can in turn help in making important decisions related to this field, and all of this computation can be done in the same cloud server. Robots, drones, and other automated vehicles are also being implemented in the field of smart agriculture. These devices are automated or semi-automated and are connected to the cloud server, which helps these devices in decision making, and the data is also sensed each time and sent over to the cloud server for further analysis. Thus, the use of IoT and the cloud service together in the field of agriculture can help in optimizing the resources as well as increasing the yield, by not only monitoring the whole system, but also by taking important decisions related to the field of smart agriculture.

## REFERENCES

[1] Y. Murakami, S. K. T. Utomo, K. Hosono, T. Umezawa, and N. Osawa, "iFarm: Development of Cloud-Based System of Cultivation Management for Precision Agriculture," in *2013 IEEE 2nd Global Conference on Consumer Electronics (GCCE)*, Tokyo, Japan, Oct. 2013, pp. 233–234, doi: 10.1109/ GCCE.2013.6664809.

[2] J. A. López-Riquelme, N. Pavón-Pulido, H. Navarro-Hellín, F. Soto-Valles, and R. Torres-Sánchez, "A Software Architecture Based on FIWARE Cloud for Precision Agriculture," *Agricultural Water Management*, vol. 183, pp. 123–135, Mar. 2017, doi: 10.1016/j.agwat.2016.10.020.

[3] A. Khattab, A. Abdelgawad, and K. Yelmarthi, "Design and Implementation of a Cloud-Based IoT Scheme for Precision Agriculture," in *2016 28th International Conference on Microelectronics (ICM)*, Giza, Egypt, Dec. 2016, pp. 201–204, doi: 10.1109/ICM.2016.7847850.

[4] S. K. Choudhary, R. S. Jadoun, and H. L. Mandoriya, "Role of Cloud Computing Technology in Agriculture Fields," *Computing*, vol. 7, no. 3, p. 8, 2016.

[5] S. Rajeswari, K. Suthendran, and K. Rajakumar, "A Smart Agricultural Model by Integrating IoT, Mobile and Cloud- Based Big Data Analytics," in *2017 International Conference on Intelligent Computing and Control (I2C2)*, vol. 5, 2017.

[6] H. Channe, S. Kothari, and D. Kadam, "Multidisciplinary Model for Smart Agriculture Using Internet-of-Things (IoT), Sensors, Cloud-Computing, Mobile-Computing Big-Data Analysis," *International Journal Computer Technology & Applications*, vol. 6, p. 10, 2015.

[7] K. E. Adetunji and M. K. Joseph, "Development of a Cloud-Based Monitoring System Using 4Duino: Applications in Agriculture," in *2018 International Conference on Advances in Big Data, Computing and Data Communication Systems (icABCD)*, Durban, South Africa, Aug. 2018, pp. 4849–4854, doi: 10.1109/ICABCD.2018.8465418.

[8] J. A. Delgado, N. M. Short, D. P. Roberts, and B. Vandenberg, "Big Data Analysis for Sustainable Agriculture on a Geospatial Cloud Framework," *Frontiers in Sustainable Food Systems*, vol. 3, p. 54, Jul. 2019, doi: 10.3389/fsufs.2019.00054.

[9] V. Saiz-Rubio and F. Rovira-Más, "From Smart Farming towards Agriculture 5.0: A Review on Crop Data Management," *Agronomy*, vol. 10, no. 2, p. 207, Feb. 2020, doi: 10.3390/agronomy10020207.

[10] Y. Tang, S. Dananjayan, C. Hou, Q. Guo, S. Luo, and Y. He, "A Survey on the 5G Network and its Impact on Agriculture: Challenges and Opportunities," *Computers and Electronics in Agriculture*, vol. 180, p. 105895, Jan. 2021, doi: 10.1016/j.compag.2020.105895.

[11] S. Balamurugan, N. Divyabharathi, K. Jayashruthi, M. Bowiya, R. P. Shermy, and D. R. G. Kruba, "Internet of Agriculture: Applying IoT to Improve Food and Farming Technology," *International Research Journal of Engineering and Technology (IRJET)*, vol. 03, no. 10, p. 8.

[12] K. Venkataramana, "A Design of Framework for AGRI-CLOUD," *IOSRJCE*, vol. 4, no. 5, pp. 01–06, 2012, doi: 10.9790/0661-0450106.

[13] A. Giri, S. Dutta, and S. Neogy, "Enabling Agricultural Automation to Optimize Utilization of Water, Fertilizer and Insecticides by Implementing Internet of Things (IoT)," in *2016 International Conference on Information Technology (InCITe) - The Next Generation IT Summit on the Theme - Internet of Things: Connect your Worlds*, Noida, Oct. 2016, pp. 125–131, doi: 10.1109/INCITE.2016.7857603.

[14] J. Kaivosoja, L. Pesonen, J. Kleemola, I. Pölönen, H. Salo, E. Honkavaara, H. Saari, J. Mäkynen, A. Rajala, "A Case Study of a Precision Fertilizer Application Task Generation for Wheat Based on Classified Hyperspectral Data from UAV Combined with Farm History Data," in *Remote Sensing for Agriculture, Ecosystems, and Hydrology XV*, Dresden, Germany, Oct. 2013, p. 88870H, doi: 10.1117/12.2029165.

[15] H. U. Leena, B. G. Premasudha, and P. K. Basavaraja, "Sensible Approach for Soil Fertility Management Using GIS Cloud," in *2016 International Conference on Advances in Computing, Communications and Informatics (ICACCI)*, Jaipur, India, Sep. 2016, pp. 2776–2781, doi: 10.1109/ICACCI.2016.7732483.

[16] S. R. Prathibha, A. Hongal, and M. P. Jyothi, "IOT Based Monitoring System in Smart Agriculture," in *2017 International Conference on Recent Advances in Electronics and Communication Technology (ICRAECT)*, Bangalore, India, Mar. 2017, pp. 81–84, doi: 10.1109/ICRAECT.2017.52.

[17] P. Zhang, Q. Zhang, F. Liu, J. Li, N. Cao, and C. Song, "The Construction of the Integration of Water and Fertilizer Smart Water Saving Irrigation System Based on Big Data," in *2017 IEEE International Conference on Computational Science and Engineering (CSE) and IEEE International Conference on Embedded and Ubiquitous Computing (EUC)*, Guangzhou, China, Jul. 2017, pp. 392–397, doi: 10.1109/CSE-EUC.2017.258.

[18] Lavanya, G, et al., "An Automated Low Cost IoT Based Fertilizer Intimation System for Smart Agriculture," *Sustainable Computing: Informatics and Systems*, vol. 28, p. 100300, 2020.

[19] P. Divya Vani and K. Raghavendra Rao, "Measurement and Monitoring of Soil Moisture Using Cloud IoT and Android System," *Indian Journal of Science and Technology*, vol. 9, no. 31, Aug. 2016, doi: 10.17485/ijst/2016/v9i31/95340.

[20] N. Pavón-Pulido, J. A. López-Riquelme, R. Torres, R. Morais, and J. A. Pastor, "New Trends in Precision Agriculture: A Novel Cloud-Based System for Enabling Data Storage and Agricultural Task Planning and Automation," *Precision Agriculture*, vol. 18, no. 6, pp. 1038–1068, Dec. 2017, doi: 10.1007/s11119-017-9532-7.

[21] M. R. Bendre, R. C. Thool, and V. R. Thool, "Big Data in Precision Agriculture: Weather Forecasting for Future Farming," in *2015 1st International Conference on Next Generation Computing Technologies (NGCT)*, Dehradun, India, Sep. 2015, pp. 744–750, doi: 10.1109/NGCT.2015.7375220.

[22] E. Kanagaraj, L. M. Kamarudin, A. Zakaria, R. Gunasagaran, and A. Y. M. Shakaff, "Cloud-Based Remote Environmental Monitoring System with Distributed WSN Weather Stations," in *2015 IEEE SENSORS*, Busan, Nov. 2015, pp. 1–4, doi: 10.1109/ICSENS.2015.7370449.

[23] L. Tan, "Cloud-based Decision Support and Automation for Precision Agriculture in Orchards," *IFAC-PapersOnLine*, vol. 49, no. 16, pp. 330–335, 2016, doi: 10.1016/j.ifacol.2016.10.061.

[24] S. Tenzin, S. Siyang, T. Pobkrut, and T. Kerdcharoen, "Low Cost Weather Station for Climate-Smart Agriculture," in *2017 9th International Conference on Knowledge and Smart Technology (KST)*, Chonburi, Thailand, Feb. 2017, pp. 172–177, doi: 10.1109/KST.2017.7886085.

[25] O. Elijah, T. A. Rahman, I. Orikumhi, C. Y. Leow, and M. H. D. N. Hindia, "An Overview of Internet of Things (IoT) and Data Analytics in Agriculture: Benefits and Challenges," *IEEE Internet of Things Journal*, vol. 5, no. 5, pp. 3758–3773, Oct. 2018, doi: 10.1109/JIOT.2018.2844296.

[26] S. Jaiganesh, K. Gunaseelan, and V. Ellappan, "IOT Agriculture to Improve Food and Farming Technology," *2017 Conference on Emerging Devices and Smart Systems (ICEDSS)*, pp. 260–266, 2017, doi: 10.1109/ICEDSS.2017.8073690.

[27] R. Reshma, V. Sathiyavathi, T. Sindhu, K. Selvakumar, and L. SaiRamesh, "IoT Based Classification Techniques for Soil Content Analysis and Crop Yield Prediction," in *2020 Fourth International Conference on I-SMAC (IoT in Social, Mobile, Analytics and Cloud) (I-SMAC)*, Palladam, India, Oct. 2020, pp. 156–160, doi: 10.1109/I-SMAC49090.2020.9243600.

[28] A. C. Bartlett, A. A. Andales, M. Arabi, and T. A. Bauder, "A Smartphone App to Extend Use of a Cloud-Based Irrigation Scheduling Tool," *Computers and Electronics in Agriculture*, vol. 111, pp. 127–130, Feb. 2015, doi: 10.1016/j.compag.2014.12.021.

[29] K. Foughali, K. Fathallah, and A. Frihida, "Using Cloud IOT for Disease Prevention in Precision Agriculture," *Procedia Computer Science*, vol. 130, pp. 575–582, 2018, doi: 10.1016/j.procs.2018.04.106.

[30] S. K. Choudhary, R. S. Jadoun, and H. L. Mandoriya, "Role of Cloud Computing Technology in Agriculture Fields," *Computing*, vol. 7, no. 3, p. 8, 2016.

[31] N. H. Kulkarni, G. N. Srinivasan, B. M. Sagar and N. K. Cauvery, "Improving Crop Productivity Through A Crop Recommendation System Using Ensembling Technique," in *2018 3rd International Conference on Computational Systems and Information Technology for Sustainable Solutions (CSITSS)*, 2018, pp. 114–119, doi: 10.1109/CSITSS.2018.8768790.

[32] K. Pradeep, J. Ratha, A. Thillai, B. Saminathan, A. Murugan, *Smart Applications of IoT*, 2021, pp. 131–151, doi: 10.1002/9781119761655.ch7.

[33] R. Shukla, G. Dubey, P. Malik, N. Sindhwani, R. Anand, A. Dahiya, and V. Yadav, "Detecting Crop Health Using Machine Learning Techniques in Smart Agriculture System," *Journal of Scientific & Industrial Research*, vol. 80, no. 08, pp. 699–706, 2021.

[34] N. Sindhwani, R. Anand, R. Vashisth, S. Chauhan, V. Talukdar, and D. Dhabliya, "Thingspeak-Based Environmental Monitoring System Using IoT," in *2022 Seventh International Conference on Parallel, Distributed and Grid Computing (PDGC)*, 2022, pp. 675–680, doi: 10.1109/PDGC56933.2022.10053167.

# 2 A Comparative Analysis on Performance Factors for Communicating Devices in MIPv6 Environment under Smart City Model

*Shweta Singh, Manish Bhardwaj, Arun Kumar Tripathi, and Shivali Tyagi*
KIET Group of Institutions, DELHI-NCR, Ghaziabad, India

## CONTENTS

2.1   Introduction ........................................................................................................21
2.2   Related Work ......................................................................................................23
2.3   Simulation ..........................................................................................................23
       2.3.1   Components Used for the Proposed Scenario ........................................24
                2.3.1.1   Home Agent ..............................................................................24
                2.3.1.2   Foreign Agent ...........................................................................25
                2.3.1.3   Mobile Node .............................................................................25
                2.3.1.4   Correspondent Node .................................................................25
                2.3.1.5   Tunneling ..................................................................................25
       2.3.2   Types of Handover Management Techniques .........................................25
                2.3.2.1   Hard Handover ..........................................................................25
                2.3.2.2   Soft Handover ...........................................................................25
       2.3.3   Proposed Scenario .................................................................................26
                2.3.3.1   Devices Communicating in Existing IoT-based Environment ...................26
                2.3.3.2   Devices Communicating in IoT-based Environment with
                            Applied Proxy ..........................................................................26
2.4   Simulation Analysis ...........................................................................................28
       2.4.1   Delay .....................................................................................................28
       2.4.2   Load .......................................................................................................28
       2.4.3   Throughput ............................................................................................29
2.5   Conclusion and Future Scope ............................................................................30
References ....................................................................................................................30

## 2.1  INTRODUCTION

With the concept of the smart city [1], networks [2] play a vital role in establishing efficient and reliable communication between the communicating nodes [3]. Usually when it comes to the smart city concept, Internet of Things (IoT) [1], the based network is encouraged. IoT is one of the trending technologies that can be implemented in any area of engineering. IoT provides reliable solutions to

DOI: 10.1201/9781003319238-2

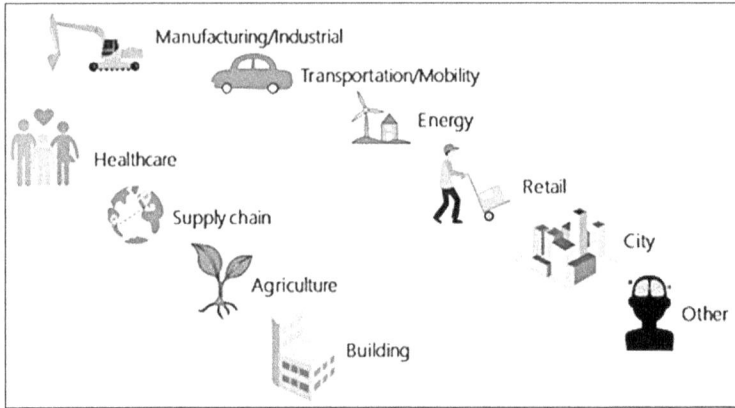

**FIGURE 2.1**  IoT technology in various sectors worldwide.

various day-to-day issues related to various fields [4]. The network in any scenario, whether it operated under IoT or a distant cloud-based network [5, 6], should be able to provide more consistent and active communication between the communicating devices. Since IoT deals with the short-range communication in a network, hence the cloud-based network is encouraged to provide more reliable ways of collecting the large volumes of data received through IoT-based devices. Generally, cloud-based architectures [7] are encouraged as cloud computing follows properties of elasticity, scalability, reliability, etc. Figure 2.1 represents the usage of IoT-based techniques all around the world in various sectors, such as manufacturing, transportation, city, etc. To operate on the network wirelessly, one must ensure that the information floating onto the network is kept secure and no information is lost while the communication is being done.

To ensure the required levels of information security, a certain set of protocols are usually applied in a network, namely TCP, HTTP, MQTT, etc. [8]. Equally, to ensure that appropriate information flow in a network, certain routing protocols are followed, namely, MIP, RIP, IS-IS, etc. [9]. Mobile Internet Protocol (MIP) [8] is the standard communication protocol for the cloud-based network. MIP allows users to roam within numerous networks while communication is established between the intended devices. For mobility management, MIP version 6 (MIPv6) [9] is applied in a network for effective communication to be done. MIP can be found operational on both wired and wireless networks. MIP also supports continuous connectivity that ensures maximum throughput and minimum interference. MIP enables the concept of tunneling which supports a consistent communication link, computing the optimal route to exchange information within a different network. Tunneling enables encryption and proxy [10] with data packets floating into the network. While proxy is utilized with tunneling, it behaves as a middleman in between Mobile Node (MN) and Correspondent Node (CN) [9]. Previously, security-based solutions have been provided by various authors concerning the improvement of the throughput in the network. In [11], the author has proposed a two-stage queuing model to determine the latency for sensor devices communicating in the IoT-based environment. In [12], the author has deployed an Access Class Barring (ACB) scheme in the IoT-based architecture to analyze the performance metrics in a network. The solutions provided by various authors need a more improved strategy to have a communication link with minimum delay and maximum throughput for an IoT-based architecture. This paper undertakes the following set of objectives:

a) To propose an IoT-based architecture as the basis for Smart City Architecture,
b) To simulate the proposed architecture to analyze the performance metrics for communication link,
c) To compare the proposed scenario with the existing one based on the performance metrics,
d) To improve the simulation analysis of the proposed scenario with applied proxy.

Further organization of the paper can be best described in the following sections: Section 2.2 represents the previous work that has been provided by various authors in terms of the performance analysis for the smart city concept in the MIPv6 environment. Section 2.3 represents the components used in the proposed scenario, along with the configuration for the proposed scenario. Section 2.4 represents a comparative analysis of a number of performance metrics, including network load, delay in exchanging information, and throughput in terms of packets/seconds in both the scenarios for communicating nodes. Section 2.5 represents the conclusion and future scope of the proposed work.

## 2.2 RELATED WORK

The performance of any network can be determined in terms of various performance metrics [13], namely delay, throughput, load, etc., since a network relies on the layered architecture, hence it becomes extremely important to coordinate the traffic flowing onto the network. The more the traffic is optimized, the higher the throughput that can be expected while communication is being done. To optimize the traffic flow in a network, some set of protocols needs to be followed along with different stages of applied proxy [14]. A proxy is responsible for providing encryption and encapsulation of information flowing in the network [15] so that no breakthrough can be performed by any of the third nodes in a network.

For any network, delay and load should be kept at minimal rates, and throughput should be maintained at higher rates. Delays in a network usually occur due to route discovery processes for information to reach the destined device. MIPv6 will help in eradicating the traditional requirements of maintaining route discovery messages, additional neighbor discovery options. [16] While operating in the MIPv6 environment, constant data session needs to be maintained. For data sessions to be accomplished, certain address types need to be maintained. The communicating devices in the MIPv6-based network are represented as Home Agent (HA), Foreign Agent (FA) [9].

New technologies are being adopted by various researchers worldwide to maximize the throughput in an IoT-based environment. In [17], the author has provided a comparative analysis based on the application layer communication protocols for devices communicating in an IoT-based environment. The author had implemented a prototype for an IoT-based microgrid communication system and recorded the experimental results for analysis. The results in the author's study concluded that the implementation of an additional set of protocols for security-related factors adds a delay to the network extensively. In [11], the author has proposed a two-stage queuing model to determine the latency for sensor devices communicating in the IoT-based environment. The author then investigates the network based on performance metrics and provided a feedback mechanism to measure channel utilization. In [18], the author has proposed the framework for IoT applications with an embedded security structure [19]. The authors analyzed the structure between Message Queue Telemetry Transport (MQTT) [20] and User-Managed Access (UMA) [21]. Results of an embedded structure proposed by the author depicted that IoT-based applications perform better when security is applied to the network [22]. In [12], the author has deployed an ACB [23] scheme in the IoT-based architecture. The author has simulated and analyzed the ACB scheme in the IoT-based architecture to maximize the throughput in the network. The author used success, collision, access delay, etc., as the performance metrics for the proposed architecture.

## 2.3 SIMULATION

This section underlines the configuration of the proposed scenario for the simulation to be done. The simulation is mainly based on two scenarios. The first scenario consists of the network configuration where IoT devices tend to communicate in an environment where a proxy is not applied in the network. The second scenario consists of the network configuration where IoT tends to communicate in an environment where a proxy is applied in the network. The simulation is based on the scenario

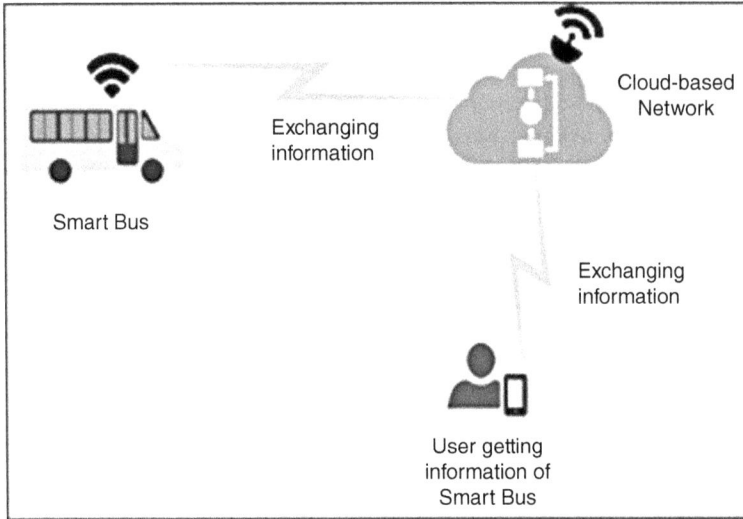

**FIGURE 2.2**   IoT-based architecture of user exchanging information with a smart bus.

consisting of a smart bus [24] and a dedicated mobile application to retrieve information via distance. Figure 2.2 represents the IoT-based architecture of information exchange with the smart bus. The common configuration for both the scenarios includes the following equipment: 07 equipment in each scenario, 03 communicating routers to act as CN & MN, 01 routers [9] to connect all the HA & FA in a scenario where a proxy is not applied in a network, and 01 IP cloud [9] for a proposed scenario with applied proxy. Each scenario consists of 01 IP Mobility Config [9]. In both scenarios, MN is allowed to communicate with other devices through the base station [25] and is allowed to travel along with other cells in a network.

### 2.3.1   COMPONENTS USED FOR THE PROPOSED SCENARIO

In any MIPv6 scenario, communicating devices are exchanging information within the scope of two different networks, i.e., a home network and a foreign network [9]. The purpose is to set up a consistent communication link with continuous data interchange between the communicating nodes. Generally, in the MIPv6 network, the process of handover is followed to establish successful communication while moving from one location to another. Handover management [9, 26] is a technique wherein HAs roam within the foreign network [27] while communication is going on. The notion of handover management is to uphold the factor of consistency of the communication link and deliver communication with the least loss of information while moving across different networks.

To every MN in a network, two addresses are attached, namely the perpetual home address and Care-of Address (CoA) [9]. These addresses help in identifying the current location of MN in both the networks (for home network and foreign network). CoA is usually maintained to detect the recent location of the MN in a network. The following entities are utilized in the MIPv6 environment for reliable communication between the communicating nodes:

#### 2.3.1.1   Home Agent

HA consists of a database comprising permanent addresses for all the MNs belonging to the same home network. Utilizing the CoA of each MN, a database including the current location of each MN will be maintained.

#### 2.3.1.2 Foreign Agent

FA consists of a database including information regarding each MN visiting the foreign network. FA uses the broadcasted CoA of MN to better locate the present location of the MN for the corresponding HA.

#### 2.3.1.3 Mobile Node

MN establishes two-way communication between coupled nodes in a network. MN in a network can roam from its home network to a foreign network with least breaking to the communication linkage so established.

#### 2.3.1.4 Correspondent Node

CN is responsible to establish reliable communication linkage between the CN and MN to exchange information. CN exchanges information to CN even when MN roams other than its home network, using the route optimization concept.

#### 2.3.1.5 Tunneling

The concept of tunneling [28] helps in switching information within MN and CN directly in a virtual manner. Tunneling is usually a virtual concept that allows working whenever CN wants to exchange information to MN while roaming within different networks. Tunneling is performed by employing broadcasted CoA of MN.

Handover is a theory of transferring ongoing data sessions between multiple channels [29], such that the devices stay connected from the home network to some foreign channel. Handover can be utilized under conditions such as:

1. When MN moves out of its coverage area while a data exchange session is accomplished and a termination of data session is to be avoided,
2. When the capacity of two connecting nodes is highly consumed and a new data session is to be established,
3. To avoid the interference caused by one of the data sessions starting up on one channel, while that particular channel is already attained for a previous data session to be exchanged,
4. To avoid the interference caused by the speed of movement of MN from one network to another network,
5. To avoid the interference caused by the MN while traveling from one channel to another channel with maximum consistency of communication.

#### 2.3.2 Types of Handover Management Techniques

Handover management is categorized mainly into subcategories, i.e., soft handover and hard handover [30]. Both the handover techniques are utilized in a network in different scenarios. Figure 2.3 represents the handover techniques that can be used in any network.

#### 2.3.2.1 Hard Handover

In a hard handover, MN is allowed to roam from its home network to the foreign network while the data exchange session resumes. The two networks or channels are separated at some geographical distance and, meanwhile, the data session has to be maintained while considering the geographical distance. For this aspect, there is some intervention caused and the data session is broken while traveling from its home network to a foreign network in a hard handover.

#### 2.3.2.2 Soft Handover

In the soft handover technique, there is no or minimal intervention caused to the data exchange session while MN roams from its home network to a foreign network. There is no breakthrough to the

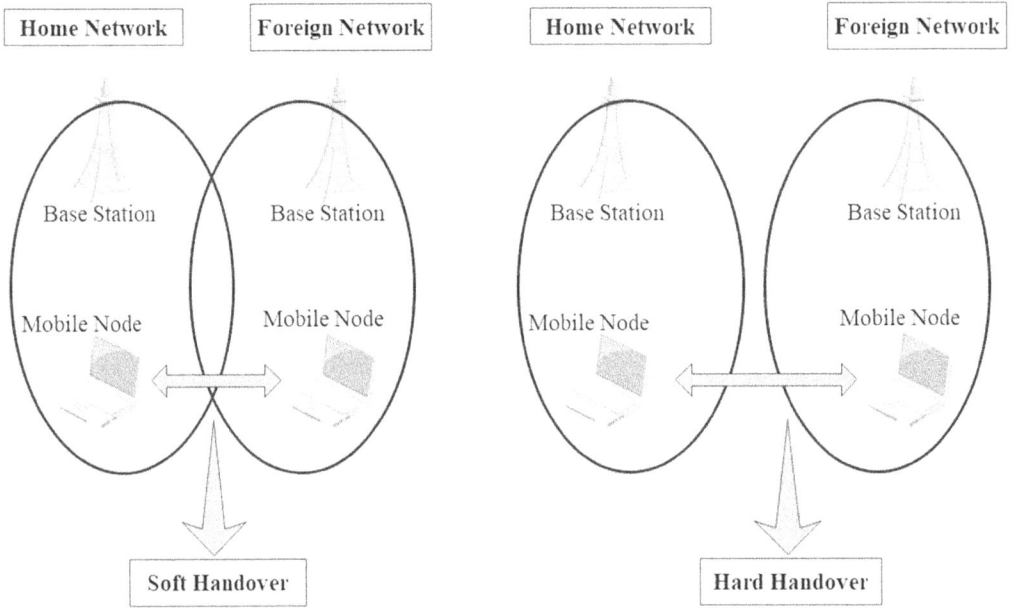

**FIGURE 2.3**  Handover management and its types.

data session as a reliable communication link is established between the parallel channels in any network in case of a soft handover. The communication link intervenes at the least possible levels.

Figure 2.3 represents the types of handover management techniques that are generally opted in two networks, i.e., home network and foreign network. For both the handover techniques, MN is allowed to roam from its home network towards the foreign network, with a purpose to minimize the intervention caused while traveling to a different network.

### 2.3.3  PROPOSED SCENARIO

This section represents the list of proposed scenarios in the IoT-enabled environment. The proposed IoT-enabled architecture is taken for comparison of devices communicating in an environment where no proxy is applied, and where devices communicate in an environment with applied proxy.

#### 2.3.3.1  Devices Communicating in Existing IoT-based Environment

This section represents the scenario where the IoT devices tend to communicate in an environment where the proxy is not applied to the network. In this scenario, the MN is considered as the smart bus traveling from one location to another, and the CN is considered as the mobile application of the passenger who is trying to retrieve the information related to the bus. Figure 2.4 represents the scenario where devices are communicating with no applied proxy in the network. The microcontroller placed in the bus and the mobile application of the passenger are communicating with each other continuously. Routers and repeaters are utilized to establish communication for farther distances [31].

#### 2.3.3.2  Devices Communicating in IoT-based Environment with Applied Proxy

This section represents the scenario where the IoT devices tend to communicate in an environment where the proxy is applied to the network. In the scenario, the MN is considered as the smart bus traveling from one location to another, and the CN is considered as the mobile application of the passenger who is trying to retrieve the information related to the bus. Figure 2.5 represents the scenario where devices are communicating with proxy in the network. In this scenario, a cloud-based network is established for reliable communication to be done.

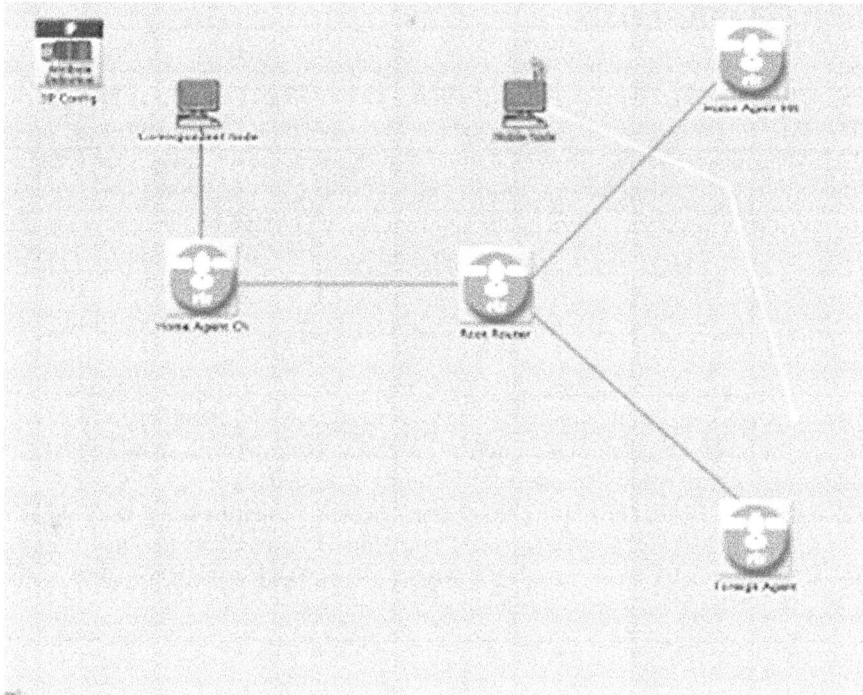

**FIGURE 2.4** Devices communicating in existing IoT-based environment.

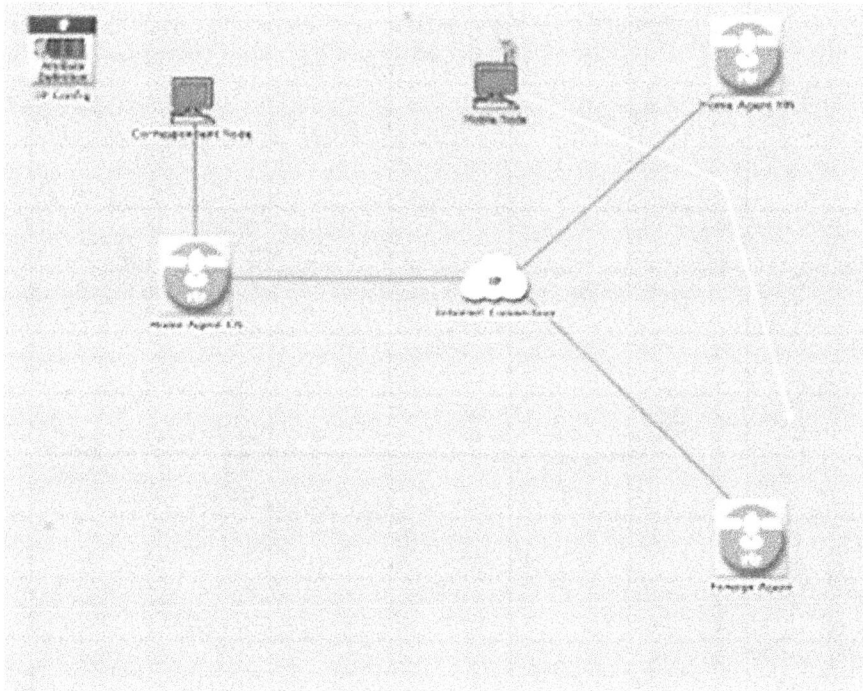

**FIGURE 2.5** Devices communicating in IoT-based environment with applied proxy.

## 2.4 SIMULATION ANALYSIS

This section represents the simulation analysis for the proposed set of scenarios. The comparison of both scenarios is being done based on the performance factors. The simulation is being performed in OPNET IT Guru Educational Version 14.5 Modeler [32]. Based on the previously set performance Discrete Event Simulation (DES) values [33], the results are drawn. The devices are communicating in MIPv6-based environment in both scenarios. The performance metrics for a network can be analyzed in terms of network delay on accessing information at HA of CN, network load, throughput at HA of CN, namely. Figures 2.6–2.8 below represent the analysis being done for the CN in both scenarios.

### 2.4.1 DELAY

Network delay is the expanse of time consumed in transferring information from one device to another [34]. For effective transmission of information, delay in a network should always be kept at a minimum. Figure 2.6 signifies network delay in terms of seconds at HA of CN.

Figure 2.6 depicts the comparison of the proposed scenario in terms of network delay received in the MIPv6 environment. It can be concluded that after cloud-based technology is utilized with applied proxy, network delay has reduced to a very low level. Devices in the proposed proxy-based environment can exchange information efficiently in a network [35].

### 2.4.2 LOAD

Network load is the information that is flowing into the network in the form of data packets [36]. Unlike the network delay, network load should be sustained within appropriate levels, if to achieve faster and more efficient communication. Figure 2.7 represents network load in terms of packets/seconds.

Figure 2.7 depicts the total load in a network that has occurred in both scenarios. It can be concluded that as the quantity of packets is increased, network load has reduced for the proposed scenario, besides cloud-based technology with applied proxy is operated in between the communicating devices.

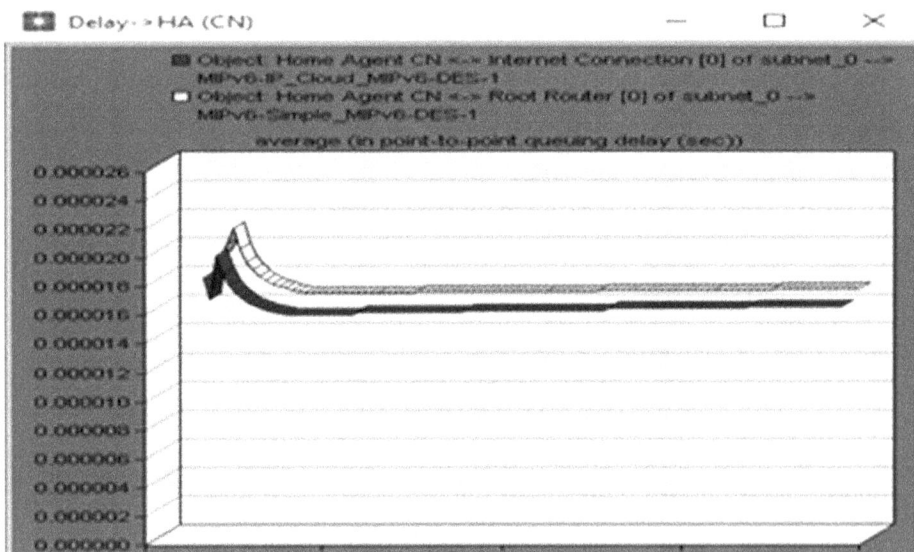

**FIGURE 2.6**   Delay measured at HA of CN (in terms of seconds).

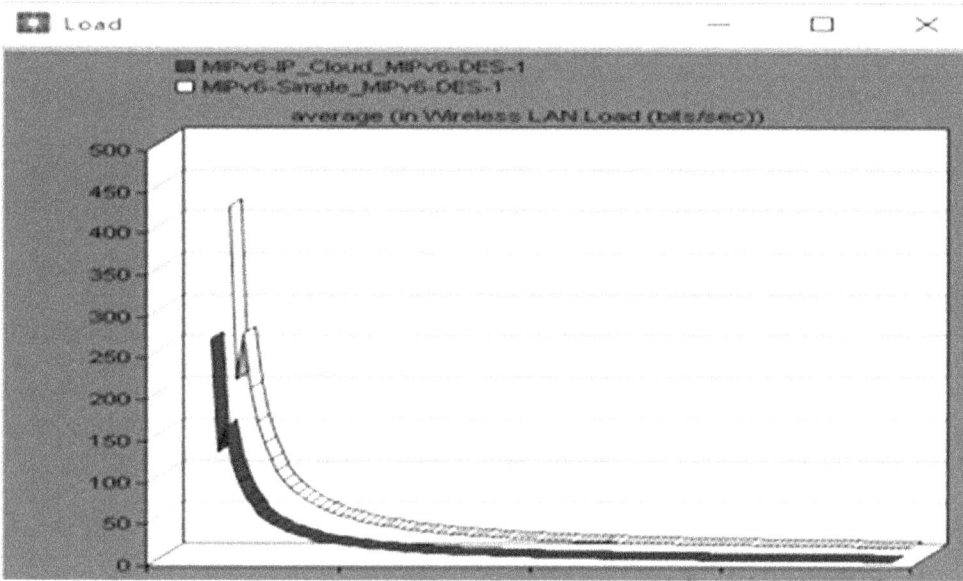

**FIGURE 2.7**    Network load in a both the scenarios (in terms of packets/seconds).

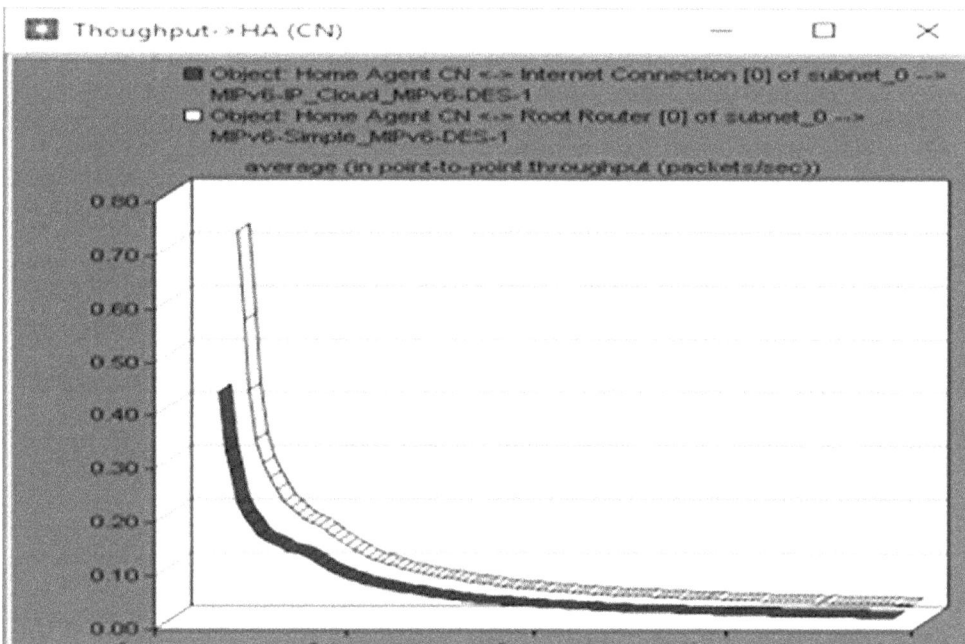

**FIGURE 2.8**    Throughput received at HA of CN (in terms of packets/sec).

## 2.4.3 THROUGHPUT

Throughput is the rate that assures that information has been successfully received at the destined node in a network [37, 38]. Figure 2.8 signifies throughput received in a network in terms of packets/sec.

Figure 2.8 depicts the throughput attained in both scenarios for HA (CN). It can be perceived that when a proxy is applied with cloud-based technology, higher throughput has been achieved in comparison to the scenario with no applied proxy.

## 2.5  CONCLUSION AND FUTURE SCOPE

With IoT-based architecture, almost every communication is performed via an Internet-based service. Hence, it becomes equally important to build the network with the least delay in exchanging information, along with improved throughput of the network [39]. In the paper, an IoT-based architecture is proposed with applied proxy at different layers of the network. In the proposed scenario, two devices – mainly MN and CN – are made to exchange information while roaming within different networks. After simulation, it can be concluded that when a proxy is applied along with cloud-based technology with the present network, more throughput is reached, with minimum delay and load in a network. The future scope of the proposed scenario is to extend the levels of security using high-end encryption techniques for information exchange in an IoT-based architecture.

## REFERENCES

[1] Zhao, Fang, et al. "Smart city research: A holistic and state-of-the-art literature review." *Cities* 119 (2021): 103406.

[2] Qureshi, Kashif Naseer, et al. "Internet of vehicles: Key technologies, network model, solutions and challenges with future aspects." *IEEE Transactions on Intelligent Transportation Systems* 22.3 (2020): 1777–1786.

[3] Jansi, K. R., G. K. Sandhia, and S. V. KasmirRaja. "A reliable cloud based framework for large scale urban sensing systems." *Smart Intelligent Computing and Communication Technology* 38 (2021): 67.

[4] Anand, R., N. Sindhwani, and S. Juneja. "Cognitive Internet of Things, Its Applications, and Its Challenges: A Survey." In *Harnessing the Internet of Things (IoT) for a Hyper-Connected Smart World* (pp. 91–113). Apple Academic Press, Oakville, Canada, 2022.

[5] Chimmalee, Benjamas, and Anuchit Anupan. "Simulation of network model on cloud technology based on mathematical framework." *2021 6th International Conference on Business and Industrial Research (ICBIR)*. IEEE, 2021.

[6] Singh, Shweta, and Arun Kumar Tripathi. "Analyzing for performance factors in cloud computing." *International Journal of Control Theory and Applications* 9 (2016): 87–96.

[7] Shakya, Subarna. "Survey on cloud based robotics architecture challenges and applications." *Journal of Ubiquitous Computing and Communication Technologies (UCCT)* 2.01 (2020): 10–18.

[8] Morzelona, Romi. "An experimental study of various network protocols and understanding its implementation in real world." *Mathematical Statistician and Engineering Applications* 71.1 (2022): 122–127.

[9] Arun Kumar Tripathi, and Shweta Singh. "Mobile IPv6 security performance metrics for WIMAX roaming environment." *International Journal of Innovative Technology and Exploring Engineering* 9.1 (2019): 68–72.

[10] Rana, Muhammad Ehsan, Mohamed Abdulla, and Kuruvikulam Chandrasekaran Arun. "Common security protocols for wireless networks: A comparative analysis." *3rd International Conference on Integrated Intelligent Computing Communication & Security (ICIIC 2021)*. Atlantis Press, 2021.

[11] Althoubi, Asaad, Reem Alshahrani, and Hassan Peyravi. "Delay analysis in IoT sensor networks." *Sensors* 21.11 (2021): 3876.

[12] Alvi, Maira, et al. "Performance analysis of access class barring for next generation IoT devices." *Alexandria Engineering Journal* 60.1 (2021): 615–627.

[13] Singh, Shweta, and Arun Kr Tripathi. "Analysis of delay and load factors in wired and wireless environment." *Second International Conference on Recent Trends in Science, Technology, Management and Social Development (RTSTMSD-15)*, IJSTM, and ISSN. No. 2321-1938. 2015.

[14] Zebari, Rizgar R., et al. "Distributed denial of service attack mitigation using high availability proxy and network load balancing." *2020 International Conference on Advanced Science and Engineering (ICOASE)*. IEEE, 2020.

[15] Liu, Lu, Rui Ma, and Zhangming Zhu. "An encapsulated packet-selection routing for network on chip." *Microelectronics Journal* 84 (2019): 96–105.

[16] Liu, Chunfeng, et al. "Kalman prediction-based neighbor discovery and its effect on routing protocol in vehicular ad hoc networks." *IEEE Transactions on Intelligent Transportation Systems* 21.1 (2019): 159–169.

[17] Kondoro, Aron, et al. "Real time performance analysis of secure IoT protocols for microgrid communication." *Future Generation Computer Systems* 116 (2021): 1–12.

[18] Alhazmi, Omar H., and Khalid S. Aloufi. "Performance analysis of the hybrid MQTT/UMA and restful IoT security model." *Advances in Internet of Things* 11.1 (2021): 26–41.

[19] A. Juneja, S. Juneja, V. Bali, V. Jain, and H. Upadhyay. "Artificial intelligence and cybersecurity: Current trends and future prospects," *The Smart Cyber Ecosystem for Sustainable Development* 27 (2021): 431–441, 2021.

[20] Hintaw, Ahmed J., et al. "MQTT vulnerabilities, attack vectors and solutions in the internet of things (IoT)." *IETE Journal of Research* (2021): 1–30. https://doi.org/10.1080/03772063.2021.1912651

[21] Pöhn, Daniela, and Wolfgang Hommel. "Proven and Modern Approaches to Identity Management." *Advances in Cybersecurity Management* (pp. 421–443) Springer, Cham, 2021.

[22] Juneja, A., S. Juneja, V. Bali, and S. Mahajan. "Multi-criterion decision making for wireless communication technologies adoption in IoT." *International Journal of System Dynamics Applications* 10.1 (2021): 1–15. https://doi.org/10.4018/ijsda.2021010101.

[23] Qiu, Gongan, Zhihua Bao, and Shibing Zhang. "Adaptive channel status based access class barring scheme in C-V2X networks." *2021 7th Annual International Conference on Network and Information Systems for Computers (ICNISC)*. IEEE, 2021

[24] Kaur, J., N. Sindhwani, and R. Anand "Implementation of IoT in various domains." In *IoT Based Smart Applications* (pp. 165–178). Springer International Publishing, USA, 2022.

[25] Kavitha, V., G. Manimala, and R. Gokul Kannan "AI-based enhancement of base station handover." *Procedia Computer Science* 165 (2019): 717–723.

[26] Mollel, Michael S., et al. "A survey of machine learning applications to handover management in 5G and beyond." *IEEE Access* 9 (2021): 45770–45802.

[27] Madhusudhan, R. "A secure and lightweight authentication scheme for roaming service in global mobile networks." *Journal of Information Security and Applications* 38 (2018): 96–110.

[28] El Idrissi, Dounia, Najib Elkamoun, and Rachid Hilal "Study of the impact of the transition from IPv4 to IPv6 based on the tunneling mechanism in mobile networks." *Procedia Computer Science* 191 (2021): 207–214.

[29] Caviglione, Luca. "Trends and challenges in network covert channels countermeasures." *Applied Sciences* 11.4 (2021): 1641.

[30] Manh, Linh Dao, Nam Vi Hoai, and Quy Vu Khanh. "Advanced handover techniques in 5G LTE-A networks." *International Journal* 9.3 (2021).

[31] Askarani, Mohsen Falamarzi, Kaushik Chakraborty, and Gustavo Castro Do Amaral. "Entanglement distribution in multi-platform buffered-router-assisted frequency-multiplexed automated repeater chains." *New Journal of Physics* 23.6 (2021): 063078.

[32] Sood, A. "Network design by using opnet™ it guru academic edition software." *Rivier Academic Journal* 3.1 (2007).

[33] Vizzarri, Alessandro "Analysis of VoLTE end-to-end quality of service using OPNET." *2014 European Modelling Symposium*. IEEE, 2014.

[34] Malandri, Caterina, et al. "Public transport network vulnerability and delay distribution among travelers." *Sustainability* 13.16 (2021): 8737.

[35] Juneja, S., A. Juneja, V. Bali, and H. Upadhyay. "Cyber Security: An Approach to Secure IoT from Cyber Attacks Using Deep Learning." In *Industry 4.0, AI, and Data Science* (pp. 135–146). CRC Press, UK, 2021.

[36] Mishra, Sambit Kumar, Bibhudatta Sahoo, and Priti Paramita Parida "Load balancing in cloud computing: A big picture." *Journal of King Saud University-Computer and Information Sciences* 32.2 (2020): 149–158.

[37] Sivaraman, Vibhaalakshmi, et al. "The effect of network topology on credit network throughput." *Performance Evaluation* 151 (2021): 102235.

[38] S. Juneja, A. Juneja, and R. Anand "Reliability modeling for embedded system environment compared to available software reliability growth models," in *2019 International Conference on Automation, Computational and Technology Management (ICACTM)*, 2019, pp. 379–382. https://doi.org/10.1109/ICACTM.2019.8776814

[39] N. Sindhwani, R. Anand, R. Vashisth, S. Chauhan, V. Talukdar, and D. Dhabliya "Thingspeak-based environmental monitoring system using IoT," in *2022 Seventh International Conference on Parallel, Distributed and Grid Computing (PDGC)*, 2022, pp. 675–680. https://doi.org/10.1109/PDGC56933.2022.10053167

# 3 Futuristic Trends in Vehicle Communication Based on IoT and Cloud Computing

*Chhote Lal Prasad Gupta and Shashank Gaur*
Bansal Institute of Engineering and Technology, Lucknow, India

## CONTENTS

3.1 Introduction .................................................................................................33
3.2 Background ..................................................................................................34
    3.2.1 Traditional VANETs ........................................................................34
    3.2.2 Cluster ..............................................................................................35
    3.2.3 V2I ...................................................................................................36
    3.2.4 V2V ..................................................................................................36
    3.2.5 Fog Computing ................................................................................37
    3.2.6 Improvement in V2V Communication Using Fog Computing .........37
    3.2.7 Dedicated Short-Range Communication (DSRC) ............................38
    3.2.8 Edge Devices ...................................................................................39
    3.2.9 AdHoc Routing Protocols for Vehicle Communication ...................39
    3.2.10 Proactive Routing Protocols ...........................................................39
    3.2.11 Reactive Routing Protocol ..............................................................40
    3.2.12 Hybrid Routing Protocol .................................................................40
    3.2.13 Dynamic Source Routing (DSR) Protocol ......................................40
        3.2.13.1 Route Discovery ...............................................................40
        3.2.13.2 Route Maintenance ...........................................................41
        3.2.13.3 Route Cache ......................................................................41
    3.2.14 AdHoc On-Demand Vector(AODV) Routing Protocol ...................44
    3.2.15 In AODV Routing Protocol-based Scenario There Are Two Phrases .....45
    3.2.16 Hierarchical Routing Protocols .......................................................45
    3.2.17 Hierarchical State Routing (HSR) Protocol ....................................45
3.3 Conclusion ..................................................................................................49
    3.3.1 Future Scope ....................................................................................49
References .............................................................................................................50

## 3.1 INTRODUCTION

Vehicle communication is a term that is used to describe the connection between the vehicles running on the road. In vehicle communication, there are two methods of communication: vehicle-to-vehicle (V2V) and vehicle-to-infrastructure (V2I). Vehicular Adhoc Network (VANET) terminology came into existence when the concept of V2V and V2I came into existence [1]. In our surroundings, everything in our day by day is getting connected with the Internet that is mobile to mobile connectivity, TVs to mobile connectivity, TVs to music system connectivity, and many more examples. The sensors which are being used in these devices are playing the main role in the communication of these

DOI: 10.1201/9781003319238-3

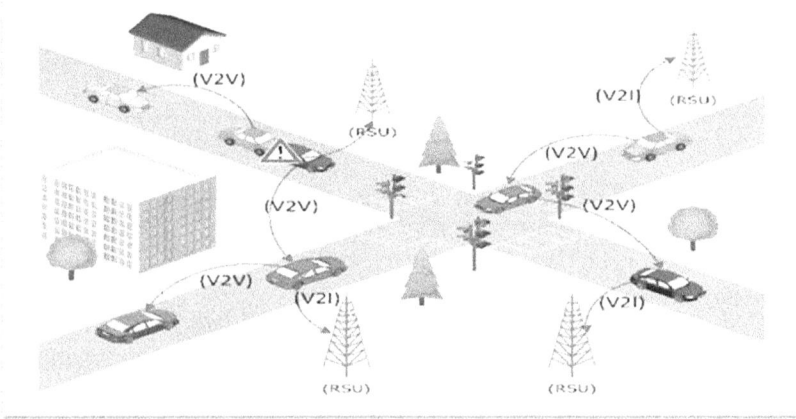

**FIGURE 3.1**    VANET Communication Architecture.

*Source*: https://www.researchgate.net/profile/Anis-Ur-Rahman/publication/330772224/figure/fig3/AS:
727739887783938@1550518118286/VANET-communication-architecture.png

devices among each other with the help of the Internet and because of this, the future aim of global connectivity of things can be achieved. Internet of things (IoT) is the emerging field of research and due to this research, our surrounding society is being facilitated by health using the Internet of Things (IoT), vehicle communication using IoT, and energy production using IoT. Industries are evolving and making a profit using IoT and producing many more benefits to society. Increasing causalities in road accidents can be reduced by developing a communication channel between the vehicles running on the road [2]. Many organizations (some of them are renowned) have predicted that as the years go by, the number of vehicles on roads also increases [3].

Vehicle communication using IoT, or in other words VANET, is an emerging research area. In this field, the main focus is on making a number of a reliable network of the vehicles running on the road according to the surroundings, and these networks can change their structure and behavior according to the current condition or we can say these networks can change their structure and nature on real- time basis. Around $210–740 billion in economic value has been estimated per year by 2025 by using Internet of Vehicles (IoV) [4].

VANET has huge potential to address road safety and can reduce the cost of communication between the vehicles on the road, but till now VANET has not been capable of implementation at a commercial level [5]. Implementing a vehicle communication network for a commercial purpose includes several issues like a fully Adhoc-based network [6], and not fully trustable Internet connectivity [7]. See Figure 3.1.

## 3.2   BACKGROUND

### 3.2.1   Traditional VANETs

A lot of research work is going on in the field of wireless communication and due to which, the concept of Intelligent Transport System (ITS) has evolved and the concept of V2V communication and Vehicle-to-Roadside (V2R) units has been introduced. In particular, the urgency of IoT technology has drawn a lot of attention to it in both academia and business [8]. In the United States each vehicle which is running on the road is equipped with a chip so that it should be identified on the Internet [9]. If we look at India, in Delhi every transportation bus, public auto, electronic vehicle, and metro rail is equipped with Wi-Fi and GPS [10]. V2V and V2R units enable vehicle communication among the vehicles that are running on the road for this, and several wireless technology-related devices – sensors, radar, camera, antenna, Global Positioning System (GPS) and Central Processing

Unit (CPU) – are being used instead within the vehicles which make the vehicle driving through vehicle communication comfortable, safe, and affordable. A number of initiatives have been taken by the European Commission for the development of the next-generation Cooperative Intelligent Transportation System (C-ITS) [11]. Much research has been performed by many renowned companies like the 'Car Play' system developed by Apple which gives full authority to the driver to use all the services of the iPhone with the help of the display of the car which also uses a voice support feature [12]. A positive response about the 'Connected Vehicles' in countries like the US, UK, and Australia has been suggested in various reports [13]. In VANET, a pre-defined communication network is established and in that network a vehicle can communicate with another vehicle with the help of a Road Side Unit (RSU) and these RSUs can be any vehicle, like a bus, which is running on a specific route and through these RSUs vehicles can communicate. In other words, we can say that VANET is made up primarily of three parts: On Board Units (OBUs), Application Units (AUs), and Roadside Units (RSUs) [14, 15].

The network architecture of VANET can be categorized in three ways:

I. Vehicle-to-Vehicle (V2V) communication with the help of RSUs which can be used as an infrastructure for transferring the message packet from one vehicle to another vehicle while running on the road.

II. Vehicle-to-Vehicle (V2V) communication with the help of cloud computing architecture; with cloud computing architecture, the vehicles running on the road can communicate with each other within a cluster and within these clusters, a cluster head helps in making a connection with the Regional Central Processing Unit (RCPU) which send back the response to the Cluster Head (CH) and these cluster heads then communicates with the vehicles that are within the cluster.

III. Vehicle-to-Vehicle (V2V) communication, with the help of fog computing architecture; with fog computing architecture, the vehicles running on the road can communicate more effectively than cloud computing architecture with each other because the vehicles communicate on the basis of fog computing architecture is more focused towards the edge devices and due to this the latency, while transmitting and receiving the message got reduces and the energy consumption at each level whether it is at cluster level or overall vehicles communication network level, the energy consumption is reduced.

If we see the market opportunity of VANET then, it has been observed and predicted that the vehicle running on the road is increasing in the world [16]. Congestion on roads is becoming a big issue when traveling from one destination to another [17]. If we analyze a scenario whereby proper monitoring of a vehicle traveling from one destination to another destination saves 5 minutes, then is has been estimated that these saved 5 minutes can generate Euro 25 billion every year by 2030 [18]. The main purpose of the Internet of Vehicles (IoVs) is to utilize time to travel more efficiently. The high market penetration rate of IoT plays an important role in the design and development of IoV [19]. The automobile industry is growing day by day and one of the factors of this increment is the IoT [20].

## 3.2.2 Cluster

A cluster can be defined as the network infrastructure where each object within the cluster will be connected and when a node within the cluster wants to send a communication packet then the node will send the communication packet to the cluster's head. See Figure 3.2.

In this example, there are three clusters that is cluster1, cluster2 and cluster3. They have cluster heads CH1, CH2 and CH3 respectively. In the cluster, there are several nodes (N11, N12, N13, N14, N21, N22, N23, N24, N31, N32, N33, N34). These nodes transmit their data packet to other nodes through cluster head (CH) and these CH then, send the data packet to the sink node to analyze the

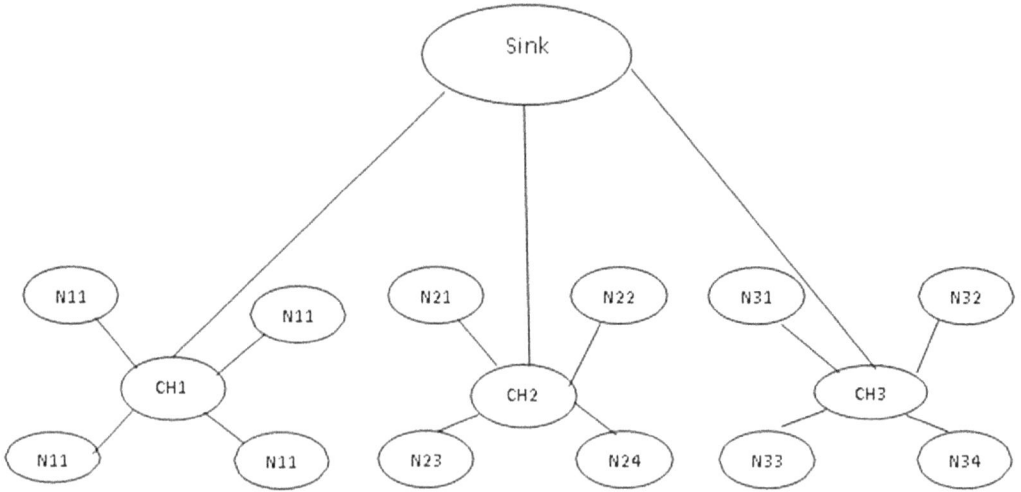

**FIGURE 3.2**  Cluster-based network architecture.

data packet and then, after analysis of these data packets sink node transmits the data packet to the cluster head (CH) of the receiver node in the cluster.

### 3.2.3  V2I

In vehicle communication, vehicles which are running on the road can communicate with the RSUs. However, external applications containing information about entertainment and road safety have used VANET in vehicle-to-infrastructure (V2I) communications, as seen in [21] and [22]. In these types of communication when a vehicle communicates with other vehicles then, an RSU acts as a mediator to transfer information from one vehicle to another vehicle. Under the consortium of the Open Automobile Alliance (OAA), Google is working with many leading automobiles and IT companies to develop an Android system for 'connected drive' [23]. These RSUs are devices that can send and receive information and due to this feature vehicles can communicate; examples of RSUs are mobile towers that have a huge infrastructure all across the globe. Suppose a vehicle wants to communicate to another vehicle that is ten meters away that the first car (V1) wants to reach the destination quickly, and for that V1 wants to speed up and want to overtake V2, and for that V1 will change the lane of the road. So, this message of V1 is transmitted to the nearby RSU and the RSU transmits the information to the V2 and in response to the message of V1 will give the status of V2's front view and this status is transmitted to V1 through a nearby RSU in the form of message. After analyzing the message received by V2, the vehicle V1 will take the decision to speed up or not or change lane or not. See Figure 3.3.

### 3.2.4  V2V

In VANET vehicle-to-vehicle communication is needed because as the vehicles move on the road they need a mediator to transfer the data between the cluster's head to understand this mechanism we have to understand this example, suppose 200 cars are moving on the road and each car which is running on the road is connected to any of the clusters and the capacity of a cluster is 50 cars that in a cluster, 50 cars can communicate with each other and from those 50 cars, a car can be a cluster head. Now, suppose a car in a cluster which is equipped with IoT sensors notices that a car that is a part of the same cluster is changing its lane then the neighboring cars who notice this change will send a communication packet to the cluster head and the communication packet the information would be

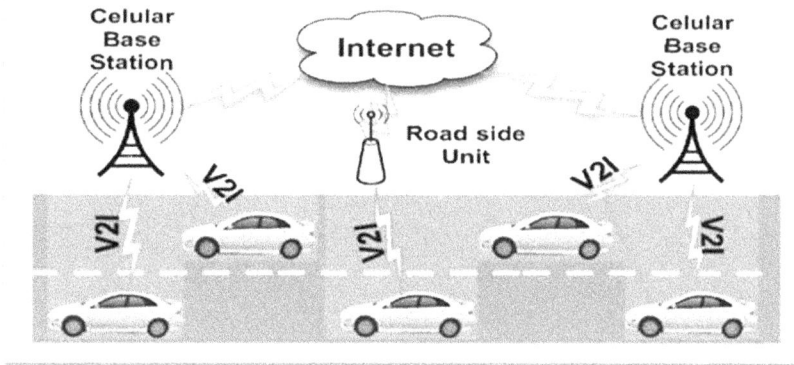

**FIGURE 3.3**   Vehicle-to-Infrastructure (V2I) communication.

*Source*: https://www.researchgate.net/profile/Kifayat-Ullah-3/publication/309546589/figure/fig1/AS:
422889376751618@1477836092182/Vehicle-to-Infrastructure-V2I-communication.png

of the overall surrounding of the cars which are in the cluster. As soon as the cluster head receives the communication packets from the neighboring cars the cluster head will send the communication packets to the cloud computing centralized unit and their the data which are sent by different cluster heads will be analyzed as soon as possible and the result will be sent to the cluster's head and according to the communication packet indexing the cluster's head then send the communication packets to the respective cars in the cluster and according to that if a car is changing its lane then the neighboring cars adjust their position, speed, and lane on the road.

### 3.2.5   FOG COMPUTING

Fog computing [24], a phrase coined by Cisco Systems, is a concept inspired from content distribution networks (CDNs). Fog computing uses an overlay network to distribute content based on end users' geographic locations through localized surrogate servers, which was developed to overcome the shortcomings of the centralized processing architecture. In other words, Fog computing is an improved version of cloud computing in which the communication packets are analyzed at the cluster's head level and the analyzed data are sent back to the neighboring cars according to the scenario. Fog computing has evolved as a system model to handle various data-intensive or delay-sensitive IoT applications [25]. Fog computing places resources closer to the edge network. In this scenario, the data did not have to be sent to the cloud centers because the Fog computing network architecture is decentralized and local. If we say that the Fog computing is local network architecture, then it means that the nodes in the Fog computing network architecture are closer to the cars that are within the cluster. In a scenario where five clusters are working on managing the communication between the cars within those clusters suppose, a cluster can have a capacity of 50 cars then if a car notice some changes in its surrounding scenario then it will send the communication packet to the cluster's head and the cluster head analyze the communication packet and send again to the source of the communication packet and according to that the neighboring cars of the cluster take the decision and change according to the scenario.

### 3.2.6   IMPROVEMENT IN V2V COMMUNICATION USING FOG COMPUTING

In the previous version of the V2V vehicle-to-vehicle communication the data is sent from the cluster head to the central command of the cloud computing center and after analyzing the data, these data are sent to the cluster's head and according to the returned data (communication packet)

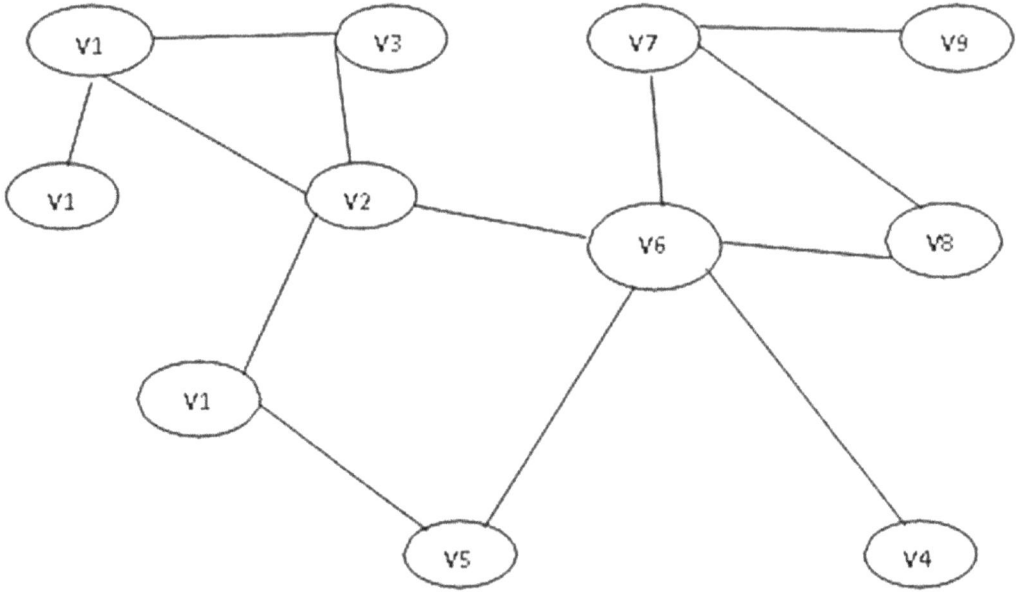

**FIGURE 3.4**  Vehicular Fog Computing.

Source: https://miro.medium.com/max/500/0*SieEmhHmOqnSqZYC.jpg

the cars within the cluster take a decision but, in this environment, an issue has been noticed that is latency (When there is a delay while receiving the data/ communication packet then that is called as latency). The current Internet infrastructure can be thought of as a three-tier distributed system, with a client layer, a cloud layer, and a communication network layer [26]. In a cloud computing-based environment all the communication packets are analyzed at the cloud center and then distributed to the cluster's head which produces a delay in the communication packet received. In the solution to this Fog computing-based networking infrastructure has been introduced. See Figure 3.4.

### 3.2.7 DEDICATED SHORT-RANGE COMMUNICATION (DSRC)

Dedicated Short-Range Communication (DSRC) is a technology that was introduced in 1999 when the United State Federal Communication Commission (FCC) introduced this technology by allocating the 5,725 MHz to 5,875 MHz radio frequency band [27]. DSRC technology is designed to work for short or medium-distance communication. In other words, in mobile vehicle safety systems, DSRC is a two-way short-to-medium range wireless communication that may convey a variety of data [29, 30]. This technology is used for the communication of V2V and V2I. Suppose a scenario of V2I communication, vehicle V1 wants to communicate with other vehicle V2 then, vehicle V1 will send the message to the nearby RSU through DSRC and then, the message from the RSU will be transmitted to the vehicle V2 using DSRC.

In another scenario we will take V2V communication where multiple vehicles are running on the road then, if a vehicle V1 want to communicate to a vehicle V5 then, by using DSRC vehicle V1 will send the message to the vehicle V2 and when vehicle V2 receives the message packet then, the vehicle V2 will analyze the message packet and transmit the message packet toward its destination vehicle that is vehicle V5 in the same way when the message packet will receive by the vehicle V3 and V4 then they will analyze the message packet and will transmit the packet to the vehicle V5 and in this way through DSRC the message packet from its origin that is vehicle V1 will transmit to its destination vehicle V5.

### 3.2.8  EDGE DEVICES

An edge device can be defined as a device that has the feature that can act as a service provider to the core network. Edge devices can configure their services in Metropolitan Area Network (MAN) and Wide Area Network (WAN) suppose, some vehicles are running over a large geographical region so, as per the concept of VANET these vehicles can communicate with each other and these communications become easier when edge devices like multiplexers, routers, router switches and many more are used within that wide geographical network. Edge devices also help in configuring a secure network and in these networks the message can only be transmitted and received by the vehicles when the edge devices authenticate those message packets.

### 3.2.9  ADHOC ROUTING PROTOCOLS FOR VEHICLE COMMUNICATION

The Ad hoc routing protocols for the communication between vehicles running on the road can be categorized in many ways:

a) Based on modifications to the routing information Ad hoc protocols may be divided into three types, which are as follows:
 I. Proactive Routing Protocol
 II. Reactive Routing Protocol
 III. Hybrid Routing Protocol

b) Based on Time information routing protocols can be categorized into two protocols, they are:
 I. History-based Routing Protocol
 II. Future-based Routing Protocol

c) Based on Topological information routing protocol which can be categorized into two protocols, they are:
 I. Flat Routing Protocols
 II. Hierarchical Routing Protocols

d) Based on the utilization of specific resources routing protocol can be categorized into three protocols:
 I. Flooding
 II. Geographical
 III. Power-Aware

### 3.2.10  PROACTIVE ROUTING PROTOCOLS

This routing protocol is based on a routing table that each node which is present in the network will maintain a table and that table will tell the activities of the node within the network. Suppose for example in a proactive based network a number of vehicles are running on a road and they all are communicating with each other through proactive routing protocols, then let's take a scenario where vehicle V1 wants to communicate with vehicle V5 but vehicle V1 cannot communicate with vehicle V5 directly, so vehicle V1 will firstly communicate to vehicle V2, then vehicle V3, vehicle V4, and lastly with vehicle V5; we can understand this scenario by the table. Some of the routing protocols come under this category, they are:

 IV. Dedicated Short-Range Distance Vector (DSDV) Routing Protocol
 V. STAR Routing Protocol
 VI. CGSR Routing Protocol

### 3.2.11   Reactive Routing Protocol

These types of routing protocols are not based on a routing table, they work on the principle of on-demand that is in a reactive routing protocol-based network when a vehicle wants to transmit the message to the destination vehicle then the initial vehicle will send the ROUTE REQUEST (RREQ) message to all the nodes within the network and in response all the vehicles within the network will respond. In other words, we can say that the vehicles find the necessary path to transmit the message it is required to send the message within the network. Some of the routing protocols that come under this category are:

VII.  Dedicated Short-Range Communication Protocol.
VIII.  Ad Hoc On-Demand Vector (AODV) Routing Protocol.

### 3.2.12   Hybrid Routing Protocol

A hybrid routing protocol is the combination of proactive routing protocol and reactive routing protocol that is in a hybrid routing protocol when a vehicle wants to transmit and receive a message then, according to the condition the hybrid routing protocol uses the best feature of both proactive routing protocol and reactive routing protocol within that network.
   Some of the routing protocols come under this category, they are:

IX.  CEDAR Routing Protocol
X.  ZRP Routing Protocol

### 3.2.13   Dynamic Source Routing (DSR) Protocol

The Dynamic Source Routing (DSR) protocol comes under the category of reactive routing protocol which means it works on the concept that when the path within the network is needed then only the path in the network will be discovered by the transmitting vehicle node in the network. So, it can be said that the transmitting vehicle node within the network will be the only node that knows the path of transmitting the data packet to the receiver vehicle node within the network; in this way we can conclude that in the DSR protocol only the source node maintains the routing path and all the intermediate paths do not maintain the information of the path from which the data packet will be transmitted from the sender vehicle node to the receiver vehicle node.
   **There are different phases in the Dynamic Source Routing (DSR) protocol, they are:**

#### 3.2.13.1   Route Discovery
The route discovery phase is the composition of two sub-phases:

1. Route Request (RREQ) packets
   AnRREQ packet is sent by the transmitter vehicle node to the destination vehicle node within the network. RREQ packets are broadcast in nature which means the transmitter sends the RREQ packet to every its neighboring node and from those neighboring node to another neighboring node; in this way the RREQ packet is received by the receiver vehicle node within the network.
2. Route Reply (RREP) packets
   An RREP packet is sent by the receiver vehicle node to the sender vehicle node within the networks. The RREP packet is sent when the route for sending the packet from the sender to the receiver has been discovered. An RREP packet is unicast which means when the route has been discovered then the receiver vehicle node will not broadcast the RREP

packet to its neighboring nodes, the receiver vehicle node sends the RREP packet on the discovered path for communication between the sender and receiver vehicle node within the network.

### 3.2.13.2 Route Maintenance

In the DSR protocol, when a sender vehicle node broadcast the packet to its neighboring vehicles nodes for sending the packet to the destination vehicle node then at some nodes in the network the packet will face route break errors like flooding in which all the vehicles send the packet to their neighboring vehicle node. Due to this at some condition, the nodes in the networks are not capable to perform the computation of the packet transmitting from different nodes within the network and there are many more errors such that the packet cannot reach its destination vehicle node then, to notify the conditions that within the network there is a route link break at a node, a Route Error (RERR) message is generated to all its neighboring nodes within the network and then, according to flooding concept, the sender vehicle node will not select the network path to transmit the packet from sender to receiver vehicle node.

### 3.2.13.3 Route Cache

In the DSR routing protocol each node maintains a piece of route information in its routing table. Each vehicle node within the network keeps track of all RREQ, RREP, and RERRs, and according to that the network will decide the best route.

Let's take a scenario to better understand the working of the DSR protocol:

In this example of DSR protocol, eleven vehicles (V1, V2, V3, V4, V5, V6, V7, V8, V9, V10, and V11) running on a road are forming a DSR protocol in which eleventh vehicle (V11) wants to transmit the data packet to the fourth vehicle (V4) within the network. See Figure 3.5 and Table 3.1.

Suppose the Unique ID (UID) of the Route Request (RREQ) packet that has been transmitted by the vehicle node (V11) is '5'. According to the Dynamic Source Routing (DSR) protocol, the sender vehicle node (V11) will transmit the data packet to its neighboring vehicle nodes (V1, V3, and V2).

**FIGURE 3.5**  Dynamic Source Routing (DSR) Protocol Scenario.

**TABLE 3.1**

**Route Request (RREQ) Packet transmit from the sender vehicle node**

| Nodes | V1 | V2 | V3 | V4 | V5 | V6 | V7 | V8 | V9 | V10 | V11 |
|---|---|---|---|---|---|---|---|---|---|---|---|
| V1 | | | | | | | | | | | [5, V11, V4, V11] Discard |
| V2 | | | [5, V11, V4, V2] | | | [5, V11, V4, V2] | | | | [5, V11, V4, V2] | [5, V11, V4, V2] Discard |
| V3 | | [5, V11, V4, V3] Discard | | | | | [5, V11, V4, V3] | | | | [5, V11, V4, V3] Discard |
| V4 | | | | | | | | | | | |
| V5 | | | | | | [5, V11, V4, V5] Discard | | | | [5, V11, V4, V5] Discard | |
| V6 | | [5, V11, V4, V6] Discard | | [5, V11, V4, V6] | [5, V11, V4, V6] | | [5, V11, V4, V6] | | | | |
| V7 | | | [5, V11, V4, V7] Discard | | | [5, V11, V4, V7] Discard | | [5, V11, V4, V7] | [5, V11, V4, V7] | | |
| V8 | | | | | | [5, V11, V4, V8] Discard | [5, V11, V4, V8] Discard | | | | |
| V9 | | | | | [5, V11, V10] | | [5, V11, V4, V9] Discard | | | | |
| V10 | | [5, V11, V10] Discard | | | Discard | | | | | | |
| V11 | [5, V11, V4] | [5, V11, V4] | | - | - | - | - | - | - | | |

The data packets within the DSR protocol network will contain:

a) Unique ID (UID): In the DSR protocol, UID is used to check the duplicity of the data packet within the network.
b) Sender ID (SID): In the DSR protocol, SID is used to represent the sender ID who wants to transmit the data packet within the network.
c) Destination ID (DID): In the DSR protocol, DID is used to represent the destination ID that wants to receive the data packet from the sender vehicle node.
d) Route Cache: In the DSR protocol, Route Cache maintains different routes to reach the destination vehicle node from the sender vehicle node in the format of [Unique ID (UID), Source ID (SID), Destination ID (DID), Route Cache].

The data packet from the sender vehicle node (V11) will be broadcast to every neighboring vehicle node (V1, V3, andV2) of the vehicle node (V11). Then, the data packet will be received by the neighboring vehicle nodes in the format:

- Vehicle Node (V1) = [5,V11,V4]
- Vehicle Node (V2) = [5,V11,V4]
- Vehicle Node (V3) = [5,V11,V4]
- Vehicle Nodes (V1, V2, and V3) will again transmit the data packet to their neighboring vehicle nodes within the network that is:
  - Vehicle node (V1) will transmit the data packet to its neighboring vehicle nodes (V11) but, vehicle node V11 will check that this data packet has the same UID as transmitted by the vehicle node V11 so vehicle node V11 will discard the data packet from vehicle node V1 as shown in table 6.1.
  - Vehicle node (V2) will transmit the data packet to its neighboring vehicle nodes (V3, V6, V10, and V11).
- When the vehicle node (V3) receives the data packet from its neighboring vehicle nodes (V2) then, it will discard the data packet as it has received the data packet with the same UID from the vehicle node V11.
- When vehicle node V6 will receive the data packet from its neighboring vehicle node V2 in the data packet format [5, V11, V4, V2] then vehicle node V6 will receive the data packet and then, transmit the data packet with the same UID to its neighboring vehicle nodes (V5, V7, V8, and V4) and does not discard the data packet.
- When vehicle node V10 will receive the data packet from its neighboring vehicle node V2 then it will transmit the data packet with same UID to its neighboring nodes (V2 and V5) in the data packet format [5, V11, V4, V10] but, this packet will be received by the vehicle node V2 then, V2 will discard the data packet because of the same UID and in the case when data packet from vehicle node V2 will be received by vehicle node V5 then, it will receive the data packet in the format [5, V11, V4, V10] and transmit the data packet to its neighboring nodes (V6 and V10) but, vehicle node V6 and V10both will discard the data packet because of the same UID in the data packet.
- Now, when vehicle node V2 has transmitted the data packet with UID '5' to vehicle node V6 then V6 will transmit the data packet to its neighboring nodes (V5, V7, V8, and V4) in data packet format [5, V11, V4, V6] but, as the vehicle node V5 will receive the data packet, it will discard the data packet because V5 has already received the data packet from V10 and in the same way in the case of V7 when it receives the data packet from V2 it will discard it. After all, V7 has already received the data packet from the vehicle node V3 with the same UID '5' but, when the data packet is transmitted from vehicle node V6 to the vehicle nodes (V8 and V4) then they will receive the data packet and transmit to their neighboring vehicle nodes (V6 and V7) and (V6) respectively but vehicle node V6 and V7 will discard

the data packet because of the same UID that they have already received from V2 and V3 respectively.

- In this way, we have got our first Dynamic Source Routing (DSR) protocol-based network route from the sender vehicle node (V11) to the receiver vehicle node (V4) that is from (from V11 to V2 to V6 to V8 and from V8 to V4).
- In this network scenario of Dynamic Source Routing (DSR) protocol there would be the following network path that will generate a link between sender vehicle node V11 and receiver vehicle node V4, they are:
  o From sender vehicle node V11 to V3 to V7 to V6 and from vehicle node V6 to receiver vehicle node V4.
  o From sender vehicle node V11 to V2 to V10 to V5 to V6 and from vehicle node V6 to the receiver vehicle node V4.
  o From sender vehicle node V11 to V2 to V6 and from vehicle node V6 to receiver vehicle node V4.
  o From sender vehicle node V11 to V3 to V7 to V6 and from vehicle node V6 to receiver vehicle node V4.

So, as we have seen in this scenario of a DSR protocol-based network there would be four best network route possibilities to reach from source vehicle node V11 to destination vehicle node V4 but, out of these network routes the route that is from sender vehicle node V11 to V2 to V6 and from vehicle node V6 to receiver vehicle node V8 is the best network route due to the following reason:

- Less hop count (Number of vehicles) on the specific network route as compared to other network routes in the scenario of the DSR protocol.
- Due to less hop count, the latency on this route will be less that the time to reach from sender vehicle node V11 to the receiver vehicle node V4 will be less as compared to other network routes.
- In the Dynamic Source Routing (DSR) protocol-based network scenario, energy consumption is a big issue but, the DSR protocol provides a basic or we can say DSR protocol provides a platform in which by assuming some of the new energy matrices and by modifying the DSR protocol-based network structure of controlling the packet we can reduce the energy consumption [30].
- Now, as the data packet has reached from the sender vehicle node V11 to the receiver vehicle node V4 a specific path has been selected with a minimum distance that is V11 to V2 to V6 and from vehicle node, V6 to receiver vehicle node V4 and in response of the data packet, vehicle node V4 will send a Route Reply (RREP) to vehicle node V11 from the chosen minimum distance path.

### 3.2.14 ADHOC ON-DEMAND VECTOR(AODV) ROUTING PROTOCOL

Ad Hoc On-Demand Vector (AODV) routing protocol is the updated version of the DSR protocol as we look at the disadvantages of the DSR protocol, the network route to reach from source vehicle node to destination vehicle node will be included the data of network route of DSR protocol into the data packet but, this concept is efficient until the network based on DSR protocol is small but, if the range of DSR protocol-based network is large the bandwidth will only be utilized to store the data packet from source vehicle node to destination vehicle node but, this problem has been resolved in the AODV Routing Protocol by adding a feature to the vehicle nodes present within the AODV routing protocol network that is each vehicle nodes that are present in the AODV routing protocol-based scenario will only maintain its previous hop count and next hop count.

Suppose the previous scenario in the DSR routing protocol here, we will take same example in the AODV routing protocol format where, a source vehicle node V11 wants to send an RREQ packet to the destination vehicle node V4 then, Vehicle node V11 will only maintain its previous hop count (Number of vehicles) and next hop count but, if we put the same scenario in the DSR routing protocol then, the sender vehicle node V11 have to maintain the whole network route for transmit the RREQ of data packet to the destination vehicle node V4.

### 3.2.15 IN AODV ROUTING PROTOCOL-BASED SCENARIO THERE ARE TWO PHRASES

➢ **RouteDiscovery**
  ✔ Route Request (RREQ)
     In the AODV routing protocol-based scenario the RREQ format for the AODV routing protocol-based network is similar to the format of RREQ of the DSR routing protocol-based network but, with some differences.
         The Route Request of the AODV routing protocol includes:
     i. Source ID
     ii. Destination ID
     iii. Recent Sequence number
     iv. Broadcast ID
     v. Hopcount
     vi. Time To Leave (TTL)
  ✔ Route Reply (RREP)
     In the AODV routing protocol-based scenario when the RREQ data packet has reached to the destination vehicle node, then the destination vehicle node generates an RREP packet for the best AODV routing protocol-based network route by which the RREP data packet will be received by the sender vehicle node in the network.
         In the AODV routing protocol-based network scenario the RREP data packet includes:
     i. Source ID
     ii. Destination ID
     iii. Destination sequence number
     iv. Hop count
     v. Life time: In the RREP of the AODV routing protocol life time will represent the time duration of RREP data packet means it tells that how much time this packet will be active.
  ✔ Route maintenance
     In AODV routing protocol-based scenario route maintenance is similar to the route maintenance format of the DSR routing protocol-based scenario.

### 3.2.16 HIERARCHICAL ROUTING PROTOCOLS

Hierarchical routing protocol is a routing protocol in which different groups of vehicle nodes combine to form a network in which these vehicle nodes communicate with each other are managed at different levels or hierarchy.

Suppose a scenario in which multiple vehicles are running on a road. As per the assumption of the hierarchical routing protocol in which a network of ten vehicles acts as a node in the network, in this way 50 vehicles are running on road and so five networks based on hierarchical routing protocol are formed.

So, now we will discuss one of the hierarchical routing protocols – the Hierarchical State Routing protocol – in more detail.

### 3.2.17 HIERARCHICAL STATE ROUTING (HSR) PROTOCOL

In an HSR protocol-based network scenario each vehicle node which wants to communicate with other vehicle nodes within the network can form a cluster that means if some vehicle nodes within the HSR protocol-based network want to transmit the data packets which have similar features then they can form a group of those vehicle nodes. So, we can say that HSR protocol-based networks form a cluster-based multi-layer network environment and these HSR protocol-based network scenarios are distributed in nature and due to the clustering nature of the HSR protocol some features are added, which are:

- In the HSR protocol-based network scenario, the resource allocation within the network increases.
- In the HSR protocol-based network scenario, after the resource allocation within the network, resource management is also performed.
- As we know that the HSR protocol has a cluster nature, then as per the concept of cluster, each cluster has a cluster head which means if some vehicle nodes with similar features are forming a network, then from those vehicle nodes a vehicle node will be selected as a cluster head through which all other vehicle nodes within that cluster will transmit the data packet to each other.
- In the HSR protocol-based scenario the nodes within the network are divided into three categories, they are:
  o Cluster head – These are those nodes which are used to build a communication channel between two or more clusters through there cluster heads nodes, these cluster heads have the following operations:
    i. If an internal node within the cluster wants to transmit data to another internal node, then the cluster head analyzes the data packet and finds the best route to send the data packet to the receiver internal node in the network.
    ii. If an internal node1 of cluster1 wants to transmit the data packet to an internal node2 which is a part of cluster2 then internal node1 will transmit the data packet firstly to the cluster head1 of cluster1 and then the cluster head1 will transmit the data packet to the cluster head2 of cluster2 and then cluster head2 will transmit the data packet to the internal node2 within the network.
  o Internal nodes – These nodes are the nodes which are defined within a cluster and cannot be categorized as cluster heads or gateways nodes.
  o Gateway nodes – These nodes are those which belong to two or more clusters, that is they are present in two or more clusters.

In the HSR protocol as we have discussed, the HSR protocol is multi-leveled, so in the HSR protocol there are two levels:

- Initial level
- Updated level

Now, let's understand these two levels by assuming a network scenario based on the HSR protocol.
Suppose eight vehicles (V1, V2, V3, V4, V5, V6, V7, and V8) running on a road want to communicate with each other through the HSR protocol and these vehicles, when they transmit the data packet on the basis of the HSR protocol, form two clusters which will come under the initial level, that are (see Figure 3.6):

- Initial level – This level includes two clusters that are:
- Cluster1 – This cluster includes vehicle nodes V1, V2, V3, and V4.
- Cluster2 – This cluster includes vehicle nodes V5, V6, V7, and V8.
- Cluster3 – This cluster includes vehicle nodes V9, V10, and V11.

In this scenario of the HSR protocol, there are eleven vehicles which are running on the road and these vehicles are categorized into three clusters so that each vehicle node within the cluster can communicate with each other.

Updatelevel3:                                    V6

Updatelevel2:              V11                    V6

Updatelevel1:         V11              V2              V6

InitialLevel:

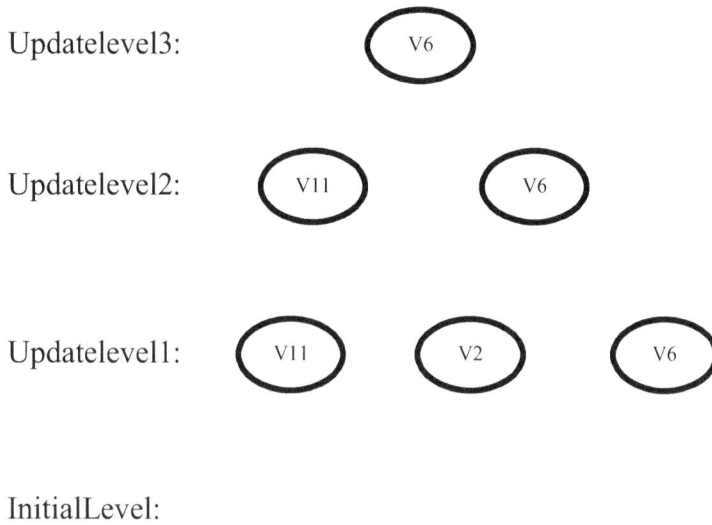

**FIGURE 3.6**   Hierarchical State Routing (HSR) protocol scenario.

In this scenario, the nodes can be categorized into cluster heads, internal nodes, and gateways as:

- Cluster head – In this scenario cluster heads are V11, V6, and V6.
- Gateway nodes – Node V2 will act as a gateway node because it is communicating with node V6 which is a node of cluster2 and V2 is also communicating with node V10 which is a node of cluster3.
- Internal nodes – All other nodes which do not come under the category of cluster heads and gateway nodes are defined as internal nodes (V1, V3, V4, V5, V7, V8, V9, and V10).

At the initial level the three clusters are defined and according to the HSR protocol, the main functionalities that are performed at the initial level is to perform communication between those vehicle nodes which have a single hop count, for example, if vehicle node V11 wants to transmit the data packet to the vehicle node V4 then, firstly, V11 will transmit the data packet to the vehicle node V2 and hop count between V11 and V2 is one, and then, vehicle node V2 will transmit the data packet to V6 and hop count between V2 and V6 is also one and in the same way, V6 will transmit the data packet to the vehicle node V4 and hop count between V6 and V4 is also one so, in this way we can see the working of the initial level of the HSR protocol-based scenario.

Now, as the initial level of the HSR protocol is cluster-based, so a cluster head is also selected; these cluster heads are selected on the following parameters:

- Based on latency from each node that is within a cluster the time to transmit the data packet from the cluster head to every node within that cluster should be low.
- Based on the energy consumption of a vehicle node while transmitting the data packet within HSR protocol-based network.

Now, as for the updated levels in the scenario based on HSR protocol when, the cluster heads are selected in the cluster then, these cluster head nodes (V11, V2, V6) have to do transmission of the data packet from their cluster to another cluster head of another cluster.

Energy consumption of the cluster head in HSR protocol-based cluster can be calculated, for that, we take an example from the journal "Energy Consumption in Wireless Sensor Network" [31], to transmit the data packet from one node to another node within the network there are several IEEE standards like IEEE 802.11, IEEE 802.11b and many more. In this example, the author has taken an IEEE standard that is IEEE 802.11b to transmit the data packet from one node to another node. Now, if we talk about IEEE 802.11b:

a) IEEE802.11b was released in September 1999.
b) IEEE802.11b is used for High Rate Direct Sequence Spread Spectrum (HR-DSSS) modulation.
c) IEEE802.11b is also used for Complementary Code Keying (CCK) modulation.
d) The data rate to transmit the data packet from the sender vehicle node to the receiver vehicle node is 11 Mbps.
e) IEEE802.11b operates at 2.4 GHz bandwidth.

So, as we come to the example IEEE 802.11b interface is used to operate the data packets using the Distributed Coordination Function (DCF) with a reservation-based scheme that is a Ready To Send (RTS)/Clear To Send (CTS) handshake.

In the reservation-based scheme, there are two types of operations – RTS and CTS:

- Ready To Send (RTS): In a network when the sender wants to transmit the data packet to the receiver node then before sending the data packet the sender node transmits an RTS to the receiver node to check whether the receiver node is in ideal condition or not.
- Clear To Send (CTS): In a network when the sender node transmits the RTS data packet to the receiver node then in response the receiver node will transmit the CTS to the same sender node that the receiver node is in ideal condition and the sender node can send the data packet to the receiver node in the network.

If we take an example of the RTS/CTS handshake model then,

Step 1:

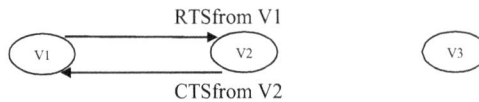

Step 2:

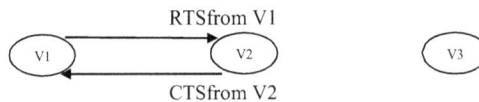

Step 3:  If node V3 want to transmit a data packet to the node V2 in the network then,

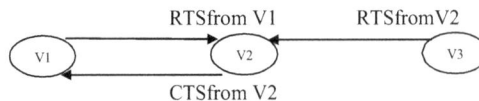

Step 4:  In response to RTS from node V3 the node V2 will transmit CTS to that V2 is not in ideal condition so, transition of data packet from sender node V2 to receiver node V3 cannot be done.

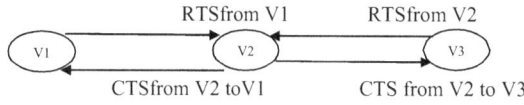

In this example the average power (*Pm*) consumption by using the interface IEEE802.11b is:

$$P_m = t_{sl}{}^* P_{sl} + t_{Id}{}^* P_{Id} + t_{Rx}{}^* P_{Rx} + t_{Tx}{}^* P_{Tx}$$

where *tsl, tId, tRx, and tTx* are the spent time in fraction by the IEEE802.11b interface in the different possible sleep, Idle, Receive, and Transmit, respectively.

Now in this example, there are two types of energy consumption of the node are considered:

- Initial energy (E) of the node within the Wireless Sensor Network (WSN).
- Power consumption (*Pm*) of the four states:
  $P_{sl}$ = Power consumption at sleep state;
  $P_{Id}$ = Power consumption at Idle state;
  $P_{Rx}$ = Power consumption at Receive state;
  $P_{Tx}$ = Power consumption at Transmit state.

$$T_v = E / P_m$$

So, by calculating average power consumption ($P_m$) and initial energy (E) of the node, we can calculate the node lifetime ($T_v$) which represents the time before the energy of the node reaches zero.

## 3.3  CONCLUSION

Vehicle communication is a growing field in which many types of research are going on, such as vehicle-to-infrastructure and vehicle-to-vehicle communication, with the help of cloud computing architecture and its advanced version fog computing architecture. By using these architectures for the communication of vehicles running on a road we can analyze and improve the latency, energy consumption, and throughput of the vehicle's communicating network.

### 3.3.1  FUTURE SCOPE

The future of vehicle communication can be the integration of VANET with some of the emerging technologies like the 5G network, Named Data Networking (NDN), and Image processing which can play an important role in improving vehicle-to-vehicle and vehicle-to-infrastructure communication. In the network where vehicles are running on the road, IoT-based 5G network integration can play an important role in improving the communication features between vehicles by using some 5G network features like millimeter WAVE (mm WAVE) which can be used in short-range communication [24]; many more aspects are being discovered in the field of VANET. In the same way, image processing with VANET can detect vehicles and objects running on the road and can reduce the probability of accidents between them. If VANET is integrated with NDN it is not based on IP address so if any packet has been lost within the VANET then the communication packet's path can be back tracked to reach to the node within the VANET from which the packet has been lost.

## REFERENCES

1 Mckinsey&Company, "Mobility of the future," 2013. Available: http://www.mckinsey.com/client_service/automotive_and_assembly/latest_thinking

2 J. Barbaresso, G. Cordahi, D. Garcia, C. Hill, A. Jendzejec, and K. Wright, "USDOT's IntelligentTransport System (ITS) ITS strategic plan 2015–2019," Report: FHWA-JPO-14-145, US Department of Transportation Intelligent Transportation System, Joint program office. Available: https://www.its.dot.gov/startegicplan/index.html

3 Mckinsey & Company, "The road to 2020 and beyond-what is driving the global automotive industry," 2013 Available: http://www.mckinsey.com/client_service/automotive_and_assembly/latest_thinking

4 Gartner, "Internet of things units installed base by category," 2014. Available: http://www.gartner.com/newsroom/id/2905717

5 M. Saini, A. Alelaiwi, and A.E. Saddik, "How close are we to realizing a pragmatic VANET solution: A meta-survey," *ACM Computing Surveys*, vol. 48, no. 2, pp. 29–65, 2015.

6 S.F. Hasan, X. Ding, N.H. Siddhique, and S. Chakraborty, "Measuring disruption in vehicular communications," *IEEE Transactionon Vehicular Technology*, vol. 60, no. 1, pp. 148–159, 2011.

7 B. Aslam, P. Wang, and C.C. Zou, "Extention of internet access to VANET via satellite receive-onlyterminals,"*International Journal of Ad Hoc and Ubiquitous Computing*, vol. 14, no.3, pp. 172–190, 2013.

8 K. Hill, "What is DSRC for the connected car?," *RCR Wireless News*, 2015. https://www.rcrwireless.com/20151020/featured/what-is-dsrc-for-the-connected-car-tag6

9 Mckinsey & Company, "Unlocking the potential of the Internet of Things," 2015. Available: http://mckinsey.com/insights/business_technology/the_internet_of_things_the_value_of_digitization_the_physical_world

10 The White House US, "National strategies for trusted identities in cyberspace (NSTIC): Enhancing online choice, efficiency, security, and privacy," 2011. Available: https://www.whitehouse.com/sites/default/files/rss_viewer/NSTICstrategy_041511.pdf

11 Transport Department, GOVT of NCT of Delhi, "Installation of GPS in buses and autos," Order No: Odr (2010)/75/8, 2010.

12 S.A. Ahmed and S.H.N. Ariffin Fisal, "Overview of wirless access in vehicular environment (WAVE) protocols and standards," *Indian Journal of Science and Technology*, vol. 6, no. 7, pp. 4994–5001, 2013.

13 European Commission, "Digital signal market strategy," 2015. Available: http://ec.europa.eu/priorities/digital-signal-market/

14 SDxCentral, *Understanding the SDN Architecture-SDN Control Plane& SDN Data Plane*, SDxCentral, 2018. Available: https://www.sdxcentral.com/networking/sdn/definitions/inside-sdn-architecture/

15 L. Zhang, X. Men, K.-K.R. Choo, Y. Zhang, and F. Dai, "Privacy-preserving cloud establishment and data dissemination scheme for vehicular cloud," *IEEE Transactions on Dependable and Secure Computing*, vol. 17, no. 3, pp. 634–647, 2018.

16 European Environment Agency, "Assessment of Global Megatrends-States and Outlook," Report, 2015.

17 Available: http://www.eea.europa.eu/soer/europe-and-the-world/megatrends

18 O. Kaiwartya et al., "Internet of vehicles: Motivation, layered architecture, network model, challenges, and future aspects," *IEEE Access*, vol. 4, pp. 5356–5373, 2016, doi: 10.1109/ACCESS.2016.2603219.

19 S. Al-Sultan, M.M. Al-Doori, A.H. Al-Bayatti, and H. Zedan, "A comprehensive survey on vehicular ad-hoc network," *Journal of Network and Computer Applications*, vol. 37, no. 1, pp. 380–392, 2014.

20 Govt. of UK Govt, "The Internet of Things: Making the most second digital revolution," Report: 2014. Available: http://www.gov.uk/goverment/uploads/system/uploads/attachment_data/file/409774/14-1230-internet-of-things-review.pdf

21 A. Akhunzada and M.K. Khan,"Toward secure software defined vehicular networks: Taxonomy, requirements, and open issues," *IEEE Communications Magazine*, vol. 55, no. 7, pp. 110–118, Jul. 2017.

22 B.A.A. Nunes, M. Mendonca, X.-N. Nguyen, K. Obraczka, and T. Turletti, "A survey of software-defined networking: Past, present, and future of programmable networks," *IEEE Communications Surveys and Tutorials*, vol. 16, no. 3, pp. 1617–1634, 2014.

23 University of Michigan, "Public opinion about self-driving vehicles in China, India, Japan, the US, the UK, and Australia," Transportation Research Institute (UMTRI), pp. 122, 2014. Available: http://deepblue.lib.umich.edu/bitstream/handle/2027.42/106590/102996.pdf?sequence=1&isAllowed=y Google, "Open Automobile Alliance," 2015. Available: http://www.openautoalliance.net/. [20]. Apple, "Car Play," 2014. Available: http://www.apple.com/ios/carplay/

24  Shashank Gaur, "Millimeter WAVE (mm WAVE) in VANET," 2021. Available: https://shashankgaur123. blogspot.com/2021/07/mmwave-in-vanet.html

25  N. Giang, V. Leung, and R. Lea, "On developing smart transportation applications in fog computing para-digm," in *Proceedings of the 6th ACM Symposium on Development and Analysis of Intelligent Vehicular Networks and Applications*, 2016.

26  H. Atlam, R. Walters, and G. Wills, "Fog computing and the internet of things: A review," *Big Dataand Cognitive Computing*, vol. 2, no. 2, 2018, p. 10.

27  J. Lin, W. Yu, N. Zhang, X. Yang, H. Zhang, and W. Zhoo, "A survey on internet of things: Architecture, enabling technologies, security and privacy, and applications," *IEEE Internet of Things Journal*, vol. 4, no. 5, pp. 1–1, 2017.

28  A.B. Nkoro and Y.A. Vershinin, "Current and future trends in applications of intelligent transport system on cars and intelligent infrastructure," in *IEEE 17th International Conference on IntelligentTransportation Systems(ITSC-2014)*, Qingdao, China, 8–11 October 2014.

29  Y.A. Vershininand and Y. Zhan, "Vehicle to vehicle communication: Dedicated short range communica-tion and safety awareness," in *2020 Systems of Signals Generating and Processing in the Field of on Board Communications*, 2020, pp. 1–6, doi: 10.1109/IEEECONF48371.2020.9078660.

30  B. Baisakh, "A review of energy efficient dynamic source routing protocol for mobile ad hoc networks," *International Journal of Computer Applications*, vol.68, no.20, pp. 06–15, 2013.

31  T.O. John, H.C. Ukwuoma, S. Danjuma, and M. Ibrahim, "Energy consumption in wireless sensor net-work," *Energy*, vol. 7, no. 8, pp. 63–67, 2016.

# 4 Towards Resolving Privacy and Security Issues in IoT-Based Cloud Computing Platforms for Smart City Applications

*Soma Saha*
VIT Bhopal University, Bhopal, India

*Bodhisatwa Mazumdar*
IIT Indore, India

## CONTENTS

4.1 Introduction .........................................................................................................54
4.2 Smart City Overview: Cloud-Based System Architecture, Security
and Privacy Issues, and Applications .................................................................56
    4.2.1 General Architecture of a Smart System .................................................56
    4.2.2 General Cloud-based Architecture of a Smart System ............................57
        4.2.2.1 Cloud Computing Deployment Models .................................58
        4.2.2.2 Cloud Computing Service Delivery Models .........................58
    4.2.3 Privacy and Security Issues in Cloud Computing Infrastructure ...........59
4.3 Privacy and Security Issues in Smart City Infrastructure ..................................59
4.4 Data Confidentiality and Security of Smartphone Devices and Services ...........60
4.5 Power Grid Systems within Smart Cities ...........................................................63
4.6 Smart Healthcare Systems within Smart Cities ..................................................64
4.7 Privacy and Security in Smart Transportation ...................................................65
4.8 Smart Environment ............................................................................................66
4.9 Security in Smart IoT Devices ...........................................................................66
4.10 Security and Privacy Frameworks .....................................................................67
4.11 Cryptography .....................................................................................................67
    4.11.1 Homomorphic Encryption .....................................................................67
    4.11.2 Zero-Knowledge Proofs ........................................................................68
    4.11.3 Secret Sharing ......................................................................................68
    4.11.4 Secure Multi-party Computation ...........................................................68
    4.11.5 Multi-Factor Authentication ..................................................................69
    4.11.6 Blockchain .............................................................................................70
    4.11.7 Side-Channel Attacks ...........................................................................71
4.12 Biometric ...........................................................................................................72
4.13 AI-ML Models for Detecting Security and Privacy Issues .................................73
4.14 Anonymity .........................................................................................................73
4.15 Security Issues, Properties, and Countermeasures Against Threats ...................74
4.16 Sustainable Security and Privacy Approaches ...................................................76

DOI: 10.1201/9781003319238-4

4.17  Conclusion ....................................................................................................76
4.18  Future Scope .................................................................................................77
References ...............................................................................................................77

## 4.1  INTRODUCTION

Smart cities have emerged radically with the advent of disruptive technologies over the past two decades. The necessity of smart cities is evident with the recent advancement in communication technologies, mobile computing, wireless sensing, and consumer electronics. This advancement along with the ever-growing world population has been more concerning in cities wherein one can find possessing multiple electronic devices and gadgets per person connected to the Internet [1, 2]. Therefore, a city's infrastructure in the present era implicitly embeds billions of devices and systems, producing excessive burden on the environment and living conditions.

With an aim to alleviate such challenges, upgrading the living conditions of citizens, generating economic developments, and governing present-day cities in a comfortable and intuitive way, a perpetually increasing count of the world's urban settlements resulted in their own smart methodologies in various technologies and products. The term "smart environment" encompasses a varied spectrum of domains, such as embedding and managing endpoint systems to public utility devices for smart lighting, managing road traffic and easy pedestrian movement, monitoring gas and water supply, managing waste collection and disposal, precision agriculture, smart transportation, smart governments, smart healthcare, smart buildings, and smart homes.

Implementing a smart system across a city involves deploying a large number of uniquely identifiable smart sensors and personal devices [3]. These sensors are connected through a wireless sensor network for data acquisition, processing, dissemination, and transmission. Smart cities generate large volumes of data at different rates from these sensors and individual objects or devices from various services and applications, which may pertain to varied social media and associated digital platforms. The requirement for storing and interpreting large volumes of data ushers in the emergence of cloud as an active platform [4].

Moreover, smart city-based individual devices are connected through multiple features of other emerging technologies comprising the Internet of Things (IoT) and cyberphysical systems. Therefore, the requirement for a unified framework for managing data generated from a growing number of devices in everyday life is critical for a smart city and all its applications [2]. The aforementioned technologies have adopted artificial intelligence (AI) techniques due to their proven abilities in processing large volumes of data intelligently while deriving learning rules. These learning rules may be generated through machine learning (ML) models, building complex unions, and the resulting predictions of aggregated physical procedures [5]. As data rates across applications are evolving, the already deployed AI and ML-induced algorithms need to be improved continually for extracting intelligent information, secured data storage, and data transmission.

The presence of different types of end users, such as government entities, citizens, and industrialcollaborators, creates different sets of requirements for smart city applications and services for maintaining the quality of service (QoS). Government agencies or public healthcare polyclinics need secure connections to cloud servers for storing data that comprise patient critical information or patient healthcare information. Citizens who adopt smart home facility may need high-speed Internet connectivity for accessing applications with high-definition streaming of multimedia files. Therefore, data management determines data distribution according to different group of end users of a smart city application. AI and ML-based techniques are constantly evolving to prioritize end-user requirements that are specific to data distribution [6]. However, security and privacy of personal and non-shareable data is one of the primary service requirements for all end users (Figure 4.1).

The deployment of smart applications may lead to wide-ranging threats to confidentiality and security aspects due to the commonly existing vulnerabilities in each layer or cross-layer of smart city systems. Figure 4.1 shows a general structure of different layers in a smart system. A variety of

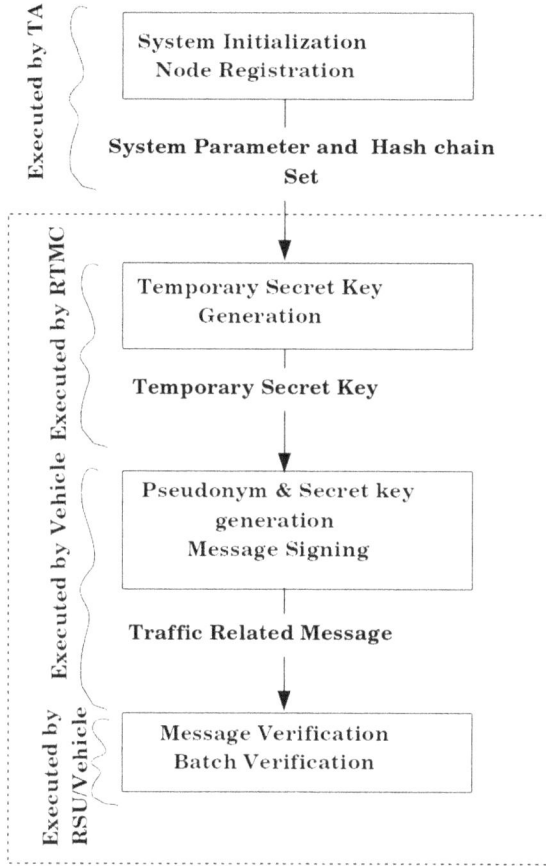

**FIGURE 4.1**    IoT-based General Architecture of a Smart System.

active as well as passive attacks, such as unauthorized access, eavesdropping, traffic analysis attack, spoofing, password pilfering, man-in-the middle, tampering, flooding, data modification, time synchronization, denial of service (DoS) or distributed denial of service (DDoS), can affect the quality of intelligent applications [2, 7–9]. For example, in 2015 hackers attacked the power grid system that affected nearly 230,000 Ukrainian citizens inducing a long period of electricity disconnection [8]. In October 2016, attackers exploited the security loopholes in IoT devices to paralyze the entire city by mounting a DDoS attack on Dyn [2]. In June 2020, a group of researchers demonstrated that Amazon's Alexa, the virtual assistant whose count can cross over 15 billion by 2025, has vulnerabilities, such as installing applications in a user's Alexa profile without the user's knowledge, extracting an information list of all applications existing in the user's Alexa account, removing an existing application without informing the user, capturing the victim's voice history from the Alexa device, thereby obtaining the victim's personal credentials without its knowledge. These vulnerabilities can be used by hackers to hack a victim's smart home system [10]. Amazon has subsequently fixed these issues after the researchers notified the company about the vulnerabilities.

The wave of AI and ML-induced real-life applications in almost every aspect of urban lifestyle has encouraged the security researchers for applying AI and ML-based techniques to prevent and detect security vulnerabilities. However, a similar wave has also encouraged intelligent minds from the attackers' perspective to induce cyberthreats enhanced by AI, such as fuzzing and ML poisoning to break the security safeguards of a smart system. In summary, AI and ML-based techniques may act as a boon for processing and extracting relevant information from large volumes of data and transferring and storing data in encrypted format, thus providing privacy, and detecting and resolving cyberthreats in the cloud. In addition, AI and ML-based techniques may assist unethical hackers to intelligently analyze the data on cloud servers. However, such techniques can also be used by malicious cloud-based services to work remotely and even breach the security in seemingly secured smart systems. Therefore, AI and ML-based techniques, while improving the everyday life of an urban smart system, may introduce privacy and security threats. Along with smart system implementation, continuous monitoring and upgrades of software, communication systems, storage and data transfer techniques, finding faults/bugs in hardware are the only way to restrict intelligent malicious minds to retain a secure smart system.

The major contribution of this chapter are as follows:

(i) This chapter presents the privacy and security issues that exist in present-day smart city infrastructures, including cloud computing-based applications.
(ii) The chapter also illustrates data confidentiality and security aspects of smartphone devices and the applications that they host.
(iii) The chapter exemplifies the security and privacy issues in protecting electronic healthcare records of patients stored on cloud platforms in smart healthcare systems.
(iv) The chapter also demonstrates different security and privacy frameworks, such as cryptography, homomorphic encryption, zero-knowledge proofs, secret-sharing, secure multi-party computation, blockchain, and multi-factor authentication, which can have their own application-specific implementation for cloud computing systems.

The organization of the chapter is as follows. We analyzed the cloud-based smart cityarchitecture, security, and privacy issues in cloud computing infrastructure, as well as in smart city architecture and smart city applications. Thereafter, we described state-of-the-art security and privacy frameworks. Later, we pointed out security issues, properties, and countermeasures against threats. Subsequently, sustainable security and privacy approaches are analyzed. We conclude with the direct future scope towards resolving privacy and security issues in IoT-based cloud computing platforms for smart city applications.

## 4.2   SMART CITY OVERVIEW: CLOUD-BASED SYSTEM ARCHITECTURE, SECURITY AND PRIVACY ISSUES, AND APPLICATIONS

Most protection methods developed for smart city security challenges and resulting requirements depend on customized scenarios of the embedded smart applications. There exists a variety of architectures suitable to specific smart systems. Therefore, a general structure of unified framework of a smart system that includes the features, design, common applications, and constraints of smart cities provides the readers an overview of basics of many smart systems. In the next three sections, we depict a general structure of a smart system along with its generic cloud-based architecture, and security threats in cloud infrastructure. Next, we explore various cloud-based applications and related constraints associated with these applications.

### 4.2.1   GENERAL ARCHITECTURE OF A SMART SYSTEM

A smart city comprises wide-ranging smart applications, devices, communication protocols, data storage, management and processing techniques, end users and their specific requirements. In addition, numerous privacy and security issues exists for these parameters. Therefore, it is difficult to propose and implement a uniform architecture of a smart system that can be strictly followed in all kinds of smart applications and achieve all privacy and security requirements [2, 8, 11–15]. However, IoT imparts a key function in the smart city as it holds the networking system accountable for collecting and computing data from a large number of distributed sensors, actuators, and RFID tags, i.e., from multiple heterogeneous sources [14]. Depending on the installation of multiple IoT endpoints in a smart system, the system architecture may be based on service-oriented architecture (SOA), cloud and fog-based architecture, social IoT (SIoT) architecture [16]. Habibzadeh et al. [11] have divided the smart city system architecture into five planes: security, data, communication, sensing, and application planes. These five planes can be categorized into different layers from the security and confidentiality aspects in smart cities as shown in Figure 4.1 [8].

The *perception layer*, also known as sensing or data acquisition layer, collects data through a set of sensors and actuators to measure real-world signals and interact with heterogeneous devices in everyday urban life. Moreover, it is also involved in transferring the obtained data to the *network layer*, which is primarily known as the core of an IoT architecture or a smart system and preprocesses and aggregates the acquired data received from perception layer for other layers. In addition, the network layer connects network devices, smart things, and servers interfaced with wireless sensor networks (WSNs), Internet, and other communication networks.

The *support layer* is responsible for retrieving information from the voluminous incoherent data. In recent past, ML algorithms are used for data processing. These processing techniques often require excessive processing power which may not be attainable using a single host. Therefore, computational power of a combined set of resources/hosts may be used to act as a single computational infrastructure, such as clusters for processing. Each resource/host may take part in executing the algorithm partially. This distributed/decentralized implementation reduces the expenses by recycling existing resources. In another aspect, the support layer may work in a centralized way through physically separated cloud-based servers while using intelligent computing methods, such as cloud, edge, and fog computing. The *application layer* or top-most layer is interfaced with the support layer and provides intelligent services and applications based on user requirements.

### 4.2.2 General Cloud-based Architecture of a Smart System

The architectural essence of cloud computing can be split into cloud computing deployment and service delivery models. The cloud, fog, and edge layers constitute the three levels of IoT combined cloud-based smart system.

The edge computing layer comprises billions of IoT elements, such as, devices, actuators and sensors in security and surveillance systems, multimedia systems, energy management systems, and healthcare systems. All these devices have limited computing resources in terms of hardware footprint, power, and speed. This layer collects large volumes of data from multiple sources and applications, subsequently transferring it to upper layers.

The fog computing layer is located between edge layer and cloud layer. It comprises routers, gateways, controllers, and servers operated by Internet Service Providers (ISPs). These nodes have larger computing resources as compared to edge computing layer. Services and processes that require high memory, power, and computational logic are run on fog nodes. These nodes are dispersed across large geographical areas inside ISP networks and contain multiple interfaces that allow them to communicate with various applications and protocols.

The cloud layer, being the high-end layer, comprises high-performance servers and storage devices. Present-day cloud computing platforms employ large amounts of resources in terms of data storage and scalable processing resources for data analytics. The service provider cloud collects data from different devices and peer cloud services. It further employs stream processing algorithms to transform data, filter data, and analyze data in real time. Subsequently, the cloud platform stores the data in various repositories wherein advanced analytics can be performed. In addition, the service provider cloud provides core IoT applications and associated services, such as process management for the entire IoT system, and creates data visualization.

#### 4.2.2.1    Cloud Computing Deployment Models

Cloud computing architecture comprises three primary deployment models: private cloud, public cloud, and hybrid cloud [17]. Private cloud primarily focuses on security by dedicating hardware and software resources exclusively to a single organization/customer. Due to its restricted accessibility, private cloud provides freedom of customization of its resources and greater control over operational and capital costs. Public cloud architecture uses the same computing resources among multiple organizations/customers in a shared manner. In general, the service and resources are provided dynamically on a demand basis. Therefore, public cloud yields the benefits of greater flexibility and scalability, lower costs, and faster access to the state-of-the-art technologies to the customer over the challenge of security.

Hybrid cloud aims to integrate the benefits of both private as well as public cloud. An organization/enterprise can keep the sensitive data and applications that cannot be easily migrated to public cloud, in its on-premises infrastructure. However, the enterprise can use public cloud for software-as-a-service (SaaS) application, storage, computation and additional platforms simultaneously depending on demand. As a result, hybrid cloud balances the workload division seamlessly between private and public cloud architectures to meet the requirements of cost, performance, and security. But the implementation of hybrid cloud is more complex due to the involved complex operations and issues in data migration between different clouds, data integration, and visibility.

A *community cloud* architecture is based on shared infrastructure as a result of collaborative effort involving several organizations from a single community with common interests such as jurisdiction, security, and compliance. For example, the virtual private cloud model employs public cloud infrastructure privately and interconnecting resources using virtual private network (VPN).

#### 4.2.2.2    Cloud Computing Service Delivery Models

Service delivery models in cloud computing can be broadly classified into Software as a Service (SaaS), Platform as a Service (PaaS), Infrastructure as a Service (IaaS), and Unified Communications as a Service (UCaaS). Before the emergence of UCaaS, cloud service delivery models were abbreviated as the Software, Platform, and Infrastructure (SPI) service frameworks. The SaaS layer provides access to various software applications such as email or office productivity tools as cloud services [18]. Well-known SaaS applications are Google Docs, Microsoft online, and Facebook. SaaS providers depend on Extensible Markup Language (XML) encryption, Web Service (WS) security, Secure Sockets Layer (SSL) and other various methods for data protection.

The PaaS layer provides a platform for customers/consumers for creating and operating their own software. Microsoft Windows Azure, Google App Engine, and Amazon S3 are few popular PaaS providers. IaaS providers allocate computing resources and storage to the end users dynamically through virtualization technology or on-demand basis. Virtualization technology comprises of virtualization layer, known as *hypervisor*, and virtualization resource layer, referred to as *virtual machines* (VMs), virtual network, and virtual storage. Leasing VMs for OS and virtual resources (e.g., memory, CPU, hard disk, etc.) is a cost-effective solution than procuring physical machines. VM's management as well as data and instructions flow control between VMs and the physical hardware are the primary responsibility of hypervisor. Google cloud storage, Amazon EC2, and Microsoft Azure are few widely used IaaS. The physical computing resources of the cloud, collectively known as the hardware layer, are implemented in the data centers. This layer is responsible for managing traffic, hardware configuration, fault tolerance, cooling resources, and power consumption.

Cloud computing provides unbounded computational and data storage resources for associated smart applications on top of the IoT infrastructure. Therefore, entire data is sent back to the cloud for processing. Sending the entire collected data to the cloud incurs high costs from various parameters such as storage, bandwidth, latency, and energy consumption for communication. To address the weaknesses of cloud computing, fog computing has emerged as a solution for processing data in an efficient way. In the next section, we investigate the threats to confidentiality and other security issues associated with various layers of the architecture.

### 4.2.3 PRIVACY AND SECURITY ISSUES IN CLOUD COMPUTING INFRASTRUCTURE

The widely accepted definition of the cloud computing model is introduced by *National Institute of Standards and Technology* (NIST) as "a model for enabling ubiquitous, convenient, on-demand network access to a shared pool of configurable computing resources (e.g., networks, servers, storage, applications, and services) that can be rapidly provisioned and released with minimal management effort or service provider interaction."[19]. This chapter presents security threats and security attacks that exist for cloud computing systems. In addition, it also mentions the mitigation steps that the system designers and engineers can adopt to thwart such threats and attacks.

Entities, such as users and services, authenticate themselves to the service providers to avail their services. Each entity is required to provide *personally distinguishable information* (PDI) that distinguishes it to a service provider. An entity, be it an individual or an object, possesses a set of unique features or characteristics referred to as an identity. Entity identifiers are used to authenticate entities to the service providers that helps the service providers to decide whether to allow the entity to avail a service. Moreover, each entity can possess multiple identities that requires an *identity management system* (IDM) for portfolio management of such identities. Typically, an IDM can establish identities that associates PDI with an entity. It can also describe or record use of identity information by logging identity activity in a system, and thereafter providing access to such logs. The IDM can also destroy an identity if the PDI is rendered unusable after an expiration date.

An IDM can be used by an *identity provider*, which issues digital identities to the entities, such as credit card service providers issue identities that enable payment for different services used. The *service provider* can use an IDM to authenticate an entity with its identity information prior to providing a service. For this purpose, it requires an identity verifier to verify any specific claim from an entity. The verifier evaluates the correctness of a claim from an entity. This entire process requires protection mechanism for confidentiality and data integrity of PDI of the entity. In the next section, we discuss the privacy and security issues that may arise for information flow in a smart city infrastructure.

## 4.3 PRIVACY AND SECURITY ISSUES IN SMART CITY INFRASTRUCTURE

IoT-based smart applications may impose vulnerability within the smart city infrastructure asphysical objects, for example, sensor-based devices directly collect data from environmentsbased on direct association within the network infrastructure and communicate to transmit data using wireless or wired connections. Therefore, IoT-based smart city applications have their vulnerabilities directly associated with those of the corresponding network paradigms [14]. The collected data which is uploaded, processed, and stored for data analytics, and is available to authorized users, can expose major vulnerabilities through various external attacks, such as DoSor DDoS attacks, remote brute force attacks, or man-in-the-middle (MITM) attacks, and internal attacks, such as data tampering, password sniffing, and Trojan codes. Therefore, storing and transferring data using the IoT infrastructure through the cross-layers of smart city architecture may lead to defective applications and a degrading quality of services, and also severely impact the security and data confidentiality in smart cities, such as paralyzing the entire city [2, 20].

The large number of smart devices, sensors, actuators, RFIDs, different type of end users and frequent data transfer may easily compromise the privacy of generated, and stored or transmitted data. Researchers have demonstrated that the user acceptance is the primary criteria for successful development of new smart city applications. The user acceptance of a smart system is primarily dependent on privacy friendliness or the level of privacy protection. In [21], the authors have briefly described the identification of various privacy threats and their possible solutions.

The physical distributive nature of a smart city network needing widely spread accessible services incorporates large number of small, smart devices and sensors, introducing privacy and security threats. On account of the constraints of limited computational power, data processing, and analyzing capabilities of cost-effective smart devices, smart city infrastructure utilizes the advantages of powerful cloud servers for data storage and processing. However, these cloud servers, if untrusted, may impose critical privacy and security threats [22].

The cumulative study by researchers from various interrelated research domains demonstrated the merits of strategic smart city security policies over the confined privacy aspects of smart data [14, 22–24]. The general concept of failure of existing smart systems lies in not including interoperable security goals for information security even though aspects of privacy, confidentiality, integrity, and availability have been considered [14, 25, 26].

Access control of data and the data communicated between network components and the IoT devices must adhere to effective risk management. Early detection and timely response to threats prevent smart city applications from entering an unsafe physical state. In [27], authors develop a prototype framework for IoT-based systems named IoTSan that uses model checking for identifying interaction level pitfalls by pinpointing the events that may induce system transition to insecure states.

In the next section, we investigate certain smart city applications based on the security and privacy of smartphones, power grid systems utilized within smart cities, smart healthcare and transportation, smart environment, protocols and algorithms, security and confidentiality policies, operational threats for smart cities.

## 4.4 DATA CONFIDENTIALITY AND SECURITY OF SMARTPHONE DEVICES AND SERVICES

Smartphones, tablets, smart watches, and other mobile devices have become an integral part of an urban citizen's everyday life. These devices provide a way to connect citizens with the smart city ecosystem, such as connected cars or smart homes. For example, the Google Cloud Print™ service allows a user to print to a printer registered to the user's Google account using a network-compatible device such as a mobile smartphone without installing the printer driver on the device. The pairing facility of these devices helps to connect citizens with industrial equipment and services of a smart city in order to achieve low-cost sustainability. The processing power of these mobile devices, and the technological advancement in wireless technology along with the advancement in security technologies such as cryptography, digital signatures, and public-key infrastructure, provide ample facility to use these devices for easing everyday life hassles. Apart from accessing basic telecommunication services, such as communicating via direct phone calls, these mobile devices can be used to connect with smart city ecosystem and access various websites for numerous web services comprising collecting information, location finding to online banking, e-commerce, e-governance, or e-administration. Over the past decade, mobile devices have become the backbone of a smart city infrastructure.

In general, these mobile devices have unique identifiers such as device ID, SIM card ID, International Mobile Station Equipment Identity (IMEI), Media Access Control (MAC) address, Mobile Equipment Identifier (MEID), and Unique Device Identifier (UDID). As the aforementioned unique IDs can be retrieved and shared by apps or indirectly recorded via bluetooth signal or Wi-Fi, those captured IDs can be utilized for tracking the phone and, by association, its possessor [21, 28]. The small form factor of these devices makes it easy to steal or be left behind by the users unknowingly when they travel. Physical access to stolen mobile devices compels experienced attackers to break the lock or password easily in order to access user data, including encrypted data. These stolen devices not only reveal personal data, they may disclose corporate data where the devices were used by corporate employees.

With the massive growth in digital technology and wireless communication, corporate officesadopt a *bring-your-own-device* (BYOD) policy. BYOD promotes use of personal mobile devices, such as smartphones, laptops, etc., to retrieve official data. A malicious corporate insider can download and store large volumes of corporate data to the secure digital (SD) flash memory card of an individual's smart mobile devices or can use their smart mobile devices to transmit data via email services to external email accounts. So, the malicious insider can misuse the corporate data, or if outsider attacker accesses that malicious insider's mobile devices, or email IDs, sensitive corporate data will be disclosed.

Data leaks from malicious corporate insider can be monitored using *data loss prevention* (DLP) technologies. Additionally, attackers may gain access to corporate services such as *virtual private network* (VPN), email, and credit card details through the passwords residing in seemingly secure places like the iPhone Keychain. Therefore, in this current fastest growing digital era, the company needs to protect its data by using *mobile device management* (MDM) to secure corporate as well as personal data information in employees' mobile devices. Complete removal of data from these devices is not possible through re-flashing the operating system or using the device's built-in factory reset. Therefore, the privacy and secrecy of sensitive data stored in these mobile devices has become a significant challenge in the presence of unknown security threats from the attackers. Additionally, these devices mostly rely on cloud computing or fog computing for compensating limited data storing and computing power. Both these challenges impose vulnerability on a smart city system as the mobile devices implicitly form the interface medium to the smart city's network infrastructure.

There exist application-based security threats, web-based security threats, and network secu-rity threats along with the above-mentioned data security and privacy challenges. Mobile application-based security threats occur when users download malicious applications that appear legitimate, however such applications steal data from used devices. Additionally, few legitimate mobile applications cause privacy threats. For example, spyware and malware steal personal as well as corporate information without the device user realizing. Moreover, legitimate *global positioning system* (GPS) applications can leak personal information about any place that a user visits. When a mobile device user visits a website that is affected, malicious content gets downloaded onto the device unnoticed by the user. Mobile networks are vulnerable because cybercriminals can steal unencrypted data while users access public Wi-Fi networks.

In addition, mobile network services such as SMS, MMS, and voice calls can be utilized for attacking mobile devices through phishing attacks. When data is transferred through a Wi-Fi communication medium between mobile devices and a Wi-Fi access point, unencrypted data can be easily intercepted by attackers, referred to as Wi-Fi sniffing. Therefore, transferring data using Wi-Fi communication medium is a potent security threat. A wide variety of threats with relevant examples, security solutions, and vulnerabilities for smartphones over the period of 2004–2011is described in [29]. In 2017, the U.S. Department of Homeland Security (DHS) in association with National Institute of Standards and Technology (NIST) released a detailed report on mobile device security that depicts an extensive account of threats and the wide range of categorization, vulnerabilities, and best practices as security solutions for mobiles devices, in general, referred as smartphones and tablets running mobile operating systems [30]. Additionally, the report pointed out the existing gaps in DHS authorities and also addressed the challenges that jeopardize the use of smartphones by the U.S. Government, as well as the mobile users of smart city ecosystem.

In general, mobile device security can be achieved using endpoint security solutions, secure web gateway (SWG), VPNs, cloud access security broker tools, and email security. Endpoint security solutions are incorporated by companies to monitor the processes and files on every corporate employee's mobile device that accesses the network. A constant monitoring via endpoint security system supports early detection of malicious behavior and prevention from massive damage through the removal of threats by security teams. As of 2021, leading endpoint security solution companies that provide security to corporate data and mobile devices comprise Microsoft defender advanced threat protection, Sophos endpoint protection, Trend Micro, ESET endpoint security, McAfee endpoint security, Crowdstrike falcon, and Avast advanced endpoint Protection, among many.

VPN refers to a connection that is encrypted over the Internet from a device to a private network. This encrypted connection ensures safe transmission of sensitive data while preventing unauthorized people from snooping on the traffic, thus allowing the mobile device user to perform remote work safely.

SWG is a cyberbarrier for keeping unauthorized traffic from entering a company's own network. With the increasing growth in sophistication for embedding web-based threat vectors into seemingly safe or professional-looking websites by cybercriminals, the SWG solution has kept its root in mobile device security domain. SWG detects and prevents elusive cyberattacks through constant monitoring and incorporation of new attack signatures into cyberattack detection proficiency, and removal of the SSL blind spot in order to expose threats hiding in encrypted traffic [31]. It also provides visibility into web traffic through monitoring, keeps a log for every web transaction, and supports compliance efforts in order to constructively control usage and apply standardized policies and guidelines aligned with regulatory requirements.

Email security is a service implementation to protect users from email attacks on mobile devices as well as all devices through which the email application can be accessed. In general, a strong email security service provides robust threat protection capabilities through its designing in order to handle the problem of social engineering and business email compromise (BEC) attacks. Social engineering is known as the psychological manipulation technique exploiting human error to gain confidential information or access the system or critical resources. Advanced email security services may allow the user to perform security controls such as warning banners within emails for sending alert to email users when a suspicious email is delivered to their email inbox.

In the past, most email accounts by default required one-factor of authentication in the form of password to grant account access along with the user posed factor 'email-ID'. With the technological advancement and intelligent minds behind the attacks, this one-factor authentication may not be strong enough for providing protection as the attacker can easily use tricks to access the email password. Therefore, the implementation of multi-factor authentication (MFA) policy has evolved as one important solution in order to protect corporate or individual mobile device users from phishing and email compromise attacks. MFA policy ensures that sensitive email accounts require an additional layer of verification in order to be accessed along with username and password. This extra factor of authentication prevents malicious attackers to access critical corporate as well as personal email accounts even though user account or password is compromised through a phishing attack. We have discussed MFA in detailearlier.

*Cloud access security broker* (CASB) is a cloud-based or on-premises software or hardware acting as an intermediary between cloud service users and cloud applications or cloud service providers and scanning all activity and enforcing security policy compliance. In general, CASB provides three primary services, namely, secured access to all user apps, app discovery and management, data protection, and threat protection. CASB has the ability to address extended security gaps across three environments: *software-as-a-service* (SaaS), *platform-as-a-service* (PaaS), and *infrastructure-as-a-service* (IaaS) environment. CASB protects the movement of data by restricting features such as access and sharing privileges, and the contents of the data through encryption.

To impart confidentiality of smartphone data in smart cities, [32] proposed three approaches based on fog computing, such as, foggy dummies for protecting the user privacy, a blind third party for protecting the user from the server provider by maintaining a trust relationship between them, and a double foggy cache for solving the trust issues between peers with cooperative methods.

*Privacy-preserving authentication* (PPA) protocols adopt a strong cryptographic approach to impart authentication and user privacy protection for mobile devices. In [33], authors proposed aPPA protocol for mobile services and proved protocol security in the random oracle model. They have also demonstrated that the proposed PPA protocol has less communication and computation overheads in comparison to other competing protocols.

## 4.5  POWER GRID SYSTEMS WITHIN SMART CITIES

Present-day electric power utilities depend solely on communications and computations to operate and analyze data from power systems. There exist a variety of power system applications that take advantage of cloud computing platforms. There are various computational and security requirements of core power system applications that can be moved to the cloud, which can be used in power exchange markets and operational planning. In typical operations of power systems, time-varying computational requirements exist for multiple applications. Some of these applications may be operated in periodic fashion or occasionally, but they require large computational power when used. On the contrary, a different category of applications may operate in continuous mode, but the number of computations may be a function of power system state, which itself is time variant. In such power grid systems, computing the price of consumed electricity is computed as a function of the demand for real-time market forecasted for the upcoming day, a mechanism called *locational marginal pricing* (LMP). In some applications, LMP is computed by solving power flow problems with the required inputs of predicted loads, power generation cost, and power transmission capacity. The criticality of the computation depends on uncertain power consumption in populated areas of the smart city. The cloud computing-based platforms can compute and provide reliable data storage involving this computation by providing functionalities of data confidentiality, data privacy, and data integrity.

Owing to accelerating and ever-growing urbanization, reliability, and availability issues, *power distribution networks* (PDNs) have gained paramount importance in these security aspects. To sustain smart city infrastructure along with increasing demand of deploying intelligent technology, an increased requirement of automation by operators in PDNs has been observed. To ensure reliable, efficient energy supply and management, solutions can be based on grid automation, demand response-balanced computation performed on cloud platforms, renewable integration, and energy storage.

Typical solutions for electric grids can comprise common information and operational technologies (IT/OT) that are typically deployed per application, such as *supervisory control and data acquisition* (SCADA), control operations, asset management and analytics. The major emphasis on automation of grid distribution is laid on prevention of power supply interruptions, which can be controlled through decision-making processes running on cloud comprising a shared set of servers. With emerging high-precision sensors, protections, and control relays, it is now feasible to pinpoint issues before problems lead to interruptions. However, in unavoidable circumstances of fault occurrence, the circuit can be isolated rapidly to restore the power supply, by using a fast network configuration feature in a remote control system.

In addition, such a control system can be put to effective use by incorporating entire equipment required in SCADA or Distribution Management System (DMS). This inclusion promotes real-time monitoring and other control solutions in existing networking systems. With the assistance of accurate fault location and DMS functions, critical disruptive situations can be resolved effectively. This enables an undisturbed power distribution system, which is essential for normal life and business operations, and also ensures a constant power supply for emergency services. Such services must have features to pinpoint fault localization accurately, along with fault isolation followed by power restoration. Moreover, real-time grid information must incorporate emerging needs of demand response, energy cost savings, and facility of charging of electric vehicles.

In smart cities, dense mesh and aging cable networks become critical to reliable performance with the passage of time. Hence, periodically monitoring the condition of cables is essential. With the system compensating for earth fault currents, the grid may be run on a single-phase earth fault developed for a limited time. This step provides adequate time to restore alternate power supply facility during which the faulty section is decoupled and repaired. The central automation system can analyze the data collected from grid node stations to continue maintenance actions. This can avoid unplanned power supply interruptions. As an integrated approach to mitigate such interruptions, present-day smart products have emerged for protection and control, measurement, monitoring, and communication. All these solutions are emerging for urban and rural power distribution networks for smart city infrastructures.

## 4.6   SMART HEALTHCARE SYSTEMS WITHIN SMART CITIES

*Smart healthcare systems* (SHSs) in smart cities are flourishing with the advances in smart networks, cloud computing, and big data analytics paradigms. However, the security and confidentiality issues become critical challenges due to the constraint of storing sensitive data in untrusted and remotely controlled infrastructure. This must ensure minimal disclosure and secure transmission of medical data to thwart tampering and forgery attacks [34, 35]. Relying on only one party for the smooth management of confidential healthcare data makes many current smart healthcare data systems vulnerable to a single point of failure. Blockchain technology can increase the reliability due to its distributive nature. In [35], authors have demonstrated a detailed survey on how blockchain technology can be used for addressing various privacy and security issues in storing, transmitting, accessing, obscuring, and securing medical data.

In real-time monitoring of patient's health for infectious diseases like COVID-19 or chronic diseases, such as diabetes, sensors in IoT-based smart infrastructures collect and transfer medical data continuously. Therefore, IoT technology is an integral part of smart healthcare in smart cities. In [36], authors have shown that how the *attribute base credential* (ABC) system's salient features, such as blind signatures and *zero-knowledge proofs* (ZKPs), help to achieve privacy goals. Authors have demonstrated experiments on various scenarios for Idemix (IBM) and U-Prove (Microsoft) ABC systems and shown that Idemix is computationally faster than U-Prove.

Electronic health records, or *electronic medical records* (EMRs), comprise health-related dataof the patients which form a major component of e-health. Effectively managing EMRs involves multidisciplinary teams that comprise telecommunication, computer science, and instrumentation to enable exchange of medical data across large geographic regions. In almost all cases, as it involves online database systems, such systems can be managed on a shared set of servers, referred to as the cloud. Cloud-based systems provide ubiquitous computing capabilities in terms of storage and processing power. The security principles of such data processing components typically define requirements for *confidentiality*, *integrity*, and *availability* (CIA principles) of EMRs. The confidentiality of EMRs is restored by defining policies to prevent malicious access of medical records to unauthorized users. The principle of integrity protects EMR data from unauthorized modification, whereas the availability principle restricts reliable access, or use, of EMR information to authorized users.

Confidentiality and integrity of patient data communication from and to the cloud, and also within the cloud, can be ensured with standard cryptographic protocols, such as *Transport layer security* (TLS) and IPSec. To ensure data confidentiality, integrity of computations, and preventing access by undesirable agents, strong guarantees of isolation between the *virtual machines* (VMs) on the cloud are required. Also, to thwart attacks from malicious insiders, such as server administrators, protection of patient-related VMs from host virtual machine managers (VMMs) is required. The resulting cloud-based infrastructures can be enhanced to provide automated redundant provisioning of medical data, failure handling of processes, and better reliability of VMs.

The underlying architecture of smart healthcare systems comprises a presentation layer, a big data processing and analytics layer, a data storage and management layer, and data connection components, all of which can be implemented on a cloud system. In addition to CIA principles, a general security framework for sharing health information of patients can be based on additional security principles. The security and privacy principles of EMRs comprise data sharing control wherein authorized legal users with essential rights are given access to patient record information. Moreover, legal measures are provided to ensure trust among multiple organizations, and personal data is processed to prevent releasing of identification of the patient information.

To mitigate security risks in smart healthcare applications, the considered features may comprise two components. The first component comprises technology enablers, such as IoT, wearable and immersive technologies, advanced ML, knowledge discovery, and sensor networks. The other component refers to security and privacy mechanisms, such as biometric authentication, attribute-based encryption-decryption, smart emergency access control, privacy-preserving data aggregation, group privacy while performing data fusion, and inference control. The technology enablers have been demonstrated to be vulnerable against wide-ranging threats, such as forgery attacks, impersonation attacks, spoofing, session key disclosure, and replay attacks. For such threats on emerging technologies, such as healthcare IoTs and wireless body area networks (WBNs), security schemes like lightweight hash-chain and forward secure authentication have been proposed. In such systems, security of communicated messages between the participating entities is of paramount importance. Hence, mutual authentication and anonymity of patients is addressed by such protection schemes.

The data communication channels over which WBANs and IoT devices communicate, areinsecure. An adversary may eavesdrop, intercept, tamper, or replay the communicated data orgain access to data stored in sensors or mobile devices that may have serious impact on the health and life of patients. In such healthcare environment, authentication and encryption schemes impart various security functionalities, such as, forward secrecy, user anonymity, irrevocability, mutual authentication, and resilience against various forms of attacks mentioned above.

## 4.7 PRIVACY AND SECURITY IN SMART TRANSPORTATION

The security and data confidentiality issues in intelligent transportation system (ITS) impose a major challenge in smooth conduction of traffic and road mobility management, and road transport in smart cities. The key component of ITS is vehicular adhoc network (VANET) known as a self-organized network comprising vehicles installed with onboard unit (OBU) which is a communication device, a roadside unit (RSU) that stands along the roadside and crossroads, and a trusted authority (TA) that manages the smart system [37].

The signature of traffic information exchanged between vehicles and their surrounding environments must be verified while protecting the actual identities of the vehicles. In [38], authors demonstrated that *conditional privacy-preserving authentication* (CPPA) schemes, which presently exist, may show critical vulnerabilities in driving safety within the VANET systems. They proposed SD2PA, an authentication scheme to overcome the non-safe driving problem that emanated from the existing group-based authentication schemes. The SD2PA scheme adapted the general hash function used in the system without involving computation-extensive bilinear pairings or *Elliptic-curve cryptography* (ECC). The authors demonstrated that SD2PA meets various security and privacy goals of VANET as well as smart transportation system, such as lightweight message authentication and data integrity, vehicle traceability, identity privacy preserving, non-repudiation and revocability, unlinkability, and resilience against modification attack, replay attack, and impersonation attack. The authors also claimed that along with security and privacy goals, SD2PA has advantages over prevalent schemes in computational cost and communication overhead.

In [37], the authors demonstrated an autonomous lightweight CPPA scheme based on ECC without relying on a hardware device to restore security and data confidentiality requirements in VANETs, as shown in Figure 4.1. The CPPA scheme relying on the strength of the elliptic-curve discrete logarithm problem (ECDLP), resolves the private key compromise problem, and imparts a countermeasure against privilege escalation. In this scheme, a trusted but curious party known as regional traffic management center (RTMC) is incorporated that produces secret session keys for automotive control. The authors demonstrated that the scheme assures the security and data confidentiality goals of VANETs and also provides resilience against adaptive chosen plain text and identity attacks in the random oracle model.

In [39], authors proposed a *Security-Aware Efficient Data Sharing and Transferring* (SA-EAST) model to secure cloud-based ITS implementations with an aim to yield secure real-time multimedia data sharing and transferring. Deploying mobile vehicular systems in the presence of unknown adversaries in a dynamic environment where high-performance communication is required, is still facing numerous challenges due to the limitations of wireless communications. A stable communication is required during the migration of data to cloud servers. However, rapid movement of objects in ITS leads to an unstable connection to cloud resources implying frequent changes in the quality of cloud services. Therefore, conducting secure and smooth operations of VANET is a challenge in ITS. The SA-EAST model is based on vehicular embedded systems that capture video data through vehicular sensors and cameras. In addition, the system dynamically picks the optimal cloud server for data transmissions.

## 4.8   SMART ENVIRONMENT

A smart environment can substantially support a sustainable smart society. Ubiquitous sensing and intelligent systems embedded in smart systems can work in conjunction for smart city applications, such as monitoring waste gas and greenhouse gas, waste collection, energy consumption, city noise, air and water pollution, structural reliability of buildings, forest conditions, weather forecasting, land movements, and disaster responsiveness [8, 22, 40]. Smart street light management, smart agriculture, smart surveillance, and smart retail store automation are also part of smart environment management. Virtual sensing is another paradigm where sensors are not located physically, however their sensing observations can be analyzed to obtain better information about our surroundings and environment. For example, real-time social media analytics, such as, Facebook posts, Twitter feeds, Tumblr, and Weibo posts can be used for behavior and sentiment analysis in order to make better governing or societal decisions [25].

Each smart city application generates tremendous volume of user data. This ever-growing data requires analysis in a strategic manner so that judicious decisions taken based on the data analytics possess a higher accuracy rate. In the past decade, the trends depict a significant amount of research on big data analytics using AI and ML algorithms for initiating intelligent decisions. Researchers also significantly used AI and ML algorithms to cope with various data security and privacy challenges for storing and transmitting data over wireless networks for smart city applications. Among many existing technologies, intrusion detection systems (IDSs) substantially depend on ML systems for restraining networks from various attacks, such as man-in-the-middle attacks, malware, and viruses [2]. In [12], authors have reviewed ML and game theory-based approaches for providing security in data sensing in wireless sensor networks (WSNs), discovering attacks in Wi-Fi networks, and discovering and preventing intrusions in WSNs. With the advancement in AI/ML technologies, adversarial attacks have also developed due to the ever-growing intelligent minds behind the attacks. To nullify the intended AI/ML algorithm, attackers can corrupt training data that are required to train ML models.

## 4.9   SECURITY IN SMART IoT DEVICES

The existence of a large number of off-shore vendors in integrated circuit (IC) design, fabrication and manufacturing process, and intentional inclusion of malfunctioning hardware by malicious agents in the entire IC supply-chain, has emerged as one of the severe private as well as national threats. In addition, with the revolutionary growth and power of hardware-based attacks, the need to address and resolve hardware security as well as IoT device security challenges has become apparent in context of smart city design and implementation, as well as in the context of data confidentiality, availability, and integrity.

*Hardware Trojans* (HTs), IC counterfeiting and overbuilding, reverse engineering and *sidechannel attacks* (SCAs) are a few widely used hardware-based attacks in hardware security literature, among many. HTs are malicious changes incorporated in ICs during design or fabrication. HTs can be used to mount denial-of-service attacks by tampering with the design functionality of ICs as well as smart devices. Reverse Engineering (RE) requires disassembling systems and devices and extracting their underlying design using destructive and nondestructive methods. Such a disclosure of designs may lead to device cloning, duplication, and reproduction. Dishonest intentions of the increasing amount of RE leads to design counterfeiting, piracy, insertion of HT, and cloning. Tremendous growth in VLSI technology over the decades results in the shrinking size and increasing complexity of ICs used in a system. The trend of assembling and fabricating these ICs globally over different geographical regions has led to a thriving unauthorized market that undercuts the market competition with counterfeit ICs and electronic systems. The problem when a factory produces more ICs than that required by the IC design house is termed IC overproduction. The factory sells the ICs in the market without authorization of the IC design house. SCAs are based on the analysis of side channels or physical parameters of a hardware or a chip.

Hardware security must be one of the inseparable goals of the smart device-based smart city-infrastructure. In the recent past, extensive research has been carried out for resolving security and privacy issues related to hardware. However, with the presence of intelligent and organized minds behind most of the security and privacy threats, and the rapid growth in VLSI technology and smart city applications, advanced research in hardware security is of the utmost importance. A continuous monitoring of possible threats as well as their early detection is required for electronics devices in use.

## 4.10   SECURITY AND PRIVACY FRAMEWORKS

In this section, we outline different tools and techniques that have been adopted in the recent past to thwart various forms of security threats and vulnerabilities affecting present-day components of smart cities, such as cryptography, homomorphic encryption, zero-knowledge proofs, secret sharing, secure multi-party computation, multi-factor authentication, blockchain, biometrics, etc.

## 4.11   CRYPTOGRAPHY

Cryptography is a powerful tool for providing security and privacy protection in smart city applications. One of the major challenges for implementing privacy protection within urban smart systems lies in extreme difficulty for handling large volume of continuously generating data that is collected by a growing number of smart devices embedded in smart cities [41]. In addition, end users use mobile devices with limited computational power and limited memory hindering the implementation of strong but computationally demanding standard encryption and decryption algorithms. The next subsections demonstrate various broadly classified cryptographic techniques for solving security and privacy protection issues can be used in smart cities.

### 4.11.1   HOMOMORPHIC ENCRYPTION

Encrypting the data in a smart city and transmitting the ciphertexts to the cloud servers may secure the data stored. Encryption also prevents untrusted cloud servers from directly accessing collected data, and processing and analyzing the collected data for providing intelligent services of smart city applications [2, 22]. In [42], the author introduced a fully homomorphic encryption (FHE) scheme that allows computations on its encrypted data without requiring any decryption. Qualitative research has been undertaken by researchers to developing an FHE scheme that is segregated among four generations of FHE. Partially homomorphic cryptosystems that either use addition or multiplication are computationally feasible in smart systems. However, fully homomorphic cryptosystems that can use both addition and multiplication are computationally expensive. Significant research has been undertaken in last few years for deploying this breakthrough technology. Various research groups all over the world have made libraries such as [HElib], [SEAL], [cuHE], [cuFHE], [PALISADE], [NFLlib], [HEAAN], and [TFHE]for general-purpose homomorphic encryption [43].

In order to adopt homomorphic encryption in health, medical, and financial sectors to protect patient data and consumer privacy, standardization is required from multiple research agencies and governing bodies. With the aim of standardization, an open industry/government/academia consortium named HomomorphicEncryption.org group is co-founded in 2017 by IBM, Microsoft, and Duality Technologies. The present standard document comprises specifications of secure parameters for the ring learning-with-error (RLWE) problem.

### 4.11.2 ZERO-KNOWLEDGE PROOFS

The process of sending the claimant's secret (password) to the verifier for password authentication is susceptible to eavesdropping by a third unauthorized party. Additionally, a dishonest verifier may reveal the passwords to others or utilize it to impersonate the claimant. Zero-knowledge proof (ZKP) or zero-knowledge protocol allows preservation of the user's privacy wherein the claimant is not required to reveal any information to the verifier [26]. Extensive research has been carried out on privacy-preserving mining of distributed data in the last decade.

In smart transportation, privacy-preserving distributed navigation scheme based on distributed RSUs for road planning in crowdsourcing allows both the querier and responding vehicles' location privacy during the querying, crowdsourcing, and navigation phases. This location and navigation response privacy can be achieved by ZKP-based Elgamal and AES encryption of the driving route [22]. To integrate smart vehicles or smart meters into hassle-free applications, such as toll gates, smart meter billing, and wireless payments at gas stations, ZKPs can be used for payments at toll gates computed on actual road usage or time-of-use of electricity, natural gas, water, and direct heating consumption [21].

### 4.11.3 SECRET SHARING

Secret sharing is a cryptographic tool for sharing a secret key between a group of participants where a share of the secret is allocated to each participant. If a secret is shared among $n$ participants, each of them having one share, at least $k$ shares are required to recover the secret; a single share cannot recover the secret. It is known as a $(k, n)$ secret sharing scheme [44, 45]. Hence, secret sharing imparts reliability and confidentiality in the distributed system [12]. For encryption, digital signatures, or login authorization for IoT devices, secret sharing of keys can be used in smart cities [46, 47]. It can be used to aggregate data received from participatory sensor networks, or smart meters, for distributed data storage, or distributed $k$-anonymity [21].

In [48], authors have proposed a multi-route data transmission scheme that encodes the message between multiple sources and communicates them via different paths aiming to strengthen the smart city high-confidence IoT-based services. Authors claim that the attacker cannot reveal the transmitted data easily in case of a $(k, n)$ secret sharing scheme. To obtain knowledge about transmitted data, the attacker has to compromise a large number of nodes which incurs high cost. Hence data security in smart cities gets improved. An adversary can modify the share values in the secret sharing scheme. Therefore, researchers worked on a verifiable secret sharing scheme, and algebraic manipulation detection (AMD) codes among many, to strengthen the secret sharing.In [49], authors have proposed a non-malleable secret sharing scheme (NMSS). They guarantee that if an adversary tampers with all of the authorized shares, then the secret recovery process can either generate the master secret key, or the master secret key is destroyed, and the recovery process generates a completely unrelated string compared to the key. Their proposed schemes and related constructions have already proven applications in cryptography as well as in smart city applications for sharing the secret over distributive heterogeneous system.

### 4.11.4 SECURE MULTI-PARTY COMPUTATION

Secure multi-party computation (SMPC or MPC) is a cryptographic method used to provide confidentiality and untraceability. MPC allows multiple participants to jointly compute a public function based on their private inputs without disclosing individual participants private inputs to others, and without depending on a trusted third party [21, 41]. MPC protocols generally use Shamir's secret sharing and additive secret sharing. Although, MPC provides weaker security guarantees in practice, it is a practically efficient approach than FHE [41]. Secure MPC can be used for various smart city applications. For example, homomorphic encryption-based secure MPC can be used for preventing the untrusted health cloud service provider from learning any private social and health-care data [22].

In [50], the author has mentioned the historical transformation of research in MPC from theoretical aspect to practical implementation and industrial acceptability. However, he has also stated that MPC needs substantial expertise to install. Additionally, the author considers still significant research breakthroughs are required to create secure computation practical on the ever-growing volume of large data sets and for complex problems, and to make it approachable to use for non-experts.

### 4.11.5 MULTI-FACTOR AUTHENTICATION

Multi-Factor Authentication (MFA) policy has evolved for mitigating the security and privacy issues usually raised by single-point failures or one-factor authentication. With the continuous advancement in digital as well as VLSI design technology, electronic devices that include chips as an integral part have become smaller, cheaper, and more powerful day by day. Additionally, the development in Internet and wireless communication technology introduces more human dependency on electronic as well as smart devices in daily life. The growth in IoT technology increases this dependency a step further towards connecting these smart devices with smart city applications to make life comfortable. For example, CCTV (closed circuit television) is one of the most useful electronic devices connected to IoT applications, such as healthcare, transportation, and surveillance systems. Therefore, securing these electronic devices as well as smart city applications requires a strong authentication policy for righteous users.

Single-factor authentication, aka SFA or one-point authentication, requires a sensitive password, or "salted" password, or a PIN to a username. SFA is simple and user-friendly. With the emerging growth in VLSI design technology, it has become easier to guess the right password by an attacker generating millions of passwords using several algorithms on highly computational devices within a fraction of time. Therefore, MFA has evolved to yield a higher level of security through uninterrupted protection of smart electronic devices and smart city services from unauthorized access by using two or more than two factors of credentials. Two-factor authentication (2FA), a specialized case of MFA requires a combination of username and password along with a one-time passcode. 2FA is extensively used in healthcare services such as data transmission between patients' wearable IoT devices and the wireless body area networks (WBANs), and providing perfect forward security, untraceability, and resilience of patients' medical data against various attacks [51]. In healthcare system where wearable IoTs are employed, 2FA is more convenient than MFA due to its comparatively higher complexity and user unfriendliness. Therefore, MFA can be used in applications wherein security is significantly prioritized than complexity.

Three major factor groups for connecting a user with established credentials are: ownership factor, knowledge factor, and biometric factor. The ownership factor refers to the things a user possesses, such as a smartcard, smartphone, or other token. The knowledge factor refers to user awareness, such as of a password, and the biometric factor refers to things that the user carries as a human being, such as behavior patterns or the sound of clearing their throat. An MFA policy for a smart city application or a smart device is a combination of more than two of the above factors. For example, a user provides a physical token – such as debit card – that represents the ownership factor, for withdrawing cash from banking ATM service. In addition, the user also requires a PIN – that represents the knowledge factor – to access their account to withdraw money. This 2FA system can be made more secure and complex using additional safety factor of generation of one-time password (OTP) after using debit card and PIN. This OTP is received by the user in their registered authorized smart device. With this MFA policy, the user can access their account after authentication by entering the correct OTP to the ATM machine. Instead of OTP, a user's facial or other biometric factors can be utilized as final layer of authentication.

Applicability of MFA policy can be classified according to the field of services, such as government services for identifying government ID, documents, passport, driver's license, and social security number; commercial services for authenticating physical access control, account login, ATM, and e-commerce; and forensic services for identifying a missing person or missing child and identifying the body. For example, MFA can be enabled for user's amazon webservice (AWS) account. In addition, MFA can be applied for controlling access to AWS service APIs. Google's cloud identity feature allows a wide variety of MFA methods, including a phone as a security key, hardware security keys, smartphone push notifications, SMS and voice calls to select user-friendly correct options among many by companies for their employees.

In general, MFA provides strong security for smart electronic devices and smart web services. However, the selection and implementation of a particular MFA method among many faces numerous challenges. Major challenges in terms of usability are deciding user preference of a specific authentication scheme due to differences in cognitive behavioral properties among users, requiring time for registration and authentication with the system, deciding the maximum number of login attempts for authentication with the system, etc. Integration of various devices and services is another major challenge to implement MFA policy. Additionally, the existence of various sensors, data storage, processing devices, communication channels, and other critical components in an MFA network as well as a smart city network is susceptible to various security attacks, such as replay attempts or adversary attacks at different levels. MFA fails to fulfill robustness criteria in a working environment when a biometric factor is considered for authentication. For example, early facial recognition techniques suffers from failure while the working or operating environment does not have adequate light support and quality camera. The MFA system is susceptible to another type of attack for capturing secret data samples. Therefore, MFA methods that use biometric data must be given improved security levels during data capture, transmission, storage, and processing phases.

### 4.11.6 BLOCKCHAIN

As the present-day world faces rapid urbanization, numerous challenges related to adequate housing and infrastructure emerge. The smart cities are facing security issues, such as confidentiality and authentication in data communication. In this context, blockchain-based technologies are emerging as important tools to augment data transparency and traceability in smart cities.

In the form of a decentralized IT infrastructure, blockchain technology can be deployed as a suitable means to regulate the growing networks in smart cities, such as monitoring product supply chains, managing and validating data trails, in addition to ensuring data authenticity and integrity. Blockchain technology, based on transparent and secure infrastructure presents an immutable and traceable communication of sensitive data between both people and machines. As a result, blockchain technology has obtained special emphasis in both private and public institutions. Cities employ blockchain to create a secure, shared public ledger to maintain real-time data in energy consumption, transportation, and utilities. The technology implementation can assist cities to streamline user interfaces with citizen devices, decrease energy consumption, and share public data with trusted third parties.

Blockchain is a shared and open distributed ledger system that is used for recording transactions between two entities, such that it is immutable and verifiable. The system comprises a shared data storage that is synchronized across multiple nodes in a network. The sole aim of blockchain technology is to establish accountability, transparency, and trust without any dependence on a single authority. This technology enables decentralization and tamper-proof data storage. In smart city ecosystems, the advent of blockchain technology is promising to resolve serious data privacy and integrity challenges. Moreover, this technology is employed in multiple smart city applications pertaining to data access and control, and information sharing for management of patient health record, financial transactions, and energy consumption.

## TABLE 4.1
## Blockchain Architectures

| Architecture classification based on ownership of data infrastructure | Architecture classification based on read, write, and commit permissions granted to the computing nodes | | |
|---|---|---|---|
| | | **Permissionless** | **Permissioned** |
| | **Public** | (i) Any node can join, read, commit, and write.<br>(ii) Hosted on public servers.<br>(iii) Anonymous, high resilience.<br>(iv) Low scalability. | (i) Any node can join and read.<br>(ii) Only authorized nodes can commit and write.<br>(iii) Medium scalability. |
| | **Private** | (i) Only authorized nodes can join, read and write.<br>(ii) Hosted on private servers.<br>(iii) High scalability. | (i) Only authorized nodes can join and read.<br>(ii) Only the network controller can write and commit.<br>(iii) Very high scalability. |

A typical blockchain architecture has the following components, consensus mechanism, time-stamped transactions, distributed nodes, and smart contract. All these components focus on increasing the system security, executing transactions between untrusted entities without any dependency on central authority, and maintaining auditable and tamper-proof records with integrity and transparency. Based on the available architectures, blockchain technologies used in smart city applications can be classified into public-private and permisionless-permissioned. The differences between these architectures are shown in Table 4.1.

### 4.11.7 SIDE-CHANNEL ATTACKS

Side-channel attacks or side-channel analysis (SCAs/SCA) are/is based on the information extracted from a chip or a system, or a hardware's implementation or physical properties, such as power consumption, electromagnetic emission, and execution time [21, 41, 52]. Secure cryptographic devices, such as smart cards or IoT devices, may show vulnerability against side-channel attacks [53]. Therefore, side-channel attacks pose a threat to privacy-preserving goals of smart city-based applications. For example, side channels such as radio-frequency fingerprinting or smartphone accelerators can be used in location inference attacks to pinpoint the location of a device or user.

In [52], the authors have depicted a detailed taxonomy of general SCAs. Broadly, the SCAs are categorized into active and passive attacks depending on how the extraction of leaked information affects the system or the chip. Passive SCAs are further classified into power analysis SCA, which depends on analyzing the power consumption of devices, electromagnetic analysis SCA which focuses on estimating electromagnetic waves emitted from integrated circuits in operation, and timing analysis SCA which depends on timing analysis of required computation for extracting the secret key or narrowing down the possible secret key subset in a secure cryptographic system. Fault injection SCA is classified under active SCA. Type of fault injection attack varies depending on the type of device on which the attack is performed and the information available to the attacker. These four classes of SCAs are further categorized into several sub-classes such as (a) simple, differential, and correlation power analysis; (b) simple and differential electromagnetic analysis; (c) time, trace, and access-driven timing analysis; and (d) voltage glitching, clock pin tampering, laser glitching, electromagnetic disturbance, respectively.

Cache-timing attacks can be performed based on the information gained from processing time of cache memory of a processor. IoT devices deployed in smart cities rely on cryptographic libraries in order to preserve privacy of collected user data, hence susceptible to cache-timing attacks [53].

In order to identify the location of a smart device – or user – location, interference attacks use side channels such as radio-frequency fingerprinting or smartphone accelerators [21]. Therefore, the design and implementation of the smart city infrastructure must consider the fact of not disclosing additional data of the system other than that required. The additional information leakage from the system/device may result in the attacker breaching the security and privacy shields of the system.

Extensive research has been done over the decades for preventing measures against side-channel attacks. A variety of existing countermeasures that are applicable at computationally powerful device setups cannot be implemented in IoT-based infrastructures within smart city environments due to small footprint and low power consumption nature of IoT devices. However, researchers from the security and privacy domain have made extensive progress on developing countermeasures for SCAs for IoT-based infrastructures including smart city applications.

## 4.12  BIOMETRIC

Biometrics play an important role in proper functioning of smart cities, especially in improving the security aspects in connected everyday life. This technology employs biological features, such as fingerprint, face, iris, etc., to identify individuals. This technology contributes largely to the area of public safety, as in many cases they enable law enforcement by improving the reliability and accuracy of verification and identification systems. With the advent of multi-modal biometrics, public safety and user experience has improved, whereas personal privacy has been effectively protected. Integration of such technologies has rendered the applications which embed them more robust against security threats and vulnerabilities.

A typical application of biometric systems comprises recognition of distinct facial feature of a student while the teacher is taking class attendance. In 2017, National Institute of Standards and Technology (NIST), USA, reported NEC's face recognition technology depicted 99.2% accuracy when it matched a person walking through airport boarding gate. As a result of biometric systems, automated access control and identity management can be implemented at ease with technology. Biometric systems in smart city applications are of two main types:

(i) User recognition, wherein a matching of an individual user to multiple entries in biometric data is performed.
(ii) User authentication, or one-to-one, wherein the goal is to establish that the person is indeed the person, who they claim to be.

In the present-day world, as online shopping is widely prevalent, a large demand for substantially strong authentication processes exists. Present techniques that use numbers printed on password-token combinations and credit cards are riddled with security vulnerabilities as many customers have lost money to global cybercrime activities.

On this note, biometrics provide a highly secure way to validate high value transactions. Forinstance, it may provide continuous authentication, wherein the user iris is monitored duringthe entire authentication exercise. A related requirement is the need for increased convenience. Unlike the OTPs, which are widely used in two-factor authentication, personal physical characteristics of the user cannot be lost.

Biometric technology prevents the hassle of possessing multiple authentication devices andremembering complex passwords. Realizing the potential for retail applications, present-dayonline payment powerhouses enter into partnerships with technology-driven companies to incorporate mobile payments on smartphones. Besides the cost factor, the user privacy issueprevents widespread adoption of biometric technologies. Despite the fact that biometric features are irreplaceable, and render the systems secure, it is imperative that such features are not used for unscrupulous activities.

In particular, concerns arise with availability of consumer biometric information from facialphotographs and fingerprints. As a result, such private information is susceptible to theft from public databases. However, biometric systems, such as fingerprint and facial scanners, encode personal characteristics embedded in a fingerprint, face, or iris as a template, instead of storing their actual images. Such systems typically employ multiple features for template construction. This strengthens the usage of a template captured on one device for authenticating another device that uses a different template construction. In addition, it is computationally unfeasible to reconstruct the original image from only the template itself.

## 4.13  AI-ML MODELS FOR DETECTING SECURITY AND PRIVACY ISSUES

The growing number of deployed smart devices for existing and extending coverage zone of smart city applications aim at increasing the economic competitiveness, strengthening sustainability efforts, and improving the quality of life of citizens. Naturally, the data captured by the sensors embedded in smart devices require efficient processing in order to make precise and correct decisions. Extensive research on developing efficient AI and ML algorithms has played a key role in easing data processing and decision-making in the last decade [11]. However, the dependency of training the ML models with historical data for better prediction in decision-making or for dealing with unforeseen data in future turns the AI-ML models vulnerable to security and privacy threats. For example, due to high-end hardware requirements, and the large quantum of time-requirement to train high-performance neural network (NN) ML models, the practice of outsourcing the training to a machine learning as a service (MLaaS) provider renders the integrity of trained NN models at risk. The attacker is often capable of adding or tampering with training data. The mixing of a few malicious samples with normal training data is traditionally known as Trojan in context of ML literature. Moreover, when a few selective malicious samples with the capability of generating a certain trigger pattern embedded in a trained NN with hidden functionality, is widely known as neural Trojan.

MLaaS training process is not transparent and may introduce neural Trojans in trained NNmodels. With an intention to reduce the accuracy and performance of learned models, tampering of normal data or adding malicious data samples with normal data is known data poisoning attack. In this attack, the attacker possesses the information about the training ML algorithm, however they do not have the knowledge about the training process. These trained models, when used in smart city applications, mean the performance of the system declines. In addition, the system and devices in use may be prone to privacy and security threats. Researchers have worked on prevention methods such as neural network verification, Trojan trigger detection, and restoring compromised NN models.

In another aspect, AI and ML models can be utilized to detect privacy and security threatsposed by a wide range of hardware-based attacks. For example, in [54], the authors have used image-processing algorithms and artificial neural network (ANN)-based ML models for the automation of inspection procedures for detecting IC counterfeiting. The benefits of using AI and ML models for solving privacy and security issues in smart city applications are enormous. However, through data poisoning of normal training data, tampering of weights used in NN-based ML models, or other kinds of attacks on ML models, can pose security and privacy threats to the systems that use such ML models.

## 4.14  ANONYMITY

Smart city-based applications depend on collecting data from IoT devices and transmitting them via communication networks. This process yields a trail for tracing back the communication paths back to the user location (such as influencing MAC address) to the service providers [47]. The presence of ever-increasing IoT devices in smart cities such as sensors poses a major challenge in designing a proficient end-to-end anonymity approach to preserve user privacy by obscuring the data communication paths. A full-proof anonymity approach for preserving privacy in different layers of the smart city framework, such as data collection, storage, and data analysis layers, is of paramount importance.

$k$-anonymity is a privacy-preserving model that aims to preserve the privacy of individ-uals/ users from public disclosure of statistical databases by imparting anonymity and untraceability [21].

$k$-anonymity implies a disclosed dataset or release data holds this property if the data for each individual included in the disclosure cannot be uniquely identified from at least $k - 1$ individuals whose data is also present in the disclosure. This traditional $k$-anonymity model is susceptible to various attacks for allowing re-identification of anonymized user data in order to trace the true owner, and their sensitive original data using an available public data matching procedure. In [21], the authors have studied many variations of $k$-anonymity models, such as the $l$-diversity model, $m$-invariance model, and $t$-closeness model.

In the recent past, a British scientist, Junade Ali [55], used the $k$-anonymity property in conjunction with cryptographic hashing to build a communication protocol aiming at anonymously verifying if a password was exposed without disclosing the searched password. This protocol has been extensively employed in various applications, such as in a public API in *Troy Hunt's Have I Been Pwned?* service, the *Authlogics Password Breach Database*. This protocol is absorbed by multiple services including browser extensions and password managers. This approach was reproduced by Google's Password Checkup feature [47, 56]. In [57], the authors have demonstrated that hash-based $k$-anonymity can provide real-time anonymity for MAC addresses. The utility of the proposed approach is extensively analyzed based on smart transportation systems.

## 4.15 SECURITY ISSUES, PROPERTIES, AND COUNTERMEASURES AGAINST THREATS

A smart city refers to an efficient consumption of city resources in the urban ecosystem, resulting in quality enhancement of lives of the residents. The smart city provides facilities in wide-ranging service domains, comprising energy, traffic, medicine, environment, safety, and education. However, smart city applications lead to numerous security and privacy challenges on account of weaknesses in the layers of a smart city system. The resulting attacks may decrease the quality of smart city facilities [58]. In 2015, approximately 230,000 Ukrainians suffered a long duration of power interruptions as intruders hacked the power grid [8]. The enormous volumes of data shared in IoT-enabled services can be accessed, and hence exploited, by malicious adversaries resulting in serious security implications in the smart city. Hence, minimizing such threats to confidentiality and other security risks through the promotion of efficient security solutions is important for smart city approaches to attain success.

Moreover, traffic analysis attacks can be performed against anonymous communication systems, such as analyzing communication system of smart meters for revealing private sensitive information like an individual's consumption and usage of electricity pattern. Traffic analysis-based attacks, such as snooping, performed against anonymous communication systems can reveal user data, such as the webpages or websites visited by the victim [41]. Therefore, snooping-based activities lead to privacy violations by analyzing the communication of intelligent vehicles, smart meters, and IoT devices, etc. To thwart such vulnerabilities, information theory-based packet padding is one of the key performers, which also include predicted packet padding and dummy packet padding.

As mentioned, usage of connected devices in everyday lives can lead to life-threateningsecurity issues. The improved functionality integrated into transportation systems, homes and power grids can be turned into malicious processes when accessed by adversaries. Numerous hacking cases over the past few years have exposed the extent of damage resulting from a critical security breach. The scenario has become worse with the widespread adoption of IoT applications that process classified personal, industrial, and government official data. The typical IoT security goals related to the devices in smart offices and smart homes comprise:

(i) **Authentication:** Authentication refers to the process of ensuring the identity of devices and data emanating from those objects. In IoT context, each object must have the ability to identify and authenticate the entire data originating in the system and the devices from which they are generated.

Ensuring authenticity regarding transactions is extremely important in smart cities which comprise smart grid and smart homes. For authentication, message integrity is guaranteed by using cryptographic hash functions against malicious modifications in the same way as checksums are used for detecting modified payload in message packets. In addition, message authentication codes (MACs), such as HMAC, are predominantly used techniques for authentication. Although similar in construction to cryptographic hash functions, MACs use secret keys in their constructions.

Such schemes, along with digital signature schemes, can be deployed in smart home devices that provide message authentication via asymmetric key encryption. In such schemes, each communicating party must possess a pair of keys, private and public. Prior to sending a message encrypted with the receiver's public key, the sender puts the message to a cryptographic hashing function, such as SHA-256. The hashed output is subsequently signed with the user's private key. The receiver decrypts the message using the private key, and subsequently inputs the message to hashing function to generate a hash value. Moreover, the receiver verifies the sender's signature by using the sender's public key, thereby obtaining the original hash. The obtained hash and the generated hash values are compared. If the hashes are same, the message integrity is proven, and hence the authentication of the sender is established as nobody other than the sender possesses the private key. In addition, non-repudiation is attained if the sender asks for the receiver's signed acknowledgement, which verifies that the message has been received.

(ii) **Integrity:** This encompasses the mechanism of ensuring the consistency, precision, and tamper-evidence of the information during its entire life cycle. In smart IoT-based environment, the alteration of data – or even the injection of invalid information – could create critical problems, for example tampering patient critical information in smart healthcare systems may lead to the death of the patient. Attacks against data integrity mostly belong to message modification. In addition, false data injection, replay attacks, and device impersonation are also considered as major threats to the integrity of smart homes and many other components of smart cities.

(iii) **Confidentiality:** This mechanism ensures that the communicated data is only accessible to authorized people. Some important issues should be considered regarding confidentiality in IoT-based smart systems. Firstly, to ascertain that the device receiving the data does not move/transfer the data to other unintended devices, and secondly, to address the data management issue. Examples of such attacks comprise reading device memory, maliciously editing parameters of the control program of smart meters, spoofing the contents of payloads of packets, and message replay attacks.

(iv) **Non-repudiation:** This process guarantees demonstrating that a task has been both performed and ensures that the performer cannot contradict this task later. Strictly speaking, the task performer cannot refrain from authenticating the transferred data. For instance, a compromised smart meter may send an incorrect reading to the utility provider, and later claim to have not done so. If the smart meter has a secret key embedded for data encryption, non-repudiation is enforced inherently, because no other entity will have a copy of the same secret key. On the contrary, the lack of a secret key-based mechanism will leave the attack undetected.

(v) **Availability:** This process establishes that the required service is available to the authorized users irrespective of their location and time. An important class of probable attacks against availability of resources in smart home or smart grid facilities comprise DoS attacks. This class of attacks are mounted in different forms, such as switching off devices, jamming communication channels which the devices use for data transfer, denial of service against domain name servers (DNS) in the corporate office network, and spoofing.

(vi) **Privacy:** This process ensures non-availability of secret information to public or malicious objects. Some components and devices, such as smart meters, form an integral component of smart cities. However, since their adoption, their deployment has led to concerns of being privacy invasive. The pattern of electric unit consumption can leak information about energy consumption details, the presence or absence of residents within a premise, or information about existing household equipment. Data privacy in such components can be ensured through:

- **Anonymization** – This process breaks the link between the transmitted data and its source so that an application can get the data required for processing, but it cannot establish the identity of data originating from a specific meter.
- **Privacy preserving** – Such techniques are distributed in applications wherein smart meters communicate with utilities. The deployed smart meters are reinforced with additional computational power to perform security computations. A typical example is using zero-knowledge proof (ZKP) protocols, wherein zero-knowledge billing protocol enables the smart meter to compute the bill and proves that the bill is in accordance with the energy consumption, which remains unknown.

## 4.16 SUSTAINABLE SECURITY AND PRIVACY APPROACHES

A wide range of surveys featured the priority of security and privacy requirements for wider acceptability of smart city services by citizens. In [21], authors have listed 24 deployed smart city applications across globe along with possible privacy issues existing in the implementation. Major smart city applications listed are free Wi-Fi access, smart mobility card, intelligent streetlights, smart grids, teaching robots, surveillance robots, eCall automatic emergency calls, and people-counting and noise-monitoring in a public gathering. The authors have also rated the status of privacy countermeasures implemented for deployed applications as not-taken or are unknown. The way smart city applications ease citizens' daily life means the psychological behavior of humans may suppress the concern of privacy and security threats. However, the sustainability of smart city frameworks and applications highly depends on the ability of providing continuous security and privacy measures against the already existing unforeseen attacks.

Research studies confirm that almost 50% of vehicle drivers are perturbed about the security of driver-assistance applications. Collision avoidance systems, automatic self-parking, and adaptive cruise control are driver-aid smart applications among many that depend on connected Internet for smooth operation. In 2015, a group of U.S. security researchers recognized a vulnerability in Jeeps built with a Uconnect vehicle connectivity system. The system allowed attackers/hackers to take control of the vehicle and drive it off the road [59]. Therefore, continuous combatting against cyber threats is a big challenge for the automobile industry, as well as all smart city applications. In reality, the majority of cyber threats go unnoticed until security researchers or intelligent hackers discover it. Thus, sustainable smart applications require continuous monitoring to provide immediate countermeasures once a security loophole is exposed. Recently, an RFID based electronic toll collection system, FASTag, that scans/reads from a distance and automatically deducts the toll at toll gates has been implemented in major highways across India. This smart city application saves waiting time in queue and fuel costs at toll gates.

## 4.17 CONCLUSION

Smart city applications involve a variety of services and technologies that comprise multiple research domains in order to ease citizens' lives using the boon of the digital revolution over the past two decades. Hence, the architecture and communication network of smart cities remain complex. Constant growth in the usage of smart devices, user data, and smart facilities leads to manifold increase in security and privacy challenges. These vulnerabilities and threats encompass various processes, such as data-collection, processing, transmitting through heterogeneous communication network, storing, and accessing in need [60]. In addition, preserving the privacy and security of user devices is also one of the biggest challenges in smart cities. We presented the general architecture used in most smart city applications and highlighted the security and privacy issues in all existing layers. We analyzed the privacy and security issues involved in varying applications, such as power grid systems, smart healthcare, smart metering systems, smart transportation, and smart environments. We categorized the security and privacy framework based on prominent techniques used as security and privacy-preserving tools. Furthermore, we highlighted common security issues, respective properties, and countermeasures against such threats. We identified several security and privacy-preserving issues in presently used technologies as the primary challenges for future research directions for implementation of sustainable and secure smart cities.

## 4.18  FUTURE SCOPE

The World Bank data statistics depicts a continuous growth in urban population with reference to total world population over the time period of 1960 to 2020 [61]. As populations increase, the volume, velocity, variety, and veracity of data generated from various urban digital applications increases. This ever-growing volume of data can be managed on cloud computing platforms while addressing smart city applications for improving the quality of urban life. Therefore, the importance of sustainable smart city applications with data storage and processing on the cloud has become evident. Sustainability of smart city applications depends on user-friendly infrastructure and applications, interoperability between different systems and shared servers in the cloud, and ease of handling critical situations while maintaining security and privacy of data generated and stored on cloud systems. As the urban population increases day by day, the fastest growth in developing sustainable smart city applications among various sectors, such as healthcare, transportation, power grid systems, environment, and e-governance is evident. With the wide acceptability of smart city applications, the core reliability and sustainability depends on privacy and secrecy of data at individual or various organization levels. The advancement in technology supports the growth in malicious intelligent minds. Therefore, smart city applications and their data are prone to be vulnerable unless continuous monitoring for unforeseen security and privacy breaches and an immediate recovery mechanism is implemented. The future scope in the domain of security and privacy research, as well as in the field of developing security and privacyprotection assistance services, is an emerging area [62]. This comprises addressing security threats using developing tools, such as secure multi-party computation, blockchain-based applications, post-quantum cryptographic algorithms, and resilience against a wide variation of side-channel attacks.

## REFERENCES

1. D. Evans, "The internet of things: How the next evolution of the internet is changing everything," Accessed: Jul. 09, 2022. [Online]. Available: https://www.cisco.com/c/dam/en_us/about/ac79/docs/innov/IoT_IBSG_0411FINAL.pdf
2. A. Gharaibeh, M. A. Salahuddin, S. J. Hussini, A. Khreishah, I. Khalil, M. Guizani, and A. A. Fuqaha, "Smart cities: A survey on data management, security, and enabling technologies," *IEEE Communications Surveys and Tutorials*, vol. 19, no. 4, pp. 2456–2501, 2017.
3. C. C. Aggarwal, N. Ashish, and A. Sheth, "The internet of things: A survey from the data-centric perspective," in *Managing and Mining Sensor Data* (C. C. Aggarwal, ed.), pp. 383–428, Boston MA: Springer, 2013.
4. C. Ma, "Smart city and cyber-security; technologies used, leading challenges and future recommendations," *Energy Reports*, vol. 7, pp. 7999–8012, 2021.
5. D. Luckey, H. Fritz, D. Legatiuk, K. Dragos, and K. Smarsly, "Artificial intelligence techniques for smart city applications," in *Proceeding 18th International Conference on Computing in Civil and Building Engineering* (E. T. Santos and S. Scheer, eds.), pp. 3–15, Cham:Springer International Publishing, 2021.
6. A. Perini, A. Susi, and P. Avesani, "A machine learning approach to software requirements prioritization," *IEEE Transactions on Software Engineering*, vol. 39, no. 4, pp. 445–461, 2013.

7. C. Glenn, D. Sterbentz, and A. Wright, "Cyber threat and vulnerability analysis of the U.S. electric sector," Technical Report, United States, 2016.

8. L. Cui, G. Xie, Y. Qu, L. Gao, and Y. Yang, "Security and privacy in smart cities: Challengesand opportunities," *IEEE Access*, vol. 6, pp. 46134–46145, 2018.

9. M. Z. Gunduz and R. Das, "Analysis of cyber-attacks on smart grid applications," in *Proceeding 2018 International Conference on Artificial Intelligence and Data Processing (IDAP)*, pp. 1–5, IEEE, 2018.

10. D. Barda, R. Zaikin, and Y. Shriki, "Keeping the gate locked on your IoT devices: Vulnerabilities found on Amazon's Alexa," Accessed: Jan. 7, 2021. Available: https://research.checkpoint.com/2020/amazons-alexa-hacked/

11. H. Habibzadeh, C. Kaptan, T. Soyata, B. Kantarci, and A. Boukerche, "Smart city system design: A comprehensive study of the application and data planes," *ACM Computing Surveys*, vol. 52, no. 2, pp. 1–38, 2019.

12. F. Al-Turjman, H. Zahmatkesh, and R. Shahroze, "An overview of security and privacy in smart cities IoT communications," *Transactions on Emerging Telecommunications Technologies*, vol. 33, no. 3, p. e3677, March 2022.

13. C. Perera, Y. Qin, J. C. Estrella, S. Reiff-Marganiec, and A. V. Vasilakos, "Fog computing for sustainable smart cities: A survey," *ACM Computing Surveys*, vol. 50, pp. 1–43, Jun. 2017.

14. E. Ismagilova, L. Hughes, N. P. Rana, and Y. K. Dwivedi, "Security, privacy and risks within smart cities: Literature review and development of a smart city interaction framework,"*Information Systems Frontiers*, vol. 24, p. 393–414, 2022.

15. D. Djenouri, R. Laidi, Y. Djenouri, and I. Balasingham, "Machine learning for smart building applications: Review and taxonomy," *ACM Computing Surveys*, vol. 52, pp. 1–36, Mar. 2019.

16. P. M. Chanal and M. S. Kakkasageri, "Security and privacy in IoT: A survey," *Wireless Personal Communications*, vol. 115, no. 2, pp. 1667–1693, 2020.

17. N. Amara, H. Zhiqui, and A. Ali, "Cloud computing security threats and attacks with their mitigation techniques," in *Proceeding 2017 International Conference on Cyber-Enabled Distributed Computing and Knowledge Discovery (CyberC)*, pp. 244–251, 2017.

18. V. Hu, M. Iorga, W. Bao, A. Li, Q. Li, and A. Gouglidis, "General access control guidance for cloud systems," 2020. Available: https://nvlpubs.nist.gov/nistpubs/SpecialPublications/NIST.SP.800-210.pdf

19. P. Mell and T. Grance, "The NIST definition of cloud computing," 2011. Available: http://csrc.nist.gov/publications/PubsSPs.html#800-145

20. A. I. Awad, S. Furnell, A. M. Hassan, and T. Tryfonas, "Special issue on security of IoT-enabled infrastructures in smart cities," *Journal of Ad Hoc Networks*, vol. 92, p. 101850, Sep. 2019.

21. D. Eckhoff and I. Wagner, "Privacy in the smart city-Applications, technologies, challenges, and solutions," *IEEE Communications Surveys and Tutorials*, vol. 20, no. 1, pp. 489–516, 2018.

22. K. Zhang, J. Ni, K. Yang, X. Liang, J. Ren, and X. S. Shen, "Security and privacy in smartcity applications: Challenges and solutions," *IEEE Communications Magazine*, vol. 55, no. 1, pp. 122–129, 2017.

23. R. Kitchin, C. Coletta, L. Evans, and L. Heaphy, "Creating smart cities: Introduction," in *Creating Smart Cities* (C. Coletta, L. Evans, L. Heaphy, and R. Kitchin, eds.), pp. 1–18, London: Routledge, 2018.

24. V. Moustaka, A. Vakali, and L. G. Anthopoulos, "A systematic review for smart city dataanalytics," *ACM Computing Surveys*, vol. 51, no. 5, pp. 1–41, 2018.

25. K. Chaturvedi and T. H. Kolbe, "Towards establishing cross-platform interoperability forsensors in smart cities," *Sensors*, vol. 19, no. 3, pp. 562–1–562–29, 2019.

26. C. Badii, P. Bellini, A. Difino, and P. Nesi, "Smart city IoT platform respecting GDPR privacy and security aspects," *IEEE Access*, vol. 8, pp. 23601–23623, 2020.

27. D. T. Nguyen, C. Song, Z. Qian, S. V. Krishnamurthy, E. J. M. Colbert, and P. McDaniel, "IoTSan: Fortifying the safety of IoT systems," *Proceeding 14th International Conference onCoNEXT '18*, pp. 191–203, New York, NY, USA: Association for Computing Machinery, 2018.

28. R. Kitchin, "Getting smarter about smart cities: Improving data privacy and data security," 2016. Available: https://progcity.maynoothuniversity.ie/wp-content/uploads/2016/01/Smart-Cities-data-privacy-data-security.pdf

29. M. L. Polla, F. Martinelli, and D. Sgandurra, "A survey on security for mobile devices," *IEEE Communications Surveys and Tutorials*, vol. 15, no. 1, pp. 446–471, 2013.

30. D. of Homeland Security, "Study on mobile device security. section 401 of the cybersecurity act of 2015 (consolidated appropriations act of 2016, div. n, pub. l. 114-113, 129 stat. 2244, 2977-78 [2015])," Accessed: Jul. 19, 2021. Available: https://www.dhs.gov/sites/default/files/publications/DHS%20 Study%20on%20Mobile%20Device%20Security%20-%20April%202017-FINAL.pdf

31. Symantec, "Three reasons a secure web gateway is vital for your security stance," Accessed: Jul. 22, 2021. Available: https://www.content.shi.com/SHIcom/ContentAttachmentImages/SharedResources/ PDFs/Symantec/Three%20Reasons%20a%20Secure%20Web%20Gateway%20is%20Vital.pdf

32. A. A. A. Sen, F. A. Eassa, and K. Jambi, "Preserving privacy of smart cities based on the fog computing," in *Smart Societies, Infrastructure, Technologies and Applications* (R. Mehmood, B. Bhaduri, I. Katib, and I. Chlamtac, eds.), pp. 185–191, Cham: Springer International Publishing, 2018.

33. J. Li, W. Zhang, V. Dabra, K.-K. R. Choo, S. Kumari, and D. Hogrefe, "AEP-PPA: Ananonymous, efficient and provably-secure privacy-preserving authentication protocol formobile services in smart cities," *Journal of Network and Computer Applications*, vol. 134, pp. 52–61, 2019.

34. L. Fang, C. Yin, J. Zhu, C. Ge, M. Tanveer, A. Jolfaei, and Z. Cao, "Privacy protection formedical data sharing in smart healthcare," *ACM Transactions on Multimedia Computing, Communications, and Applications*, vol. 16, pp. 1–18, Dec. 2020.

35. N. Tariq, A. Qamar, M. Asim, and F. A. Khan, "Blockchain and smart healthcare security: A survey," *Procedia Computer Science*, vol. 175, pp. 615–620, 2020. The 17th International Conference on Mobile Systems and Pervasive Computing (MobiSPC), The 15th International Conference on Future Networks and Communications (FNC), The 10th International Conference on Sustainable Energy Information Technology.

36. J. M. de Fuentes, L. Gonzalez-Manzano, A. Solanas, and F. Veseli, "Attribute-based credentials for privacy-aware smart health services in IoT-based smart cities," *Computer*, vol. 51, no. 7, pp. 44–53, 2018.

37. S. O. Ogundoyin, "An autonomous lightweight conditional privacy-preserving authentication scheme with provable security for vehicular ad-hoc networks," *International Journal of Computers and Applications*, vol. 42, no. 2, pp. 196–211, 2020.

38. S. A. Alfadhli, S. Lu, A. Fatani, H. Al-Fedhly, and M. Ince, "SD2PA: A fully safe drivingand privacy-preserving authentication scheme for vanets," *Human-centric Computing and Information Sciences*, vol. 10, pp. 1–25, Sep. 2020.

39. K. Gai, L. Qiu, M. Chen, H. Zhao, and M. Qiu, "SA-EAST: Security-aware efficient data transmission for ITS in mobile heterogeneous cloud computing," *ACM Transactions on Embedded Computing Systems*, vol. 16, pp. 1–22, Jan. 2017.

40. F. V. Paulovich, M. C. F. D. Oliveira, and O. N. Oliveira(Jr.), "A future with ubiquitoussensing and intelligent systems," *ACS Sensors*, vol. 3, no. 8, pp. 1433–1438, 2018.

41. S. Yu, "Big privacy: Challenges and opportunities of privacy study in the age of big data," *IEEE Access*, vol. 4, pp. 2751–2763, 2016.

42. C. Gentry, "Fully homomorphic encryption using ideal lattices," in *Proceeding 41st Annual ACM Symposium on Theory of Computing, STOC '09*, pp. 169–178, New York, NY, USA: Association for Computing Machinery, 2009.

43. M. Albrecht, M. Chase, H. Chen, J. Ding, S. Goldwasser, S. Gorbunov, S. Halevi, J. Hoffstein, K. Laine, K. Lauter, S. Lokam, D. Micciancio, D. Moody, T. Morrison, A. Sahai,and V. Vaikuntanathan, "Homomorphic encryption security standard," Technical Report, Toronto, Canada: HomomorphicEncryption.org, Nov. 2018.

44. A. Shamir, "How to share a secret," *Communications of the ACM*, vol. 22, p. 612–613, Nov. 1979.

45. J. Kurihara, S. Kiyomoto, K. Fukushima, and T. Tanaka, "A new (k,n)-threshold secretsharing scheme and its extension," in *Information Security* (T.-C. Wu, C.-L. Lei, V. Rijmen, and D.-T. Lee, eds.), pp. 455–470, Berlin, Heidelberg: Springer Berlin Heidelberg, 2008.

46. L. Bu, M. Isakov, and M. A. Kinsy, "A secure and robust scheme for sharing confidential information in IoT systems," *Ad Hoc Networks*, vol. 92, p. 101762, 2019. Special Issue onSecurity of IoT-enabled Infrastructures in Smart Cities.

47. M. Sookhak, H. Tang, Y. He, and F. R. Yu, "Security and privacy of smart cities: A survey, research issues and challenges," *IEEE Communications Surveys and Tutorials*, vol. 21, no. 2, pp. 1718–1743, 2019.

48. B. Yuan, C. Lin, H. Zhao, D. Zou, L. T. Yang, H. Jin, and C. Rong, "Secure data transportation with software-defined networking and k-n secret sharing for high-confidence IoTservices," *IEEE Internet of Things Journal*, vol. 7, no. 9, pp. 7967–7981, 2020.

49. V. Goyal and A. Kumar, "Non-malleable secret sharing," in *Proceeding 50th Annual ACM SIGACT Symposium on Theory of Computing, STOC 2018*, pp. 685–698, New York, NY, USA: Association for Computing Machinery, 2018.

50. Y. Lindell, "Secure multiparty computation," *Communications of the ACM*, vol. 64, pp. 86–96, Dec. 2020.

51. M. Fotouhi, M. Bayat, A. K. Das, H. A. N. Far, S. M. Pournaghi, and M. Doostari, "A lightweight and secure two-factor authentication scheme for wireless body area networksin health-care IoT," *Computer Networks*, vol. 177, p. 107333, 2020.

52. "Chapter 8 - Side-channel attacks," in *Hardware Security* (S. Bhunia and M. Tehranipoor, eds.), pp. 193–218, Morgan Kaufmann, 2019.

53. S. Takarabt, A. Schaub, A. Facon, S. Guilley, L. Sauvage, Y. Souissi, and Y. Mathieu, "Cache-timing attacks still threaten IoT devices," in *Codes, Cryptology and Information Security* (C. Carlet, S. Guilley, A. Nitaj, and E. M. Souidi, eds.), pp. 13–30, Cham: Springer International Publishing, 2019.

54. N. Asadizanjani, M. Tehranipoor, and D. Forte, "Counterfeit electronics detection usingimage processing and machine learning," *Journal of Physics: Conference Series*, vol. 787, pp. 12–23, Jan. 2017.

55. J. Ali, "Validating leaked passwords with k-anonymity," 2018. Accessed: Jul. 20, 2020. Available: https://blog.cloudare.com/validating-leaked-passwords-with-k-anonymity/

56. L. Li, B. Pal, J. Ali, N. Sullivan, R. Chatterjee, and T. Ristenpart, "Protocols for checking compromised credentials," in *Proceeding 2019 ACM SIGSAC Conference on Computer and Communications Security, CCS '19*, pp. 1387–1403, New York, NY, USA: Association for Computing Machinery, 2019.

57. J. Ali and V. Dyo, "Practical hash-based anonymity for mac addresses," in *ICETE*, 2020.

58. N. Sindhwani, R. Anand, M. Niranjanamurthy, D. C. Verma, and E. B. Valentina (Eds.). (2022). *IoT Based Smart Applications*. Cham, Switzerland: Springer Nature.

59. S. Mehra, "Smart transportation - transforming Indian cities," 2016. Available: https://www.grantthornton.in/globalassets/1.-member-firms/india/assets/pdfs/smart-transportation-report.pdf

60. R. Anand and P. Chawla, "Bandwidth optimization of a novel slotted fractal antenna using modified lightning attachment procedure optimization," in *Smart Antennas: Latest Trends in Design and Application*, pp. 379–392. Cham: Springer International Publishing, 2022.

61. United Nations Population Division, "World urbanization prospects: 2018 revision," Accessed: Aug. 20, 2022. Available: https://data.worldbank.org/indicator/SP.URB.TOTL.IN.ZS

62. R. Raghavan, D. C. Verma, D. Pandey, R. Anand, B. K. Pandey, and H. Singh, "Optimized building extraction from high-resolution satellite imagery using deep learning," *Multimedia Tools and Applications*, vol. 81, no. 29, pp. 42309–42323, 2022.

# 5 Challenges and Opportunities Toward Integration of IoT with Cloud Computing

*Manoj Kumar Patra, Anisha Kumari, Bibhudatta Sahoo, and Ashok Kumar Turuk*
National Institute of Technology, Rourkela, India

## CONTENTS

5.1 Introduction ...................................................................................................82
    5.1.1 Internet of Things .................................................................................82
    5.1.2 Cloud Computing ..................................................................................83
        5.1.2.1 Platform as a Service .............................................................83
        5.1.2.2 Infrastructure as a Service .....................................................83
        5.1.2.3 Software as a Service .............................................................84
    5.1.3 Middleware for Internet of Things ........................................................84
5.2 Architecture of Cloud-Integrated IoT ...............................................................85
5.3 Benefits of Integrating IoT and Cloud .............................................................85
    5.3.1 Storage ..................................................................................................86
    5.3.2 Processing .............................................................................................87
    5.3.3 Data Transmission ...............................................................................87
    5.3.4 Modern Capacities ...............................................................................87
5.4 Existing Solutions for Integrating IoT and Cloud ...........................................87
    5.4.1 OpenIoT .................................................................................................87
    5.4.2 Nimbits ..................................................................................................88
    5.4.3 AWS IoT ................................................................................................88
        5.4.3.1 Device Software .....................................................................88
        5.4.3.2 Connectivity and Control Services ........................................89
        5.4.3.3 Analytics Services ..................................................................90
    5.4.4 CloudPlugs IoT ......................................................................................91
    5.4.5 ThingSpeak ...........................................................................................91
5.5 Application of Cloud-Integrated IoT .................................................................91
    5.5.1 Smart City .............................................................................................92
    5.5.2 Smart Environment Monitoring ...........................................................92
    5.5.3 Smart Home ..........................................................................................92
    5.5.4 Smart Healthcare ..................................................................................92
    5.5.5 Smart Mobility ......................................................................................93
    5.5.6 Smart Surveillance ...............................................................................93
    5.5.7 Towards Serverless Computing .............................................................93
5.6 Issues and Research Challenges .......................................................................94
    5.6.1 Security and Privacy .............................................................................94
    5.6.2 Unnecessary Data Communication ......................................................94
    5.6.3 Deployment of IPv6 .............................................................................94

DOI: 10.1201/9781003319238-5

5.6.4  Resource Allocation and Management ................................................................95
5.6.5  Interoperability ..................................................................................................95
5.6.6  Service Discovery ..............................................................................................95
5.6.7  Scaling ................................................................................................................96
5.6.8  Energy Efficiency ..............................................................................................96
5.7  Conclusion ..................................................................................................................96
References ..............................................................................................................................96

## 5.1  INTRODUCTION

As a result of advancements in information technology, connecting various electronic devices to the internet is now much more straightforward. Acquiring and analyzing information is also much more convenient. Cloud computing and the Internet of Things (IoT) are two technologies that have helped us individually, but the combination of these two technologies has provided us with various new opportunities. Even though these two technologies are distinct, they will become an essential component of the not-too-distant future. The usage of cloud computing gives customers the ability to access their software and apps from any location they want. Computing in the cloud makes it easy to access a wide variety of software solutions and infrastructure, servers, memory, and databases by using the Internet (Awotunde, Jimoh, Ogundokun, Misra, & Abikoye, 2022).

### 5.1.1  INTERNET OF THINGS

The Internet of Things, or IoT, is the term used to describe the process of connecting things to the Internet. The term things used here refers to any device connected to the Internet, such as a mobile phone, medical sensor, intelligent door lock, surveillance camera, smart bicycle, or even a car. In the context of IoT, a "thing" is defined as anything that has the capacity to send and receive data across a network and an Internet Protocol (IP) address. This includes humans who have heart monitors implanted in their bodies, agricultural animals with biochip transponders, and autos with sensors incorporated into the vehicle that monitor the pressure in the tires. More electronics gadgets will join this list as the IoT expands in the coming years. Any device connected to IoT that can be operated and controlled remotely is termed an IoT device. It can be a sensor, actuator, electronic gadget, piece of equipment, or machine that has been configured for a specific use and is capable of transmitting data via the Internet. In IoT, devices and gadgets with built-in sensors are connected to an IoT platform, aggregating data from various devices and using analytics to extract information and share the most helpful information with people or apps adapted for particular needs. IoT systems allow IoT devices to connect with humans and other intelligent IoT devices by sending information over the Internet (Salih, Rashid, Radovanovic, & Bacanin, 2022).

The Internet of Things is composed of many devices responsible for data collection. Because they are all linked to the Internet somehow, each has its unique IP address. In terms of functionality, they range from autonomous automobiles that transfer products around industrial facilities to fundamental sensors that monitor the temperatures inside such buildings. These advanced IoT systems can determine which data is essential and which may be safely discarded. This information extracted from data may be used to identify patterns and trends, make recommendations, and identify potential issues before they arise. Now you might be thinking, what is IoT data? Every IoT device, such as a smart speaker that can listen to commands, a sensor that can measure a room's temperature, or an industrial drone that can gather data on agriculture conditions, can collect data from its surroundings. The information that has been acquired is then sent to the corporate systems, where it may be used in a number of different ways. Hundreds of millions of integrated Internet-enabled IoT devices worldwide provide a vibrant collection of data that businesses can utilize to gather information about company operations' safety, monitor assets, and automate procedures (Borges, Ramos, & Loureiro, 2022). A system dependent on humans' participation to function correctly is not as efficient as one that uses connected devices that can report their status in real time. The IoT seeks to eliminate this need for human intervention. The integration of cloud computing with IoT can solve the problem of data management, data analysis, information extraction, etc., with better efficiency (Li et al., 2022). The detail of integrating the cloud with IoT is presented in Section 5.2.

The IoT is being developed with the purpose of making our daily life easier. The following are some examples: Voice-activated digital assistants capable of automatically placing your typical takeaway order will make it much simpler to get freshly prepared meals sent to your home; devices for tracking your home's surveillance, turning lights on and off automatically as you move around the house, and streaming video so you can keep an eye on things even when you're not there, may all be found in smart homes (Babangida, Perumal, Mustapha, & Yaakob, 2022). Two distinguishing characteristics defining the IoT concept are automation and connectivity. The term "automation" refers to the over-arching concept that IoT entails a direct connection between distinct devices, apparatuses, and other pieces of technology without a person's involvement. The term "connectivity" refers to the improved links inside a single network to provide simple access to a broad range of information on a global scale. Many operational tasks, including inventory management, delivery monitoring, and energy management, may now be fully automated thanks to the widespread use of smart devices in daily life (Rahmani, Bayramov, & Kiani Kalejahi, 2022). For example, equipment and commodities may be tracked using RFID tags and a sensor network. Networked sensors may increase resource efficiency by automating the scheduling and monitoring of resources, such as electricity and water. Motion sensors are one of the many methods that can be used to cut down on the amount of energy and water that is used, which may help businesses of all sizes become more efficient and friendly to the environment (Hossein Motlagh, Mohammadrezaei, Hunt, & Zakeri, 2020; Liu, Hu, Li, & Lai, 2020).

## 5.1.2 Cloud Computing

Cloud computing refers to providing a variety of services by utilizing the Internet (Patra, Sahoo, Sahoo, & Turuk, 2019). These services include computing resources, software solutions, platforms, data storage, processors, database management systems (DBMS), and networking. The use of cloud computing is becoming more popular, both among people and companies, for a variety of reasons, including financial savings, increased productivity, improved efficiency and quality, improved performance, and increased privacy.

### 5.1.2.1 Platform as a Service

Platform as a Service, more often abbreviated as PaaS, gives users access to an extensive ecosystem in the cloud to create and deploy applications. PaaS gives you access to computing resources, enabling you to provide customers with a wide range of services, from the simplest cloud-based apps to sophisticated business applications run on the cloud platform. Cloud computing allows you to acquire the necessary resources on a pay-as-you-go basis from a cloud provider and then access those computing resources through a safe connection to the Internet. PaaS makes developers a platform upon which they may build to create a cloud-based software solution (Van Eyk et al., 2018). PaaS enables developers to construct applications by using pre-existing software components, which is analogous to creating a macro in Excel. There is a reduction in the amount of code that developers must do due to the incorporation of cloud capabilities such as scalability, high availability, and the potential to support many tenants. Tools that are offered as a service via PaaS make it possible for businesses to do data analysis and mining, discover insights and trends, and make predictions to enhance forecasting, product design choices, investment return rates, and other economic decisions (Modisane & Jokonya, 2021).

### 5.1.2.2 Infrastructure as a Service

There are many different cloud computing services, but one of the most common is "Infrastructure as a Service" (IaaS), one of four categories of cloud services which also includes PaaS, SaaS (software as a service), and serverless (no servers). You may avoid the expense and hassle of purchasing and operating actual servers and the complicated architecture of data centers by using IaaS. Each resource is provided as a standalone service component, and you will be charged only for that specific resource for the amount of time you use it. When it comes to moving an application or task to the cloud, this is the way that is both quickest and most cost-effective (Konjaang & Xu, 2021). You may expand the size and performance, improve the security, and lower the expenses of operating an application without redesigning the underlying infrastructure.

### 5.1.2.3   Software as a Service

The software as a Service (SaaS) model offers a comprehensive software solution obtained from a CSP on a pay-per-use basis. You may rent the usage of an application for your company, and then your employees can connect to it through the Internet, often using a web browser. The cloud data center owned and operated by the service provider houses all of the underlying architecture, middleware, application software, and application data (Loukis, Janssen, & Mintchev, 2019). The service providers are responsible for managing the hardware and software. With a suitable service contract, they will assure the availability of the application and the data you store on it, as well as its confidentiality. Using SaaS, your company will be able to get an application up and operate faster while incurring just a modest initial financial investment.

### 5.1.3   MIDDLEWARE FOR INTERNET OF THINGS

There are often a large number of heterogeneous user devices in IoT deployments. These devices have a variety of capabilities and behaviors and support several programming languages. To accomplish a smooth integration with everything, it is vital to have an abstraction layer that can abstract the variability of the environment. The cloud users and software applications can access the information and IoT devices from a network of linked items using middleware, which hides the communication and low-level acquisition components of the process. In this section, we'll look at a variety of IoT middleware. Our comparison includes web service protocols, DPWS (Device Profile for Web Services), and CoAP (Constrained Application Protocol). Though they are not middleware in the traditional sense, they have many features in common with middleware, which are critical to advancing the IoT toward Web of Things (WoT) (Palade et al., 2018).

Global Service Network (GSN) started in 2005 as a platform to process wireless sensor network data streams (Veeraiah et al., 2022). It allows flexible incorporation and deployment of different WSNs. GSN's fundamental idea lies on virtualized sensor, that specifies sensors in XML files and abstracts from actual implementation specifics of sensor data access and stream formats (Mesmoudi et al., 2018). Container-based design of GSN includes multiple layers: the virtualized sensor manager handles virtual sensors and fundamental technologies; the query manager is responsible for parsing and executing SQL queries. This layer also offers customizable alert management and an application interface for microservices and web access. VMS controls device connections via wrappers, including TinyOS, device, and available wrappers (Hammoudi, Aliouat, & Harous, 2018).

OASIS has a web services definition known as DPWS (Device Profile for Web Services), specifically for embedded applications and other devices with limited resources. DPWS uses Service-Oriented Architecture to abstract devices as services. DPWS offers a protocol stack based on SOAP 1.2, WSDL 1.1, XML Schema, and several protocols for identification, privacy, communication, and executing. DPWS services may be configured by using the schema of XML, WSDL, and the WS-Policy somewhat similar to web services that are found using just plug-and-play WS-Discovery protocols. Services are detected by type, scope, or can be both. WS-Eventing offers a publish/subscribe protocol that allows devices to register for device-related messages. WS-Security provides device authentication and secrecy. DPWS has been implemented in C/C++, Java, and Microsoft's Windows Vista and WindowsServer2008. DPWS integration in embedded systems could be too heavy owing to the protocol stack. Thus, DPWS-compliant gateways may encapsulate the underlying IoT elements while using DPWS compatibility and semantics.

LinkSmart is the result of a European research project to create network embedded system middleware based on a service-oriented design. To address the issue of interoperability between proprietary protocols and devices, LinkSmart has been developed. Due to this project, they've created an IoT gateway which allows us to connect digitally with IoT devices on the ground. Applications may control, monitor, and also manage the hardware via the gateway in LinkSmart by using weakly associated OSGi-based internet services enabled by LinkSmart components which may be deployed and controlled without needing a reboot thanks to OSGi's components system. Because it doesn't require a reboot, LinkSmart is the only one that lets you dynamically install and control components while running. It is possible to use the LinkSmart Network Manager, which is preinstalled on the device. To register and discover LinkSmart Network services and to communicate directly between LinkSmart nodes and additional components to provide a topic-based publish/subscribe service node discovery, you have to ensure user security and data encrypted communication systems (El-Hasnony, Mostafa, Elhoseny, & Barakat, 2021). If you want to run LinkSmart on an embedded device, you'll need at least 256 Mb of RAM.

For the latter, an abstract representation of various protocols for communication and different interfaces, such as ZigBee, and Bluetooth, has been implemented via the incorporation of a proxy component. The publish/subscribe method, which is supplied by the component in Event Manager, is another important part that facilitates the exchange of XML messages and is appropriate for use in asynchronous settings. In addition, tools for controlling and managing the IoT resources that have been deployed are now available for iOS, Android, and Windows (Diaz, Martin, & Rubio, 2016).

With a brief introduction to IoT, cloud computing, and middleware for IoT and cloud in Section 5.1, an architecture for integrating IoT and cloud computing has been presented in Section 5.2. Section 5.3 illustrates the several benefits of integrating the cloud and IoT. Section 5.4 presents a detailed description of various existing solutions to integrate IoT and cloud. Section 5.5 describes the possible applications of cloud-integrated IoT, including smart city, smart environment monitoring, smart home, smart healthcare, smart mobility, smart surveillance, and serverless computing. In Section 5.6, various issues and research challenges related to integrating cloud and IoT are presented and, finally, a conclusion is drawn in Section 5.7.

## 5.2 ARCHITECTURE OF CLOUD-INTEGRATED IOT

In the presented architecture in Figure 5.1, the three main components are the IoT layer, cloud layer, and service layer. The main objective of the IoT layer is to collect different types of data from the environment and send those data for further processing and analysis to the cloud. The cloud layer consists of servers, storage, and other computing resources required for analyzing the row data collected using IoT devices. On top of the cloud layer, the service layer is present where different services are availed by the users.

## 5.3 BENEFITS OF INTEGRATING IOT AND CLOUD

The IoT environment includes storage devices, network devices, and actuators. Because of its smaller processing and memory capacities, the IoT system cannot store a significant proportion of the data created by sensors, and its ability to process this data also falls short of expectations. As a result, the IoT system requires some assistance to get over these constraints. The cloud computing model offers a pool of resources, including an endless storage capacity, enormous computational power, and network bandwidth, in addition to a significant number of other resources, which can assist an IoT system in resolving the problems that have been previously discussed. The resources available through cloud computing are elastic in nature; they can grow or shrink in response to the

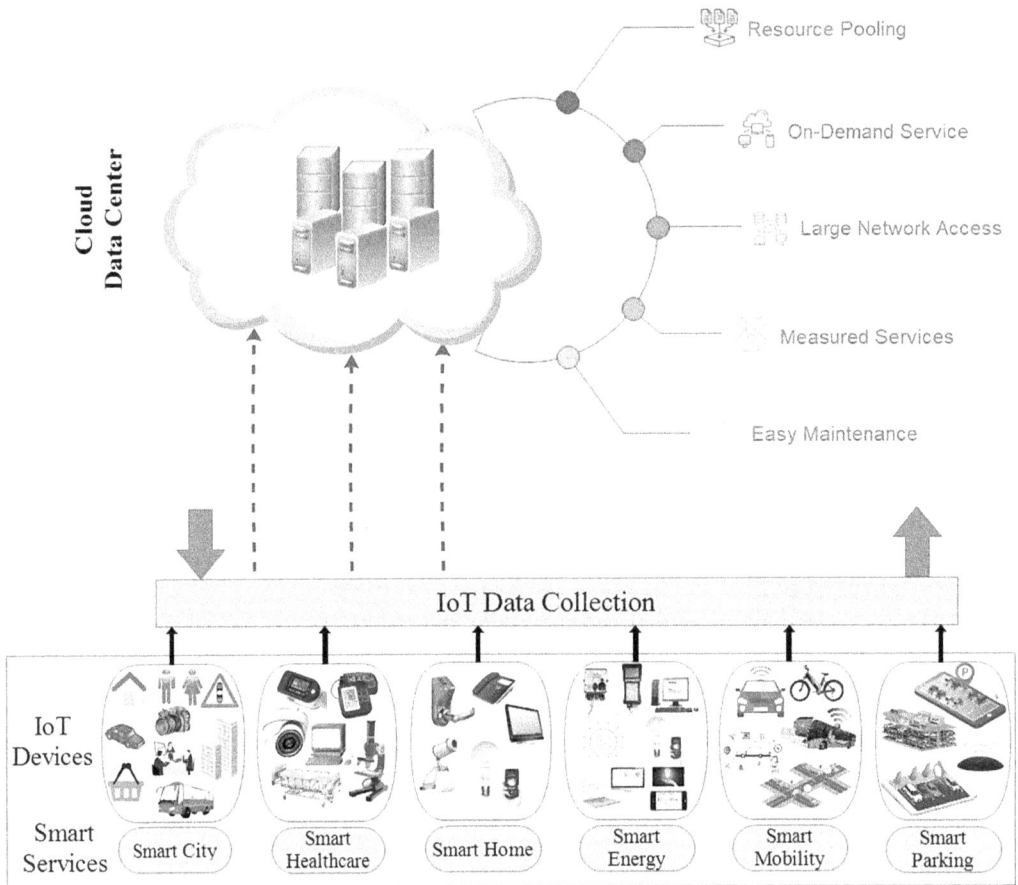

**FIGURE 5.1**    Architecture for cloud-integrated IoT.

requirements of the IoT environment. The analysis of sensor data is another application that can benefit from cloud-based big data analytics. Integrating cloud computing with IoT systems enables applications based on IoT to become more effective and trustworthy. There are a lot of advantages to combining cloud computing and IoT, which will now be discussed (Ansari, Ali, & Alam, 2019).

### 5.3.1  STORAGE

The IoT ecosystem is characterized by the presence of a massive number of interconnected devices and sensors, all of which are responsible for the continuous generation of a vast quantity of data. The local storage capacity of IoT devices is not enough to hold such a vast amounts of data (Wang et al., 2018), and such devices can also produce data in an unstructured or semi-structured format. Traditional database management systems do not have the capability to store data in such a diverse assortment of different forms. Cloud computing provides a solution to the problem of inadequate data storage that is inherent to IoT devices. The concept of cloud computing uses a distributed network of low-cost, all-purpose computers equipped with a vast amount of data storage space already installed. The data produced by IoT devices may be uploaded to the cloud, where it is made available to users in any place that has access to the Internet (Yang, Xiong, & Ren, 2020). This enormous data storage may also be used for analytical reasons, as well as for the purpose of making modifications to the system.

### 5.3.2 Processing

The computing power of IoT devices is extremely constrained. As a result, they cannot process the enormous amounts of data generated by the millions of connected smart gadgets. The IoT relies on cloud computing to supply the necessary computational capacity. Cloud computing makes use of virtualization, a method that splits a single physical machine into multiple virtual machines. The execution of IoT devices' applications can be outsourced to virtual computers that can be hired on a pay-per-use basis. When integrated, cloud computing and the IoT provide application users with low-cost processing power while simultaneously increasing income.

### 5.3.3 Data Transmission

The concept of cloud computing for the IoT enables the interchange of data. Applications that take advantage of the IoT are capable of transporting data from one data node to another, communicating and sharing data among nodes at a very low cost. Cloud computing is the technology that provides a cost-effective and more efficient solution that suits the connection, control, and transfer of data well by using integrated applications.

### 5.3.4 Modern Capacities

The IoT makes use of a wide variety of different kinds of linked devices, each of which utilizes different protocols and methods. As a result, coordinating the operation of all of these different kinds of equipment is quite challenging. In addition, achieving maximum reliability while simultaneously maximizing efficiency can be a difficult task. The elastic and user-friendly nature of cloud computing can help find solutions to the problems caused by the variety of devices. The integration of cloud computing and the IoT provides applications for users with increased reliability, scalability, security, and efficiency.

## 5.4 EXISTING SOLUTIONS FOR INTEGRATING IOT AND CLOUD

This section presents a summary of many current integration concepts for the IoT and cloud computing. The ideas include research initiatives, corporate products, and open-source projects in a variety of domains; as a result, they include a diverse collection of solutions that are already in use in this sector.

### 5.4.1 OpenIoT

OpenIoT is a collaborative project being worked on by several well-known open-source contributors to allow a new variety of open big-scale intelligent IoT applications delivered via a utility cloud computing paradigm. Everyone is aware that other technology solutions are in competition with the IoT due to this technology's platform-independent and resilient qualities. Most notably, the IoT is helping the cloud computing platform because of its scalability and quick access to advanced infrastructure. OpenIoT is seen as a logical extension of cloud computing solutions, which will provide access to new and more essential IoT-based computing resources and capabilities (Hu, Chen, He, Li, & Ning, 2019). This perception is supported by the fact that OpenIoT is now gaining popularity. In particular, OpenIoT will conduct research and offer the methods for developing and controlling environments containing IoT assets. These environments can supply on-demand consumable IoT services, like sensing, as a service. OpenIoT will also provide the environments themselves and bridges the gap between semantics and data computation for IoT programs through cloud interfaces. OpenIoT employs the W3C SSN ontology to unify IoT systems semantically. Its technology gathers and semantically annotates data from any I/O device. OpenIoT offers a function that links related data sets. It can constantly handle a data stream suitable for sensing devices, even without external interfaces. It provides GUI tools to deploy cloud-based IoT apps with a little scripting. In addition, OpenIoT has built a collection of cloud-based apps that allow on-the-fly description of service calls to the OpenIoT system, data visualization, and administration and management components over sensors and OpenIoT applications.

### 5.4.2 NIMBITS

Nimbits is a platform that uses open-source software to link devices in the cloud to one another. Downloading and installing Nimbits on a private network is now an option in addition to utilizing a public cloud version of Nimbits that has been set up on Google App Engine (da Cruz, Rodrigues, Al-Muhtadi, Korotaev, & de Albuquerque, 2018). There are two main components to the Nimbits platform: a web server that stores and analyses geo and time-stamped data, and a Java library that is used for creating and integrating new apps to the platform. The webserver records and processes the data by recording and processing the geo and time-stamped data. In addition, Nimbits comes with a library that provides support for Arduino and an Android app that can manage and display all linked data prints. The JSON format is used for data transmission by the clients. When new data is captured, users may arrange data points to respond in a variety of cascading ways, which will generate a variety of triggers and warnings depending on the circumstance. Even though they do not come with an open-source license, Nimbits' fundamental components may now be downloaded for free. Although open-source licenses are included in the distribution of many Nimbits components, they do not apply to the core components.

### 5.4.3 AWS IoT

Internet of Things (IoT) devices can communicate with AWS cloud services through the cloud services provided by AWS IoT. The device software that AWS IoT provides may assist in integrating IoT devices with the services that AWS IoT delivers. AWS IoT will be able to link IoT devices to the cloud computing services that AWS offers if the equipment can connect to AWS IoT. AWS IoT gives you the ability to pick the most appropriate and updated technologies for your product (Kurniawan, 2018). These protocols are supported by AWS IoT Core so that you may more easily manage and maintain the IoT devices you have deployed.

AWS provides services and solutions for the IoT which allows for the connection and management of billions of devices, data collection, storage, and analysis for IoT applications. Some of the services offered by AWS IoT are:

#### 5.4.3.1 Device Software

**FreeRTOS:** an open-source real-time operating system that was designed specifically for use with microcontrollers. The process of developing, installing, securing, connecting, and controlling small edge devices with less power consumption is made more accessible as a result of this. FreeRTOS is open source and distributed freely by the MIT open-source license and is made up of a core and an ever-expanding library of software that may be used in various markets and contexts because of its adaptability.

This includes the capability to connect your low-power devices securely to AWS Cloud services such as AWS IoT Core or AWS IoT Greengrass. The FreeRTOS is developed with the focus placed on dependability and the convenience of use, and it provides the predictability of long-term support release dates.

**AWS IoT Greengrass:** Construct intelligent IoT devices more quickly by employing premade or bespoke modular components. These components, which can be easily added or removed to regulate your device's software footprint, may be purchased separately or can be constructed from scratch. Without having to update the firmware, you may remotely deploy and manage device software and settings at scale. It brings the computing and logic of the cloud closer to edge devices so that they can function even with a sporadic internet connection. It is possible to configure an IoT device to provide only high-value data, which simplifies the process of providing rich insights at a lesser cost.

**AWS IoT ExpressLink:** A wide variety of AWS Partners' hardware modules, including those created and provided by Espressif, Infineon, and u-blox, are powered by AWS IoT ExpressLink. The connection modules come with software that has been approved by AWS, which makes it much simpler and more expedient for you to connect your devices safely to the cloud and interact with a variety of AWS services in a hassle-free manner.

**AWS IoT EduKit:** IoT from Amazon Web Services by combining a reference hardware kit with a set of educational tutorials and example code that is simple to understand and implement. EduKit makes it simple for developers of all experience levels, from students to seasoned professionals, to receive hands-on expertise in building end-to-end IoT applications. The sample hardware kit offered by AWS IoT EduKit is jam-packed with on-board functionality that makes it possible to run a wide variety of IoT applications right out of the box. The abilities of programmers may be expanded to address more use cases with the help of hundreds of extension options that are plug-and-play. It delivers up-to-date material and sample code to typical IoT applications. Developers have access to a wide selection of free information via the AWS IoT EduKit tutorials, which allows them to acquire skills in the areas of designing and administering IoT applications utilizing AWS services.

### 5.4.3.2 Connectivity and Control Services

This module will protect, manage, and control your devices when they are connected to the cloud.

**AWS IoT Core:** It keeps data and connections between devices safe by using mutual cryptographic authentication and encryption from end to end. It makes it easy and reliable to connect, maintain, and grow your fleet of devices without setting up or maintaining servers.

**AWS IoT Device Defender:** The Amazon Web Services (AWS) IoT Device Defender is a fully managed solution that assists you in protecting the IoT devices in your inventory. Your IoT setups will be subjected to constant scrutiny by AWS IoT Device Defender to make certain that they do not deviate from accepted safety standards. A configuration is nothing but a set of technical control procedures that you may design to keep your information safe when the devices are interacting with one another and the cloud. You can create these controls. Confirming device identification, authentication and authorization of devices, and encryption of device data are some of the IoT settings that can be easily maintained and enforced with the help of AWS IoT Device Defender's user-friendly interface. It also makes it possible to easily manage and enforce IoT setups. Continuously analyze the IoT settings on your devices to ensure they adhere to a set of preset best possible security practices. AWS IoT Device Defender provides this service. Imagine that security problems exist in your IoT setups, such as authenticity credentials being used on multiple devices simultaneously or a device attempting to connect to AWS IoT Core. In contrast, its authenticity credential has been revoked. These are both examples of potential security risks. In such a scenario, AWS IoT Device Defender would send you an alert informing you of the situation.

**AWS IoT Device Management:** It is very necessary to keep track of, monitor, and manage connected device fleets due to the fact that many IoT installations consist of hundreds of devices. After you have deployed your IoT devices, you need to check that they continue to function normally and securely. In addition to this, you will need to manage software and firmware upgrades, monitor the health of your devices, identify and remotely resolve any issues that arise, and secure access to your devices.

AWS IoT Device Management makes it simple to safely register IoT devices at scale, then organize, monitor, and take remote control of their management. You may register each of your interconnected devices one by one or at the same time with AWS IoT Device Management. Additionally, you can quickly adjust permissions to ensure that your devices continue to maintain their level of safety. Using a fully managed web application, you can organize all your devices, monitor them and troubleshoot their operations, check the condition of any IoT device in your fleet, and send over-the-air firmware upgrades. Because AWS IoT Device Management is independent of both the operating system and the kind of device being managed, you can use it to govern anything from restricted microcontrollers to linked automobiles. AWS IoT Device Management gives you the ability to increase your fleets while simultaneously lowering the price and amount of work required to manage big and varied IoT device installations.

**AWS IoT FleetWise:** Near-real-time data collection and transmission to the cloud enables automobile manufacturers to leverage analytics and machine learning techniques to enhance the quality, safety, or autonomous capabilities of their vehicles, as well as the cost-effectiveness of doing so.

### 5.4.3.3  Analytics Services

It works more quickly with data from the IoT devices to extract meaningful information from IoT data.

**AWS IoT SiteWise:** It is a managed service that makes the process of gathering, arranging, and analyzing data from mechanical components simpler. In order to improve efficiency across locations, it is necessary to organize sensor data streams coming from numerous manufacturing lines and facilities. Perform performance monitoring on production lines, assembly robots, and other industrial equipment in order to identify chances for improvement and take action on those opportunities. Remote asset monitoring, which draws on data from the past as well as the present and near real-time, enables quicker problem prevention, detection, and resolution.

**AWS IoT Events:** It's a fully managed service that detects and responds to IoT sensor and app events. Events are data patterns indicating more complex than anticipated events, such as equipment adjustments when a belt gets jammed or motion detectors activating lights and security cameras. Before IoT Events, you had to construct expensive, proprietary apps to gather data, use decision logic to recognize an event, and trigger another app to respond. IoT Occurrences make it simple to recognize activities occurring across hundreds of IoT sensors delivering telemetry data, such as the real-time temperature data of a freezer, the moisture of a respirator, and the speed of a motor belt. Choose the data sources you want to consume, specify the reason for every event by using "if-then-else" statements and then choose a particular type of alert or any personalized action you wish to fire whenever an event takes place. IoT Events performs data analysis on information gathered from various IoT sensors and different applications. It interfaces with AWS IoT Core and AWS IoT Analytics to detect and investigate each event as soon as it occurs. IoT Events initiates warnings and actions based on your logic to fix problems promptly, save maintenance charges, and boost overall operational efficiency.

**AWS IoT Analytics:** With AWS IoT Analytics, full-service analytics on vast amounts of IoT data is now possible. This managed service eliminates the need to construct an IoT analytics platform and reduces the time and expense of doing so. This is made possible by the fact that the service is completely outsourced to AWS. It is the simplest method available for analytics on data collected by the IoT and obtaining insights to help make smarter and more appropriate choices for IoT applications and several machine learning use cases.

**AWS IoT TwinMaker:** The AWS IoT TwinMaker tool makes it simpler for software developers to construct digital replicas of physical systems, such as buildings, factories, pieces of manufacturing equipment, and assembly lines. You will be provided with the tools necessary to construct digital twins via the use of AWS IoT TwinMaker. These digital twins will aid you in optimizing your business's operations, increasing your production levels, and improving your equipment's performance. You can now harness digital twins to create a more comprehensive view of your operations more quickly and with less effort than ever before. This is made possible by using existing data from multiple sources, creating virtual representations of any physical environment, and combining data from existing 3D models with data from the real world.

### 5.4.4 CloudPlugs IoT

CloudPlugs IoT is a next-generation, unified, multi-tenant Industrial IoT platform that offers all-in-one solutions for the administration of devices, applications, and data. It was designed to have a high level of speed and scalability so that it could link a large number of different edge assets and take in a significant amount of data. The system may be deployed on-premises, as an SaaS or dedicated SaaS service, or purchased as a perpetual license for use off-site. Service providers may use CloudPlugs IoT to give IIoT services to their client base, and businesses may utilize CloudPlugs IoT to manage their assets across one or more locations.

The open design of the platform makes it possible to integrate upstream with hyper scalers, systems for analytics, machine learning, artificial intelligence, and business and operation support. A robust edge stack makes it possible to integrate downstream with any field asset. IT and OT users may use customized dashboards, and teams responsible for operations can build, integrate, and deploy new generations of apps to monitor and improve asset usage and performance across many sites.

### 5.4.5 ThingSpeak

ThingSpeak is an IoT analytics platform service (Sindhwani et al., 2022) that enables users to collect, display, and do cloud-based analysis on live data streams. ThingSpeak enables quick visualizations of the data uploaded to the platform by the devices connected to it. Because of ThingSpeak's capability of executing MATLAB code, users can do real-time online processing and analysis of data as it is being collected. ThingSpeak has often been used in the prototype and proof of concept phases of IoT system development where analytics are involved.

ThingSpeak has a number of important capabilities, some of which are as follows:

- Configuring devices to communicate data to ThingSpeak using common IoT protocols is made simple with ThingSpeak,
- Real-time visualization of sensor data,
- On-demand data aggregation from different sources,
- Works well with MATLAB to use IoT data,
- Automate the execution of your IoT analytics depending on predefined schedules or triggers,
- Build working prototypes of IoT systems without the need to configure servers or write web applications,
- Act automatically based on your data and interact by using services provided by other parties.

## 5.5 APPLICATION OF CLOUD-INTEGRATED IoT

The combination of cloud and IoT can be used in many areas. A few of them are discussed below.

### 5.5.1 SMART CITY

With the continuous increase in population in our cities, the shortage of natural resources, and the worries about the environment, services and infrastructure need to be more easily accessible, more interactive, and practical to combat these issues. The goal of developing "smart cities" is to enhance our standard of living in various spheres, including public safety, tourism, transit, urban consumerism, and several others. With the cloud-integrated IoT, it is feasible to create a new generation of services and apps that are competent to interact with the neighboring environment and thus provide new geo-awareness chances. Open data is utilized in different smart cities to organize some of the data that has been gathered. Because of this, it is necessary to provide a QoS level, which includes all time availability, automatic scaling, workload balancing, better security, and confidentiality. The infrastructure of cloud computing may be able to fulfil some of these demands on their own. As a result, the most typical problems and difficulties are connected to the dependability of privacy and security, diversity, size, and durability, as well as real-time communications (Ari et al., 2020).

### 5.5.2 SMART ENVIRONMENT MONITORING

When it comes to environmental monitoring, the cloud-integrated IoT may assist in deploying a faster information communication system that connects different types of sensors and actuators that are installed in the system under consideration with the body responsible for monitoring. Monitoring water quality levels, pollution sources, air quality and gas concentrations in the air, soil humidity, and lighting conditions are just some of the applications that might benefit from this technology. These monitoring periods may go on for an extended period and be continuous.

### 5.5.3 SMART HOME

The smart home is an excellent example of how the cloud may be connected with the IoT. Users are more active inside the setting of their home networks. These days, there are many smart things that we can discover in our houses. These intelligent items have been adapted to meet our requirements. Many IoT applications may be found in domestic places thanks to the proliferation of heterogeneous embedded devices that are connected to cloud computing. These apps make it possible to automate routinely performed tasks within the home. Several of the smart home applications that have been described in published publications include the use of sensor and actuator networks. These applications also link intelligent devices to the Internet in order to enable remote management, monitoring, or control of the devices. The intelligent control of energy use is one example of this. Other examples are smart metering and billing, smart lighting, smart monitoring of room temperature, smart air conditioning, and smart lighting.

### 5.5.4 SMART HEALTHCARE

The current trend in healthcare, which can be seen in many nations today, strives to reduce the number of required hospital resources by relocating certain medical services to patients' homes. This is being done to save money and time (Diaz et al., 2016). For example, one of these services is the patient's physical examination. With the advent of digital health, further research and development should be carried out on the development of algorithms and designing of models that make use of data and help in the decision-making processes of diagnosis and medical treatment. Several health applications and health services, which were before not executable on low-powered devices but are now, can now be executed in the cloud thanks to the Cloud of Things (see later). Patients may thus be remotely watched, and appropriate responses can be made in a timely manner if they are required. In the field of healthcare, many challenges have been researched. These include the consumers' lack of faith in privacy protection, security, connectivity, streaming quality of service, legal concerns, and the question of how to expand storage capacity continuously (Kuo et al., 2011).

### 5.5.5 SMART MOBILITY

The automobile industry is transforming due to the sector's increasingly diverse client base. Shared mobility choices are supplanting the private ownership of a vehicle to achieve both environmental sustainability and financial effectiveness. The advent of "smart mobility" is attempting to bring in an adaptable transportation system that will include driverless vehicles and the electrification of vehicles. Therefore, customers need to be allowed to change the prices of their mobility depending on their specific requirements. A whole new paradigm, known as Mobility as a Service (MaaS), has evolved and it has the potential to cut costs by a significant amount. Cloud-integrated IoT can assist in these transitions by providing potential answers to problems with transportation networks and vehicle services (Mfenjou, Ari, Abdou, Spies, et al., 2018). These possible solutions are known as Intelligent Transportation Systems (Mfenjou et al., 2019).

### 5.5.6 SMART SURVEILLANCE

When it comes to safety and security, intelligent video surveillance has become an extremely useful instrument for a variety of applications. Because of the complexity that can be enabled in video analytics, it is vital to employ cloud-based solutions, such as Video Surveillance as a Service, when it comes to storing, managing, and processing the video content collected from video cameras. This is the only way to satisfy the criteria that must be met more effectively. In the same context as before, the cloud-integrated IoT also results in the automated extraction of knowledge from situations. In addition, the systems that have been presented for intelligent video surveillance can transmit video streams to a sizable number of user devices connected to the internet.

### 5.5.7 TOWARDS SERVERLESS COMPUTING

Serverless computing, popularly known as Function as a Service (FaaS), is a type of cloud computing that enables software developers to design, implement, and deploy software applications as a composition of stateless functions. FaaS is another name for serverless computing (Cassel et al., 2022). Serverless computing does not mean that there are no longer any physical servers since the servers are still in operation, but it follows that developers do not need to be concerned with the administration of servers due to this fact. The primary advantage is that developers do not need to handle the underlying services and OS since this is taken care by the FaaS platform, which also offers transparent scaling methods. This frees up the developers to focus on other aspects of the application. When there is no function being processed by the platform, the platform will automatically scale down to zero, preventing the waste of resources. In turn, the IoT is another rising idea that is closely related to ubiquitous computing. This kind of computing makes it possible for actual physical things to communicate with each other while sharing information over the internet. Although it was first conceived for the cloud, serverless computing may be a fantastic ally for the IoT. This is because FaaS platforms can be positioned physically closer to edge and fog layers, where sensors and actuators are located, to conduct operations with reduced latency. A serverless architecture has been presented in (Kumari & Sahoo, 2022) and to optimize the quality of service, different serverless framework are evaluated in (Kumari, Sahoo, Behera, Misra, & Sharma, 2021). Containerization is another technology growing with serverless that can be used in cloud-IoT for better service provisioning. In (Patra, Sahoo, & Turuk, 2022), the author proposed a model for containerized IoT for smart healthcare service provisioning.

## 5.6  ISSUES AND RESEARCH CHALLENGES

Cloud computing and the IoT will not be as straightforward as they first appear. If we want a brighter future for humanity using this Cloud of Things (CoT), we must first deal with critical challenges that must be addressed. Cloud computing has to consider more than just data storage and computing power to integrate the cloud with IoT successfully. In the future, CoT will open up additional economic prospects, which will make it a more significant target for cyber-attacks. The preservation of one's identity in the CoT is especially critical regarding security and privacy. In the cloud-integrated IoT, multiple communication networks will support various kinds of data and services. In order to handle all different types of data, the network must be able to provide Quality of Service (QoS). Some of the critical issues and their research perspectives are described below.

### 5.6.1  SECURITY AND PRIVACY

In the context of a company, privacy refers to the safeguarding of the personal information of consumers and, more importantly, their proper application of that data. The expectations of consumers should be satisfied with this utilization. When building IoT infrastructure, ensuring data safety and privacy is a top priority. IoT devices are often linked with limited equipment, making them more susceptible to intrusions and threats than other devices. On the other hand, many IoT devices utilize confidential information and vital infrastructures in many situations to perform their functions. The privacy and security of networks, cloud infrastructure, and IoT devices are crucial aspects. As a result, the CoT places a high value on protecting user privacy and the integrity of their data. CoT security threats include issues like preserving the confidentiality of users' information and the IP of manufacturers, finding malicious activity, and figuring out how to stop it. The CoT involves transferring data from the physical world to the cloud. Giving relevant authorization regulations and guidelines while guaranteeing that only authorized users have access to sensitive data is a crucial problem that has not yet been solved.

### 5.6.2  UNNECESSARY DATA COMMUNICATION

The volume of data generated by IoTs would be too much for power-constrained IoTs to process on their own. This is where cloud computing comes in. It is possible to realize the visions of the IoT and the future Internet by integrating IoTs with cloud computing in the Cloud of Things. It's not easy to connect the IoT to the cloud. Cloud as IoT backend solutions may have their advantages; however, the decentralized nature of the IoT does not match well with the centralized cloud. Sending IoT data randomly to the cloud for storage and processing, then returning it to consumers may lead to unacceptable delays (Xavier et al., 2020). There are several difficulties to overcome. Among the problems to be solved is the pruning of data. Because it puts a strain on both the cloud's data center and the network's core, the data may be pre-processed and reduced before being sent to the cloud. Even if IoT devices constantly transmit data to the cloud, it isn't always essential. In this case, either the device has to be stopped from producing data, or the gateway device needs to take action when it is necessary to suspend transferring the data. This will ensure that the network and cloud resources are appropriately utilized. It will also contribute to the effective utilization of power use. The gateway device, which connects IoT to the cloud, should have the ability to do some processing before transferring the data to the Internet and finally to the cloud. The gateway must consider the input from the application to determine the time and kind of application development on it. This might also form the basis for future research in this area.

### 5.6.3  DEPLOYMENT OF IPv6

A huge number of devices will be connected to the cloud-integrated IoT. For uniquely identifying the IoT devices, a proper addressing technique has to be adopted. The currently used IPv4 protocol will not be sufficient to address all IoT devices in the CoT. In order to support the current and future expansion of IoT, IPv6, the successor to IPv4, will offer a huge address space. Because IPv6 can effortlessly manage the ever-increasing number of intelligent sensors that are interconnected to the internet, IPv6 is often considered a crucial enabling technology for the CoT. The combination of IPv4 and IPv6 could also be a possible solution to handle huge IoT devices in the CoT. A network's internal nodes might be immediately accessible from the public Internet if IPv6 is implemented without proper safeguards in place. For example, on-packet attacks like the IPv6-based ping of death might be rapidly leveraged against IoT devices. As a result, protocols that have been designed to function on a secured local network may mistakenly run on the potentially malicious public Internet. Coexistence between IPv4 and IPv6 will not be beneficial until an appropriate, standardized, and efficient method is implemented. A lot of research challenges are there on IPv4–IPv6 integration and migration towards IPv6 to address this issue.

### 5.6.4 Resource Allocation and Management

The primary purpose of IoT devices is to collect data, which is transformed into valuable information by using data analytics. It also delivers valuable information to end-users. But, in a cloud-integrated IoT, many IoT devices require computing resources from the cloud. With limited available resources, satisfying thousands of requests from IoT devices is difficult because it is challenging to decide the exact amount of resource that an IoT device will require. Once the IoT device completes its task, the allocated resources need to be de-allocated and allocated to other IoT devices. So, allocating and managing resources in the CoT is a new research challenge. IoT resource allocation and management are the most important constraints to assure the quality of the end-user experience.

### 5.6.5 Interoperability

Interoperability is a major challenge in IoT because of the enormous number of devices and technologies. However, interoperability is required in other cases like Fog Computing, where the IoT and cloud integration are not enough. Interoperability in the CoT allows users to utilize the same techniques, servers, and other applications across several cloud service providers and different platforms. There are issues and research challenges to consider when deciding whether or not to transfer an IoT application across clouds. These include: rebuilding the cloud-targeted application and its corresponding stack, re-establishing a network to enable IoT applications in their new cloud environment, taking care of the IoT application that runs in the new cloud, and managing the transportation of data as well as encrypting it both while it is in transit and after it has arrived at its destination cloud.

### 5.6.6 Service Discovery

The automatic identification of devices and available services in a network is referred to as service discovery. In a cloud-integrated IoT environment, service discovery involves finding IoT devices as well as the services they provide on their own. To use service discovery, you'll need a service provider, a service consumer, and something called a service registry. When a service provider joins or exits the system, it registers or deregisters itself with the service registry. Using the service registry, the Service Consumer finds a service provider's address and then connects with it. The network addresses of service instances are stored in the Service Registry database. Clients must be able to access network locations retrieved from the service registry through a highly accessible and up-to-date service registry. Replication protocols are used to keep the service registry up-to-date across a cluster of servers. In a CoT, many IoT devices are mobile. So, it would be very difficult for a cloud service provider to discover, manage and keep track of so many mobile IoT devices. Managing the state of IoT nodes, tracking mobile nodes, and keeping track of existing and newly added nodes may need the use of an IoT manager for big IoT systems. In order to accomplish this goal, a standardized method of service discovery is necessary.

### 5.6.7 SCALING

Increasing the scale of IoT puts a strain on the strategy used by companies and the current architecture. It will take more than just adding more sensor nodes to more devices. Administrators of the IoT must guarantee that their platform and architecture can manage the increasing number of connected mobile devices and the stream of data. Companies using IoT systems must be able to scale up and down in response to the volume of data that their network of connected devices is generating. You have to ensure that the infrastructures, cloud, and connection layers can all grow with your operation. Increasing the number of interconnected devices is always a potential security risk when deploying an IoT solution on a large scale. As more devices are added to a network, the attack surface becomes larger and larger. Ensuring that devices and the network remain secure as you scale up the deployment of the IoT may be difficult. Another tough obstacle to overcome is managing the IoT devices once they have been discovered and deployed. Another problem that is inherently present with device management is integrating the architectures with cloud and connection providers. When companies seek to combine their hardware, software, and networking, the process may be complex and time-consuming due to the complexity and difficulty involved. It is possible that ensuring network connectivity for an IoT deployment could also be challenging, mainly when distributing devices over large geographic regions.

### 5.6.8 ENERGY EFFICIENCY

Because of the spread of the IoT networks and increased connection with cloud storage, there will be a great deal of data exchange, which will result in a significant increase in the amount of power required. Typical components in an IoT network device include sensors, processors, transceivers, and power units. Power is an important factor in the processes of video sensing, video encoding, and video decoding. Energy optimization of IoT devices in the CoT is one major issue. There is a lot of research going on for estimating energy consumption in the CoT, and many try to address it using optimization approaches (Anand & Chawla, 2016).

## 5.7 CONCLUSION

Computing in the cloud and integration of IoT devices are harbingers of the subsequent quantum jump in the world of the internet. The Internet of Things Cloud is the name given to the brand-new category of applications that emerge as a result of the integration of these two concepts. This chapter provides an overview of the various facets of the cloud and the IoT and the benefits and problems associated with the establishment of a synergistic approach. In addition to this, it provides in-depth information regarding the incorporation of cloud and IoT together for a wide range of applications. The IoT in the cloud is creating new opportunities for business and research. We have high hopes that the combination of these two factors will lead to the discovery of a new paradigm for the future of numerous networks and an open service platform for users.

## REFERENCES

Anand, R., & Chawla, P. (2016, March). A review on the optimization techniques for bio-inspired antenna design. In *2016 3rd International Conference on Computing for Sustainable Global Development (INDIACom)* (pp. 2228–2233).

Ansari, M., Ali, S. A., & Alam, M. (2019). A synergistic approach for internet of things and cloud integration: Current research and future direction. arXiv preprint arXiv:1912.00750.

Ari, A. A. A., Ngangmo, O. K., Titouna, C., Thiare, O., Mohamadou, A., Gueroui, A. M., et al. (2020). Enabling privacy and security in cloud of things: Architecture, applications, security & privacy challenges. *Applied Computing and Informatics*. https://doi.org/10.1016/j.aci.2019.11.005

Awotunde, J. B., Jimoh, R. G., Ogundokun, R. O., Misra, S., & Abikoye, O. C. (2022). Big data analytics of iot-based cloud system framework: Smart healthcare monitoring systems. In *Artificial intelligence for cloud and edge computing* (pp. 181–208). Springer.

Babangida, L., Perumal, T., Mustapha, N., & Yaakob, R. (2022). Internet of things (iot) based activity recognition strategies in smart homes: A review. *IEEE Sensors Journal*, 22 (9), 8327–8336.

Borges, J. B., Ramos, H. S., & Loureiro, A. A. (2022). A classification strategy for internet of things data based on the class separability analysis of time series dynamics. *ACM Transactions on Internet of Things*, 3 (3), 1–30.

Cassel, G. A. S., Rodrigues, V. F., da Rosa Righi, R., Bez, M. R., Nepomuceno, A. C., & da Costa, C. A. (2022). Serverless computing for internet of things: A systematic literature review. *Future Generation Computer Systems*, 128, 299–316.

da Cruz, M. A., Rodrigues, J. J. P., Al-Muhtadi, J., Korotaev, V. V., & de Albuquerque, V. H. C. (2018). A reference model for internet of things middleware. *IEEE Internet of Things Journal*, 5 (2), 871–883.

Diaz, M., Martin, C., & Rubio, B. (2016). State-of-the-art, challenges, and open issues in the integration of internet of things and cloud computing. *Journal of Network and Computer Applications*, 67, 99–117.

El-Hasnony, I. M., Mostafa, R. R., Elhoseny, M., & Barakat, S. I. (2021). Leveraging mist and fog for big data analytics in iot environment. *Transactions on Emerging Telecommunications Technologies*, 32 (7), e4057.

Hammoudi, S., Aliouat, Z., & Harous, S. (2018). Challenges and research directions for internet of things. *Telecommunication Systems*, 67 (2), 367–385.

Hossein Motlagh, N., Mohammadrezaei, M., Hunt, J., & Zakeri, B. (2020). Internet of things (iot) and the energy sector. *Energies*, 13 (2), 494.

Hu, P., Chen, W., He, C., Li, Y., & Ning, H. (2019). Software-defined edge computing (sdec): Principle, open iot system architecture, applications, and challenges. *IEEE Internet of Things Journal*, 7 (7), 5934–5945.

Konjaang, J. K., & Xu, L. (2021). Meta-heuristic approaches for effective scheduling in infrastructure as a service cloud: A systematic review. *Journal of Network and Systems Management*, 29 (2), 1–57.

Kumari, A., & Sahoo, B. (2022). Serverless architecture for healthcare management systems. In *Handbook of research on mathematical modeling for smart healthcare systems* (pp. 203–227). IGI Global.

Kumari, A., Sahoo, B., Behera, R. K., Misra, S., & Sharma, M. M. (2021). Evaluation of integrated frameworks for optimizing qos in serverless computing. In *International Conference on Computational Science and its Applications* (pp. 277–288).

Kuo, M.-H., et al. (2011). Opportunities and challenges of cloud computing to improve health care services. *Journal of Medical Internet Research*, 13 (3), e1867.

Kurniawan, A. (2018). *Learning aws iot: Effectively manage connected devices on the aws cloud using services such as aws greengrass, aws button, predictive analytics and machine learning.* Packt Publishing Ltd.

Li, X., Liu, H., Wang, W., Zheng, Y., Lv, H., & Lv, Z. (2022). Big data analysis of the internet of things in the digital twins of smart city based on deep learning. *Future Generation Computer Systems*, 128, 167–177.

Liu, X., Hu, S., Li, M., & Lai, B. (2020). Energy-efficient resource allocation for cognitive industrial internet of things with wireless energy harvesting. *IEEE Transactions on Industrial Informatics*, 17 (8), 5668–5677.

Loukis, E., Janssen, M., & Mintchev, I. (2019). Determinants of software-as-a-service benefits and impact on firm performance. *Decision Support Systems*, 117, 38–47.

Mesmoudi, Y., Lamnaour, M., Khamlichi, Y. E., Tahiri, A., Touhafi, A., & Braeken, A. (2018). Design and implementation of a smart gateway for iot applications using heterogeneous smart objects. In *2018 4th International Conference on Cloud Computing Technologies and Applications (Cloudtech)* (pp. 1–7).

Mfenjou, M. L., Ari, A. A. A., Abdou, W., Spies, F., et al. (2018). Methodology and trends for an intelligent transport system in developing countries. *Sustainable Computing: Informatics and Systems*, 19, 96–111.

Mfenjou, M. L., Ari, A. A. A., Njoya, A. N., Mbogne, D. J. F., Abdou, W., Spies, F., et al. (2019). Control points deployment in an intelligent transportation system for monitoring inter-urban network roadway. *Journal of King Saud University-Computer and Information Sciences*, 34 (2), 16–26.

Modisane, P., & Jokonya, O. (2021). Evaluating the benefits of cloud computing in small, medium and micro-sized enterprises (smmes). *Procedia Computer Science*, 181, 784–792.

Palade, A., Cabrera, C., Li, F., White, G., Razzaque, M. A., & Clarke, S. (2018). Middleware for internet of things: An evaluation in a small-scale iot environment. *Journal of Reliable Intelligent Environments*, 4 (1), 3–23.

Patra, M. K., Sahoo, B., & Turuk, A. K. (2022). Smart healthcare system using containerized internet of medical things. In *Handbook of research on mathematical modelling for smart healthcare systems* (pp. 261– 278). IGI Global.

Patra, M. K., Sahoo, S., Sahoo, B., & Turuk, A. K. (2019). Game theoretic approach for real- time task scheduling in cloud computing environment. In *2019 International Conference on Information Technology (ICIT)* (pp. 454–459).

Rahmani, A. M., Bayramov, S., & Kiani Kalejahi, B. (2022). Internet of things applications: Opportunities and threats. *Wireless Personal Communications*, 122 (1), 451–476.

Salih, K. O. M., Rashid, T. A., Radovanovic, D., & Bacanin, N. (2022). A comprehensive survey on the internet of things with the industrial marketplace. *Sensors*, 22 (3), 730.

Sindhwani, N., Anand, R., Vashisth, R., Chauhan, S., Talukdar, V., & Dhabliya, D. (2022, November). Thingspeak-based environmental monitoring system using IoT. In *2022 Seventh International Conference on Parallel, Distributed and Grid Computing (PDGC)* (pp. 675–680).

Van Eyk, E., Toader, L., Talluri, S., Versluis, L., Uta, A., & Iosup, A. (2018). Serverless is more: From paas to present cloud computing. *IEEE Internet Computing*, 22 (5), 8–17.

Veeraiah, V., Anand, R., Mishra, K. N., Dhabliya, D., Ajagekar, S. S., & Kanse, R. (2022, November). Investigating scope of energy efficient routing in Adhoc network. In *2022 Seventh International Conference on Parallel, Distributed and Grid Computing (PDGC)* (pp. 681–686). IEEE.

Wang, T., Zhou, J., Liu, A., Bhuiyan, M. Z. A., Wang, G., & Jia, W. (2018). Fog-based computing and storage offloading for data synchronization in iot. *IEEE Internet of Things Journal*, 6 (3), 4272–4282.

Xavier, T. C., Santos, I. L., Delicato, F. C., Pires, P. F., Alves, M. P., Calmon, T. S., Amorim, C. L. (2020). Collaborative resource allocation for cloud of things systems. *Journal of Network and Computer Applications*, 159, 102592.

Yang, P., Xiong, N., & Ren, J. (2020). Data security and privacy protection for cloud storage: A survey. *IEEE Access*, 8, 131723–131740.

# 6 Multi-variant Processing Model in IIoT

*N. Ambika*

St.Francis College, Bangalore, India

## CONTENTS

6.1 Introduction ....................................................................................................99
6.2 Applications of IIoT ........................................................................................100
    6.2.1 Supply Chain Management ................................................................100
    6.2.2 Automotive Manufacturing ..............................................................101
    6.2.3 Remote Power Grid ...........................................................................101
    6.2.4 Recycling and Sorting System ..........................................................102
6.3 Literature Survey ............................................................................................102
6.4 Previous Work ................................................................................................106
    6.4.1 Drawbacks of the Previous System ..................................................106
6.5 Proposed System ............................................................................................106
6.6 Analysis of Work ............................................................................................106
6.7 Future Scope ..................................................................................................107
6.8 Conclusion ......................................................................................................108
References ..............................................................................................................109

## 6.1 INTRODUCTION

IoT [1–3] is a shrewd get-together of gadgets fit for detecting the climate. The climate, accelerometer, and dampness detectors are underemployment for catching information from congregations. The IoT detector-based [4, 5] information communicates to the stockpile host. It is subsequently put away in the information stockpiling component of the Hadoop Dispersed File Scheme in gigantic dimensions. A characterization model [6] makes concerning the Support Vector Machine (SVM) classifier for foreseeing issues regarding occasions occurring ordinarily or unusually because of ends.

Observing homegrown apparatuses through execution pointers empowers various open doors in lessening costs with compelling upkeep and expenses with the energy squandered in homes. The issue discovery process includes checking the framework's present status through detectors introduced on the apparatuses. It contrasts with the typical activity design. Shortcoming discovery uses data enrolment, acknowledgment, and a sign of abnormalities in the way of behaving of the framework.

The proposed edge calculation pipeline [7] has two phases. The authors first gather the fall-curves for all non-broken detectors utilized in an IoT arrangement in the pre-sending stage, then find the best element vectors to address a fall-curve. These component vectors are improved to infer an element word reference for detectors stacking into the microcontroller for detector proof and shortcoming location. The work records the fall-curves of the non-broken sensors and their relating detector mark, then fits a multinomial bend to each fall-curve period sequence, utilizing the relating multinomial coefficients as the component route, then complete bunching on these multinomial elements for every detector to distinguish the novel highlights. Considering the asset and power limitations of the IoT gadgets, the work upgrades a bunch of hyper-boundaries. The subsequent element word reference alongside the picked hyperparameters stacks onto the IoT gadgets for continuous fall-curve investigation. In the organization phase, they concentrate on the polynomial elements of another fall-curve, then, at that point, find its nearest neighbor from the component word reference obtained during the pre-sending stage. If the closest neighbor distance is inside a specific limit, the recommendation groups the fall-curve as working and allots the comparing detector mark or arranges the detector as defective and sends the fall-curve to the entryway/stockpile for additional handling. The general exactness recognizes the sensors and identifies detector deficiencies.

The recommendation lists the liable list of the industrial equipment. This mapping is compared with the input to detect the deficiencies in the devices. If the detection turns out to be accurate, the work transmits the message to the respective executive staff. The work is simulated using Python. The suggestion finds the defects at an initial phase by 15.8%, equating to an earlier recommendation.

The work is divided into eight sections. The Introduction (Section 6.1) is followed by Section 6.2 which discusses different applications of Industrial Internet of Things (IIoT). Section 6.3 outlines the literature survey while Section 6.4 discusses previous work and its drawbacks. The proposed work is described in Section 6.5 and the analysis of the work is explained in Section 6.6. Future work is discussed in Section 6.7 and the work concludes in Section 6.8.

## 6.2 APPLICATIONS OF IIoT

The modern Internet of Things (IoT) [9] offers interconnection and knowledge to modern frameworks through detecting gadgets and actuators with omnipresent systems administration and registering capacities, which is a critical part of representing things in modern frameworks.

### 6.2.1 SUPPLY CHAIN MANAGEMENT

An excellent item fabricating framework [9] is crucial and beneficial. The PC's motherboard or the circuit inside a cell phone illustrates smart items. These are exceptionally slender, modern, and delicate things that require careful handling. These items make smart devices. Each stage in the creation framework is critical, and there are many stages to making them great. Intelligent machines are more than human work. Smart items need a superior mechanical apparatus framework. One of the elements of smart creation is the utilization of talented workers.

The work [11] is a novel multimodal correspondence model incorporating continuous cultivating creation frameworks in various climates. The architecture practices the Information Distribution Provision middleware to empower correspondence between heterogeneous creation frameworks to perform cultivating activities in an organized way. The robotization level 0 comprises detectors and actuators to amass information from the climate and play out an activity individually. Switch level 1 has machinery that has Programmable Logic Controller (PLC) or a corresponding subsidiary incorporated in it. In a plant climate, this could be a solitary creation line. Executive level 2 synchronizes and controls all devices or creates shapes in an incorporated way. The Manufacture Preparation and Regulator level involves the assembling execution framework that effectively controls the preparation and activities of the whole plant from unrefined components to the completed item. The Enterprise Resource Planning level is the administration level in the modern plant.

The work [12] utilizes RFID innovation at all stages. The item connects with the RFID tag. Esp8266 (which is a low-cost Wi-Fi microchip) checks the contents at each phase of Supply Chain Management (SCM). The label ID transfers data to the information base. The provider logs into the framework and supplements all data connecting with the item or administration. This data is saved in the framework. The director logs into the framework and can acquire all necessary data about the provider. He sends the item status and official conclusion to the framework. The content appends the Tag ID and sends the same to the information base. Utilizing RFID innovation, the data about items will be accessible. The work shares data across inventory network stages by using Esp8266. The suggestion utilizes a few advances connected with IoT. It will upgrade the information assortment process. These data divides among providers' modules. The chief and provider can get item data from the framework database. The manager will assess the provider's item and select the high-quality items.

The provider can access the framework at this stage by entering a username, secret phrase, and track item status. The framework ought to be accessible for the appropriation process. The work uses GPS, and administrators effectively utilize Esp8266 via a minimal expense Wi-Fi.

### 6.2.2 Automotive Manufacturing

Auto production is a complex, energy-intensive process that uses numerous unrefined components and water. The car fabricating process includes a complex storage network. It incorporates natural substances, creating parts and subsystems. The area influences and is impacted effectively by other energy-intensive products, like steel, glass, and petrol-related compounds such as plastic and elastic. The monetary and ecological worries for the auto industry will accordingly generally impact these areas.

The suggestion [13] is to structure the coordinated Fuzzy Fault Tree Model with Bayesian Network (FFTA-BN) model to accomplish a dependable upkeep program for the intricate liquid filling frameworks thinking about every vulnerability. The development of FTA and BN models, master judgment cycles, and support enhancement are three vital elements in the proposed system. The work performs a grouping of the calculations into subjective and quantitative levels. The FTA structure decides the main driver of the top occasion impacted by the middle-of-the-road occasions. The consequences of the FFTA are information characteristics in the BN design. The steadfast quality capacity anticipates the unique connection between occasions. At long last, the aftereffects of the consolidated FFTA-BN prototype are involved in the enhancement model to predict the ideal upkeep spans.

This work [14] explores the reception of manufacturing object-of-Things in the automatic segment fabricating SMEs in an agricultural nation. It inspects test-size prerequisites in this examination model. The organized survey was regulated to respondents to gather the essential information. The work picks SMEs deliberately. The poll overview visits the organizations after getting a positive answer on the telephone. The work rejects nine organizations, and this study endeavors to feature the reception period of the IIoT reception/dispersion process. The objective respondents were the chiefs, proprietors, and ICT officials of the ACM SMEs. The estimation properties determine the dormant developments. The variable loadings were more than 0.5. The external thing loads and tries unwavering composite quality. The typical fluctuation removed was decided to test the united legitimacy.

### 6.2.3 Remote Power Grid

Communication assumes a part in the activity of current power frameworks. It allows continuous observation, mediating, and management of the communication and dispersion of electrical fuel. The advanced framework develops toward an extended dependence on correspondence frameworks for the insurance, metering, and observing of information procurement for planning. It is necessary to grasp the test in the administrations' framework communication and their effect on the continuous inventory of electrical power. Communication delays are one of the difficulties that could influence the exhibition of the fuel framework and guide to control misfortunes and hardware harm. It is vital to explore the reasons and the choices for alleviation.

Engineering [16] has two subsystems. The Base Station (BS) speaks with versatile units and the backhaul network. It comprises various handsets with a powered speaker that intensifies the information power, a radio-recurrence little transmission handset segment, a baseband for framework handling and coding, a DC energy reserve, and a cooling framework. Solar-based chargers are accountable for retaining shortwave irradiance and converting sunlight into natural flow electricity.

Wind Turbine Generator (WTG) is responsible for switching wind energy over to a controlled fuel associated with DC-power transport. It has a controller charge. The bank stores power for future utilization. An inverter switches a low DC voltage into an operating 220 V AC voltage.

### 6.2.4 RECYCLING AND SORTING SYSTEM

The work [1] comprises two principal components: partner incorporation and maintainability appraisal of the development at the intial phase. The recognition of partners and a survey of innovation followed. Studios acquired experiences on apparent drivers and limitations of Tracer-based Sorting (TBS). In the primary studio, members recorded drivers and boundaries of TBS and bunched them into various classes. Every member then evaluated the main driver and boundary class. The process gathered driver and boundary classifications during the primary studio. The newcomers assessed the significance of these classifications. The second partner studio incorporates partner ability in application thoughts for TBS. The procedure created first agendas in a studio with project accomplices utilizing the offer and intent of action materials considering fundamental economy standards. The reasonable pursuit of action designed from the writing is adapted. The incentive and chosen plan of action designs were approved by leading eight master interviews.

The plan [19] is an instrumented mass material molecule including a movement detector. The suggestion utilizes a 9-hub outright direction detector Bosch BNO055, which comprises an accelerometer, gyrator, and tri-pivotal geomagnetic sensing elements. The sensing element gauges the outright directive coordinated towards the world's attractive domain. It permits yielding the direction as division quaternions which are mathematically ideal over other models. The sensing element gives an alignment mode to improve the accuracy of the estimations. The combination calculations work out the outright direction values with the thought of the world's attractive field.

## 6.3 LITERATURE SURVEY

The proposed edge calculation pipeline [7] has two phases. The authors first gather the fall-curves for all non-broken sensing elements utilized in an IoT arrangement in the pre-sending stage, then find the best element vectors to address a fall-curve. These component vectors are improved to infer an element word reference for detectors stacking into the microcontroller for sensing element proof and shortcoming location. The work records the fall-curves of the non-broken sensing elements and their relating sensing element mark. Then fit a polynomial bend to each full-curve time series and utilize the relating polynomial coefficients as the component vector. To distinguish the novel highlights, perform bunching on these polynomial elements for every sensing element. Considering the asset and power limitations of the IoT gadgets, the work upgrades a bunch of hyper-boundaries. The subsequent element word reference alongside the picked hyperparameters stacks onto the IoT gadgets for continuous fall-curve investigation. In the organization phase, the work concentrates on the polynomial elements of another fall-curve. Then, it finds its nearest neighbor from the component word reference got during the pre-sending step. If the closest neighbor space is inside a specific limit, the recommendation groups the fall-curve as working and allots the comparing sensing element mark or arranges the sensing element as defective and sends the fall-curve to the entryway/stockpile for additional handling. The general exactness recognizes the sensing elements and identifies sensing element deficiencies [19].

It has a shortcoming identification module utilizing an autoencoder that gets as information the sensing element estimations used to screen the interaction. The results of the autoencoder are the reproduced values from the details. By limiting the root mean square error of the recreated values, the model learns a portrayal of the information and channels the clamor. By preparing the model on issue-free information, the autoencoder will get familiar with the examples of ordinary working circumstances. It takes on a profound brain network that delays the sensing element estimations used to screen the interaction. The information attribution module depends on a Generative Adversarial system prototype that gains the relationship between the information from the information layer to substitute absent sensing element dimensions. The work supplants the missing qualities so the short-coming location and characterization modules can work accurately – the legal needs both defective and non-flawed information during preparation.

It is an original protector string discovery [20] and deformity acknowledgment structure. In the initial step, the encasing string distinguishes coarsely, and its potential bearings are assessed. The picture and assessed headings communicate back to the neighborhood server. In the subsequent advance, the recommendation uses an adjusted Faster Region-based Convolutional Neural Network (CNN) calculation for the protector string's location, and the work uses Up-Net for semantic division of the cover string. In the third step, a robust computation identifies the deformities of the cover strings.

Industrial Control System (ICS) [21] connects with the actual environment. The readings from the climate by the sensing elements are the input. An ICS isolates into dual principal areas – the domain and the governance tiers. The field layer comprises sensing elements and administrators, and the management slab is the framework's chief. The sensing elements in the field layer initially get data from the atmosphere and send it to the control layer. The management tier processes the data from the climate and gets a gauge of the framework. Based on the assessed framework condition, the relating orders are shipped off to the administrators by the control layer. Actuators in the field layer influence the climate given those orders. After carrying out the coordinated automata in this framework, the information varieties are the information in the grouping. The quality of the property read from the information changes pursue a specific direction, and these progressions are in various groups.

The proposed model [6] deals with the capacity of IoT-based sensing element information and processes it on the Big information Real-Time Engine. It forecasts a model for recognizing issues in modern gatherings. It associates IoT-based sensing elements with the recent get-together pipelines for detecting the information. The work uses temperature, accelerometer, and dampness sensing elements to obtain information from congregations. The IoT sensing element-based information is sent to the stockpile server. It is subsequently put away in the Big Data stockpiling unit of the Hadoop Distributed File System in huge volumes. The exception is to put away sensing element information separately with the assistance of a bunching strategy for Density-based spatial grouping of uses with commotion where the issue forecasts are taken by applying the AI arrangement method of Support Vector Engine. The information in HDFS is handled constantly with the service of Apache Storm. Anomaly identification finishes with the assistance of volume-based spatial grouping of utilizations with commotion for a shared database. It uses Support Vector Machine to detect deficiencies. It prepares a grouping model.

The created IoT gadget [22] detects forces of a pressure machine and holes between the parts and connected ones. The previous guide recognizes the power signals utilizing the attractive-based force sensing element. The force sensing element estimates the attractive field and its certifications in a pressure engine. It impacts the connected domain. The gadget identifies the degree and transmits it to the stockpile server. Essentially, the microlevel vicinity sensing element recognizes the hole. The pre-owned vicinity sensing element can distinguish the distance from 40 m to 5,000 m as the spot goes from 100 m to 2,000 m. Assuming that the hollow between the two rollers is broadening over the permitted width, the sensing element distinguishes the crater and conveys the message to the stockpile server. Also, the detector identifies the hole between the roller and the embedded link part. The sensing device distinguishes it and conveys the message to the control server. The IoT reference point has various LEDs and a cautioning ringer for showing alarms. The gadget is associated with the organization centers utilizing IEEE 802.11ac [23]. The guide identifies the assigned signals and sends them to the stockpile control server. The accompanying area makes sense of the elements of the control server utilizing the stockpile climate.

The proposition is an IoT framework design [24] for wind farms. It has a microcontroller with the structure, fringe-sensing elements, and accelerometer Micro Electro-Mechanical System (MEMS).

It links with the electric generator of the breeze turbine and records the vibration. The details give insight into the status checking framework. The implanted framework contains the Feature Extractor and the Classifiers. They address a dedicated decentralized application since they can distinguish the running status of the breeze turbine generator. This data about the condition of the well-being of the framework is viewed online by dependable experts. It gives security and precision in the preparation of prescient. It is feasible to decrease working expenses, exorbitant disappointments, limit removals, and pointless upkeep groups in seaward establishments. It is an astute decentralized application from the information base obtained by assessing the presence of various classifiers on various element extraction procedures, fully intent on recognizing the best blend of extractor/classifier. The structure of shortcoming discovery is implanted in the framework that mimics a wind turbine.

The suggestion [25] is to display sensing elements incentive for Outliers, Spikes location, and the application view for (occasional) discovery. The learning element has identification engineering for four modules. Every detail is stored in the Application database. It includes the sensing element's edge and the verifiable measurement result. The gathering data serializes in this data set. This database is the point of interaction for high-layer applications. It can get the sensing element shortcoming expectation and information on the client's criticism. Recognition Thread is a foundation administration and contains the primary identifying process. A progression of recognition strings functions string pool and a connected status line. The functioning series dispatcher awakens strings to deal with it. One self-learning string involves the OS clock as a driver. The series reviews the client input. It reconsiders the pattern vectors. The Group-based Fault Detection (GbFD) Algorithm begins by instating worldwide boundaries. It launches the two center cycles.

The RF calculation [25] consolidates the bootstrap accumulation proposed by Breiman and the irregular subspace calculation. An incorporated classifier goes with choices utilizing various choice trees. The choice tree can freely involve preparing tests. Arbitrarily chosen highlights isolate the hub in the preparation test. The choice tree portrays high fluctuation and a deviation. The RF calculation model coordinates the aftereffects of every powerless classifier. The RF calculation takes on bootstrap examination. It is a basic irregular inspecting strategy with substitution. The work excludes 1/3 of the examples in preparing the subset. It is alluded to as cash-on-hand information. Breiman demonstrated that this information is exceptionally significant in research and can supplant the cross-approval technique of informational indexes. The Dataset Accuracy Weighted Random Forest (DAWRF) calculation is a weighted democratic strategy that joins prescient test informational indexes. The work uses cash-based information testing exactness in the choice phase of the calculation. The preparation tests are isolated into two parts. It prepares the choice tree and the prescient test informational collection to assess the primer grouping impact of the choice tree. The visionary test rate X decides the proactive test collection arrangement rules. The choice tree's fundamental characterization impact is the informational index's grouping precision. A choice tree with high order exactness has a decent characterization impact and colossal weight.

The modern creation line [26] has machines. Each appliance has two states: solid and nonsound. A solid-state implies that the engine is working appropriately. It has three tiers. The IoT level acknowledges readings from machines and sends the information to their entryway hubs. These hubs communicate with their entryway hubs in remote mode and have restricted computational abilities and energy holds. The fog layer comprises entryway hubs. It stands on the highest point of every machine. It combines and totals the order results from the IoT layer. It performs calculations, characterizes and sends, and gets information from the stockpile server. The work trades guidelines on the Asynchronous altering Direction Method of Multipliers (ADMM) algorithm. The decision layer comprises the stockpile server liable for all independent directions. It deals with requests for computational errands and applies the proposed enhancement technique to decide the legitimate machine status result.

The recommendation [27] comprises five layers. The information securing layer is liable for obtaining information from sensing elements, or the database can be mass stacked to the system. This tier isn't answerable for tracing or assessing information. The information layer passes the information to the concentrate change load layer. The concentrate change load layer parses the information and creates semantically explained information in the asset description framework configuration. It designs documents for logic rule thinking and the information entities in the knowledge slab. For huge information stockpiling a NoSql data set is used to store sensing element knowledge. The semantic authority thinking layer processes. Its information concurs with the rules and delivers deduced information. The learning layer pre-processes the approaching information, separates significant highlights, and applies AI techniques. AI calculations are lined up in circulated PC servers to handle information quickly. The final slab is the activity tier, which is answerable for assessing the outcomes delivered in the understanding slab.

The intelligent framework [28] depends on one AI calculation uncommonly picked for defect finding and analysis. It improves on the shortcoming recognition for the treatment of factors simultaneously. It has four primary parts. Homegrown Devices screens object and can be any homegrown apparatus. Accentuation in machines is without implanted web associations like old age coolers, climate control systems, and other conventional homegrown appliances. The data collection in real-time device is a minimal expense miniature-controlled gadget liable for the assortment of information connected with the activity of the homegrown gadget in time. The device's electrical extent obtains through suitable harmless associated sensing elements. The acquired details will form the informational collection to recognize the present status of the homegrown gadget and will be accessible at the intelligent design host. Savvy System Server has three modules. An Machine-to-Machine (M2M) connector is responsible for laying out the programmed correspondence between the information gatherer gadget and the AI module. The Machine Learning Module comprises two principal calculations: the Pre-processing and the defect finding and analysis calculation. Pre-processing is responsible for gathering information to eliminate or limit issues like unsound numeric information, clean details, and flaws like an inaccurate, conflicting, copy or missing qualities. The arrangement of information will dissect and sift. It will decide the framework's computational proficiency and exactness. The producer typically distributes the specialized determinations of the observed machine. Different information subsets have the information gathered by the sensing elements introduced on the DCRT and is called web-based information assortment. The Feature Driven Development (FDD) interaction includes checking the present status and the correlation with a reference of an activity pattern. The issue is analyzed using the classifier and the FDDs chief, which will recognize the level of significance of the shortcoming by contrasting it with the producers' data put away in the information base and the Fault Data History data set.

SVM [29] removes the shallow attributes of the log record. In the subsequent stage, the work inputs the error information into Bi-directional Long Short-term Memory Network (LSTM) to dissect the particular shortcoming to decide the exact area of the issues. The main stage has to begin with characterizing all examples into typical and atypical ones. BiLSTM distinguishes deficiencies. The proposal makes an Internet observing continuous framework. The model recognizes and orders the memorable information to look for the issue that has happened. The gadgets display in the Station boundary, with the electronic retail location in the lower-left corner. The petroleum engine is justified by the EPOS. The grease tank is justified by the OM, the Edge Internet of Things Gateway in the center, and the edge server in the right corner. The framework gathers information from Hangzhou, YinChan, and Shanghai stations. The data is transferred to the stockpile and examined by SVM-BiLSTM to track down strange gadgets on the Fault Location Edge. To keep up with the ordinary activity of the GS-IoT framework, when the OM status doesn't transform, it transmits general inquiry order through the EPOS and broker to the ES in the hexadecimal configuration. Ali ECS utilizes ubuntu16.04 as the working framework on the left is the information stockpiling focus in the stockpile layer.

The model gadget [30] is savvy in observing the issue location arrangement of an independent photovoltaic framework, utilizing the cyberspace of object approach. An electronic detecting load is an online application that screens the information continuously. The explored Stand-Alone Photo Voltaic System (SAPVS) is in the Renewable Energy Laboratory at Jijel University.

It comprises 18 PV modules, 12 capacity batteries, two controllers, and one inverter. It has three PV modules associated in equal, two batteries in series, one accuses controller of MPPT, and a heap. The securing load comprises an Arduino Mega 2560 microcontroller that recognizes information from the identification circuits. A 16x4 LCD shows the framework status and checks information progressively. ESP-01 Wi-Fi is a module for information transmission to a host PC. As the ESP-01 works at 3.3 V, a Diatone's BEC-MINI-3.3 V voltage controller diminishes the inventory voltage to 3.3 V.

The strategy [31] is hands-off on a calculation executed on an entrenched framework. This inserted framework adds to the PV framework to empower shortcoming location and identification. The implanted framework has the IoT breaker connector framework and an issue conclusion framework with sensing elements. The PV string was associated with the FDSS framework and electrical burden through a regulator/inverter framework. The Internet of Things Breaker Connector (IoTBC) and the FDSS worked as a team over a Wi-Fi association to complete a tree search fluffy NARX brain network issue. The web association permits the framework to observe and be constrained by a far-off gadget. The change condition of a breaker division was the state during which the unit was idle and becoming [8] dynamic.

## 6.4   PREVIOUS WORK

The proposed edge calculation pipeline [7] has two phases. The authors first gather the fall-curve for all non-broken sensing elements utilized in an IoT arrangement in the pre-sending stage, then find the best element vectors to address a fall-curve. These component vectors are improved to infer an element word reference for detectors stacking into the microcontroller for sensing element proof and shortcoming location. The work records the fall-curve of the non-broken sensing elements and their relating sensing element mark. Then fit a polynomial bend to each fall-curve time series and utilize the relating polynomial coefficients as the component vector. To distinguish the novel highlights, perform bunching on these polynomial elements for every sensing element. Considering the asset and power limitations of the IoT gadgets, the work upgrades a bunch of hyper-boundaries. The subsequent element word reference alongside the picked hyperparameters stacks onto the IoT gadgets for continuous fall-curve investigation. In the organization phase, the work concentrates on the polynomial elements of another fall-curve. Then, it finds its nearest neighbor from the component word reference got during the pre-sending step. If the closest neighbor space is inside a specific limit, the recommendation groups the fall-curve as working and allot the comparing sensing element mark or arranges the sensing element as defective and sends the fall-curve to the entryway/stockpile for additional handling. The general exactness recognizes the sensing elements and identifies sensing element deficiencies.

### 6.4.1   Drawbacks of the Previous System

- The recommendation finds the deficiencies of homogenous datasets.

## 6.5   PROPOSED SYSTEM

The suggestion considers streaming of various datasets. It uses classification and clustering methodology to recognize the deficiencies. Table 6.1 portrays the algorithm to detect the deficiencies in the input.

## 6.6   ANALYSIS OF WORK

The proposed edge calculation pipeline [7] has two phases. The authors first gather the fall-curve for all non-broken sensing elements utilized in an IoT arrangement in the pre-sending stage, then find the best element vectors to address a fall-curve. These component vectors are improved to infer an

## TABLE 6.1
### Algorithm 1 to Detect the Deficiencies in the Input

Step 1 – List the industrial equipment's considered to find defects.

Step 2 – Analyze the various parameters in the input to be considered.

Step 3 – Store the defects liable to be detected.

Step 4 – Accept the input equipment.

Step 5 – Compare the relative parameters list with the liable defects in the input.

Step 6 – If defect is available, mark the serial number, vendor number, defect and other details.

Step 7 – Send the details to the respective executive for further action.

## TABLE 6.2
### List of Simulation Parameters Considered

| Parameter List | Description |
| --- | --- |
| Number of pieces of equipment considered | 5 |
| Number of properties considered | 5 variants per instrument |
| Number of IoT devices considered | 5 |
| Total number of pieces of equipment considered | 40 (8 similar devices * 5) |
| Number of images taken | 40* 5 angles |

element word reference for detectors stacking into the microcontroller for sensing element proof and shortcoming location. The work records the fall-curve of the non-broken sensing elements and their relating sensing element mark. Then fit a polynomial bend to each fall-curve time series and utilize the relating polynomial coefficients as the component vector. To distinguish the novel highlights, perform bunching on these polynomial elements for every sensing element. Considering the asset and power limitations of the IoT gadgets, the work upgrades a bunch of hyper-boundaries. The subsequent element word reference alongside the picked hyperparameters stacks onto the IoT gadgets for continuous fall-curve investigation. In the organization phase, the work concentrates on the polynomial elements of another fall-curve. Then, it finds its nearest neighbor from the component word reference got during the pre-sending step. If the closest neighbor space is inside a specific limit, the recommendation groups the fall-curve as working and allot the comparing sensing element mark or arranges the sensing element as defective and sends the fall-curve to the entryway/stockpile for additional handling. The general exactness recognizes the sensing elements and identifies sensing element deficiencies.

The recommendation lists the liable list of the industrial equipment. This mapping is compared with the input to detect the deficiencies in the devices. If the detection turns out to be true, the message is sent to the respective executive staff. The work is simulated in Python. Table 6.2 contains the list simulation parameters considered. The suggestion finds the defects at an early stage by 15.8% compared to previous work. The same is represented in Figure 6.1. Figure 6.2 represents the defects located in the considered variants.

## 6.7   FUTURE SCOPE

The recommendation lists the liable list of the industrial equipment. This mapping is compared with the input to detect the deficiencies in the devices. If the detection turns out to be accurate, the work transmits the message to the respective executive staff. The work is simulated using Python. The suggestion finds the defects at an initial phase by 15.8%, equating to an earlier recommendation.

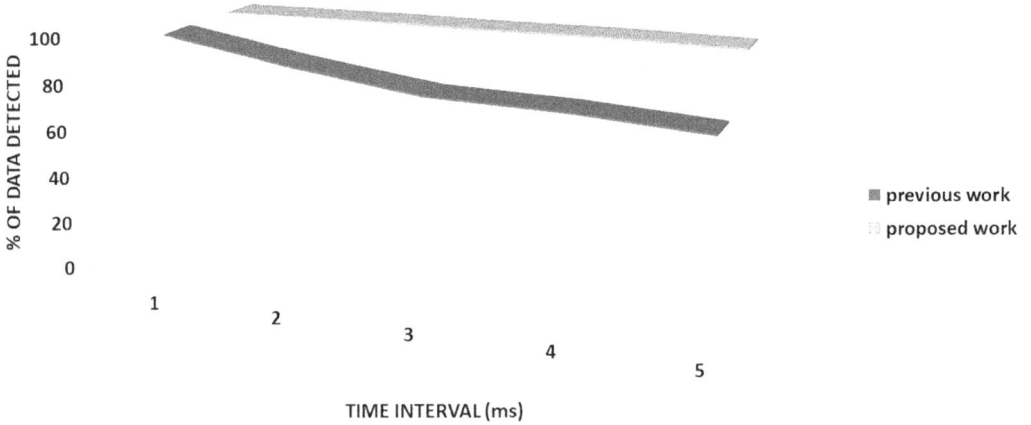

**FIGURE 6.1**   Detection of data.

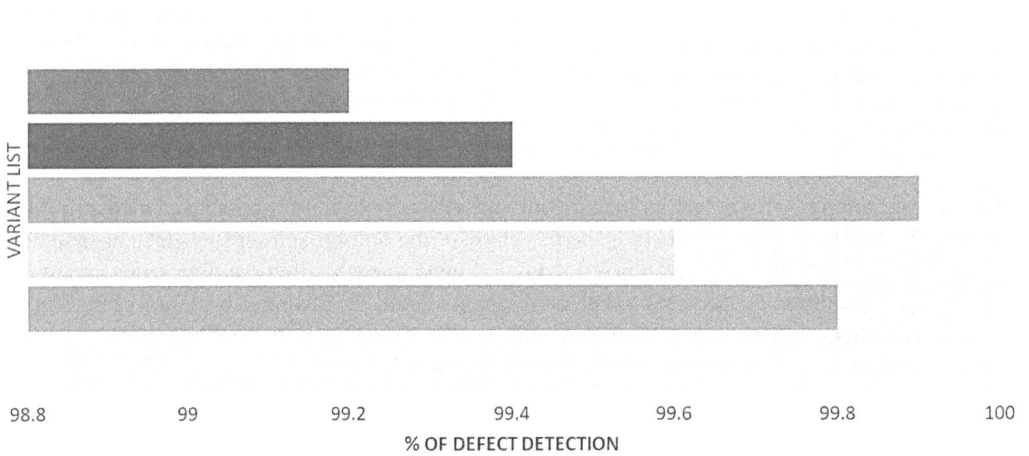

**FIGURE 6.2**   Different variants considered in the industrial equipment.

Future systems can make suggestions regarding alternative industrial equipment available for replacement.

## 6.8   CONCLUSION

IoT is an exceptionally complete innovation which has disciplinary applications cutting across a few regions like communication, software engineering, sensing device innovation, and interchanges [32, 33]. IIoT is an organization of frameworks, actual articles, applications, and stages that has installed innovation to share knowledge, conveying it to associates and outside the organization. It empowers the coordination and communication of information and its examination between associated modern frameworks. It acts as an impetus to work on current performance. IIoT is described by distributed computing, security, universal information, information investigation, clever machines, and consistent client experience [34].

The proposed edge calculation pipeline has two phases. The authors first gather the fall-curve for all non-broken sensing elements utilized in an IoT arrangement in the pre-sending stage, then find the best element vectors to address a fall-curve. These component vectors are improved to infer an element word reference for detectors stacking into the microcontroller for sensing element proof and shortcoming location. The work records the fall-curve of the non-broken sensing elements and their relating sensing element mark. Then fit a polynomial bend to each fall-curve time series and utilize the relating polynomial coefficients as the component vector. To distinguish the novel highlights, perform bunching on these polynomial elements for every sensing element. Considering the asset and power limitations of the IoT gadgets, the work upgrades a bunch of hyper-boundaries. The subsequent element word reference alongside the picked hyperparameters stacks onto the IoT gadgets for continuous fall-curve investigation. In the organization phase, the work concentrates on the polynomial elements of another fall-curve. Then, it finds its nearest neighbor from the component word reference got during the pre-sending step. If the closest neighbor space is inside a specific limit, the recommendation groups the fall-curve as working and allot the comparing sensing element mark or arranges the sensing element as defective and sends the fall-curve to the entryway/stockpile for additional handling. The general exactness recognizes the sensing elements and identifies sensing element deficiencies [35].

The recommendation lists the liable list of the industrial equipment. This mapping is compared with the input to detect the deficiencies in the devices. If the detection turns to be true, the message is sent to the respective executive staff. The work is simulated in Python. The suggestion considers multiple variants to provide early detection by 15.8%.

## REFERENCES

[1] Ambika, N. (2020). Tackling Jamming Attacks in IoT. In A. M., S. K., & K. S. (eds), *Tackling Jamming Attacks in IoT*. (pp. 153–165). Cham: Springer. https://doi.org/10.4018/978-1-6684-7132-6.ch011

[2] Nagaraj, A. (2021). *Introduction to Sensors in IoT and Cloud Computing Applications*. UAE: Bentham Science Publishers.

[3] Hassan, W. H. (2019). Current research on Internet of Things (IoT) security: A survey. *Computer Networks*, *148*, 283–294.

[4] Dian, F. J., Vahidnia, R., & Rahmati, A. (2020). Wearables and the Internet of Things (IoT), applications, opportunities, and challenges: A Survey. *IEEE Access*, *8*, 69200–69211.

[5] Balakrishna, S., Solanki, V. K., Kumar, R., & Thirumaran, M. (2020). Survey on Machine Learning-Based Clustering Algorithms for IoT Data Cluster Analysis. In Solanki, V., Hoang, M., Lu, Z., & Pattnaik, P. (eds), *Intelligent Computing in Engineering* (pp. 1195–1204). Singapore: Springer.

[6] Rashid, M., Singh, H., Goyal, V., Ahmad, N., & Mogla, N. (2022). Efficient Big Data-Based Storage and Processing Model in Internet of Things for Improving Accuracy Fault Detection in Industrial Processes. In Khosrow-Pour, M. (ed), *Research Anthology on Big Data Analytics, Architectures, and Applications* (pp. 945–957). US: IGI Global.

[7] Chakraborty, T., Nambi, A. U., Chandra, R., Sharma, R., Swaminathan, M., Kapetanovic, Z., & Appavoo, J. (2018). curve: A novel primitive for IoT fault detection and isolation. *16th ACM Conference on Embedded Networked Sensor Systems* (pp. 95–107). Shenzhen China: ACM.

[8] Choudhary, K., Gaba, G., Butun, I., & Kumar, P. (2020). MAKE-IT—A lightweight mutual authentication and key exchange protocol for industrial Internet of Things. *Sensors*, *20*, 5166.

[9] Houlihan, J. B. (1985). International supply chain management. *International Journal of Physical Distribution and Materials Management*, *15*, 22–38.

[10] Bhuniya, S., Pareek, S., Sarkar, B., & Sett, B. ( 2021). A smart production process for the optimum energy consumption with maintenance policy under a supply chain management. *PRO*, *9*, 19.

[11] Almadani, B., & Mostafa, S. M. (2021). IIoT based multimodal communication model for agriculture and agro-industries. *IEEE Access*, *9*, 10070–10088.

[12] Abdel-Basset, M., Manogaran, G., & Mohamed, M. (2018). Internet of Things (IoT) and its impact on supply chain: A framework for building smart, secure and efficient systems. *Future Generation Computer Systems*, *86*, 614–628.

[13] Soltanali, H., Khojastehpour, M., Farinha, J., & Pais, J. (2021). An integrated fuzzy fault tree model with bayesian network-based maintenance optimization of complex equipment in automotive manufacturing. *Energies*, *14*, 7758.

[14] Sivathanu, B. (2019). Adoption of industrial IoT (IIoT) in auto-component manufacturing SMEs in India. *Information Resources Management Journal (IRMJ)*, *32*(2), 52–75.

[15] Muyizere, D., Letting, L., & Munyazikwiye, B. (2022). Effects of communication signal delay on the power grid: A review. *Electronics*, *11*, 874.

[16] Alsharif, M. H., & Kim, J. (2016). Hybrid off-grid SPV/WTG power system for remote cellular base stations towards green and sustainable cellular networks in South Korea. *Energies*, *10*(1), 9.

[17] Gasde, J., Woidasky, J., Moesslein, J., & Lang-Koetz, C. (2021). Plastics recycling with tracer-based-sorting: Challenges of a potential radical technology. *Sustainability*, *13*, 258.

[18] Maier, G., Pfaff, F., Bittner, A., Gruna, R., Noack, B., Kruggel-Emden, H., … Beyerer, J. (2020). Characterizing material flow in sensor-based sorting systems using an instrumented particle. *at-Automatisierungstechnik*, *68*(4), 256–264.

[19] Dzaferagic, M., Marchetti, N., & Macaluso, I. (2021). Fault detection and classification in Industrial IoT in case of missing sensor data. *TechRxiv*.

[20] Song, C., Xu, W., Han, G., Zeng, P., Wang, Z., & Yu, S. (2020). A cloud edge collaborative intelligence method of insulator string defect detection for power IIoT. *IEEE Internet of Things Journal*, *8*(9), 7510–7520.

[21] Moradbeikie, A., Jamshidi, K., Bohlooli, A., Garcia, J., & Masip-Bruin, X. (2020). An IIoT based ICS to improve safety through fast and accurate hazard detection and differentiation. *IEEE access*, *8*, 206942–206957.

[22] Lee, H. (2017). Framework and development of fault detection classification using IoT device and cloud environment. *Journal of Manufacturing Systems*, *43*, 257–270.

[23] Sindhwani, N., Anand, R., Vashisth, R., Chauhan, S., Talukdar, V., & Dhabliya, D. (2022). Thingspeak-based environmental monitoring system using IoT. *Seventh International Conference on Parallel, Distributed and Grid Computing* (pp. 675–680). IEEE.

[24] Liu, Y., Yang, Y., Lv, X., & Wang, L. (2013). A self-learning sensor fault detection framework for industry monitoring IoT. *Mathematical Problems in Engineering*, *2013*. https://doi.org/10.1155/2013/712028

[25] Zhang, W., Wang, J., Han, G., Huang, S., Feng, Y., & Shu, L. (2020). A data set accuracy weighted random forest algorithm for iot fault detection based on edge computing and blockchain. *IEEE Internet of Things Journal*, *8*(4), 2354–2363.

[26] Xenakis, A., Karageorgos, A., Lallas, E., Chis, A. E., & González-Vélez, H. (2019). Towards distributed IoT/cloud based fault detection and maintenance in industrial automation. *The 10th International Conference on Ambient Systems, Networks and Technologies (ANT 2019) / The 2nd International Conference on Emerging Data and Industry 4.0 (EDI40 2019) / Affiliated Workshops.151*, pp. 683–690. ELSEVIER.

[27] Onal, A. C., Sezer, O. B., Ozbayoglu, M., & Dogdu, E. (2017). Weather data analysis and sensor fault detection using an extended IoT framework with semantics, big data, and machine learning. *IEEE International Conference on Big Data (Big Data)* (pp. 2037–2046). Boston, MA, USA: IEEE.

[28] Seabra, J. C., Costa, M. A., & Lucena, M. M. (2016). IoT based intelligent system for fault detection and diagnosis in domestic appliances. *IEEE 6th International Conference on Consumer Electronics-Berlin (ICCE-Berlin)*, (pp. 205–208). Berlin, Germany.

[29] Jiahao, Y., Jiang, X., Wang, S., Jiang, K., & Yu, X. (2020). SVM-BiLSTM: A fault detection method for the gas station IoT system based on deep learning. *IEEE Access*, *8*, 203712–203723.

[30] Mellit, A., Hamied, A., Lughi, V., & Pavan, A. M. (2020). A low-cost monitoring and fault detection system for stand-alone photovoltaic systems using IoT technique. In *ELECTRIMACS 2019* (pp. 349–358). Cham: Springer.

[31] Natsheh, E., & Samara, S. (2020). Tree search fuzzy narx neural network fault detection technique for PV systems with iot support. *Electronics*, *9*(7), 1087.

[32] Ambika, N. (2021). A Reliable Blockchain-Based Image Encryption Scheme for IIoT Networks. In *Blockchain and AI Technology in the Industrial Internet of Things* (pp. 81–97). US: IGI Global.

[33] Kaur, J., Sindhwani, N., Anand, R., & Pandey, D. (2023). Implementation of IoT in Various Domains. In *IoT Based Smart Applications* (pp. 165–178). Springer, Cham.

[34] Raghavan, R., Verma, D. C., Pandey, D., Anand, R., Pandey, B. K., & Singh, H. (2022). Optimized building extraction from high-resolution satellite imagery using deep learning. *Multimedia Tools and Applications*, *81*(29), 42309–42323.

[35] Anand, R., & Chawla, P. (2022). Bandwidth optimization of a novel slotted fractal antenna using modified lightning attachment procedure optimization. In *Smart Antennas: Latest Trends in Design and Application* (pp. 379–392). Cham: Springer International Publishing.

# 7 Mobile Health

## *Roles of Sensors and IoT in Healthcare Technology*

*Amit Kumar Singh*
VSB Engineering College, Karur, India

*Anshul Gaur*
Uttarakhand Technical University, Uttarakhand, India

*Maya Datt Joshi*
GLA University, Mathura, India

*M. Marieswaran*
National Institute of Technology, Raipur, India

## CONTENTS

7.1 Introduction .................................................................................................. 111
7.2 Roles of Sensors in mHealth ........................................................................ 113
    7.2.1 Roles of External Sensors Used in mHealth ..................................... 114
    7.2.2 Roles of Sensors of the Smart Mobile in mHealth ............................ 116
7.3 Software Feature of the Smart Mobile ......................................................... 119
7.4 IoT Role, Benefits, and Customers in mHealth ........................................... 121
7.5 Role of the Cloud in mHealth ...................................................................... 124
7.6 IoT in Healthcare: Challenges and Future Trends ....................................... 125
7.7 Future Role of IoT and Cloud in mHealth .................................................. 126
7.8 Conclusions .................................................................................................. 127
References ............................................................................................................. 127

## 7.1 INTRODUCTION

Due to the developing populace, the patient to doctor ratio is increasing nowadays, which causes many illnesses and problems to be left undiagnosed in the international scenario. We can say that currently the world's diseases are under-diagnosed due to lack of availability of clinical facilities. There are many causes for this deficiency, indeed one of them is the higher value of the diagnostics and the clinical professionals, as most international locations don't have a free or government-based healthcare system. So, to resolve the above troubles, the engineering generation performs an important role in this area. One of these spheres is the Internet, which has changed the sector and entered human life in the usual and unusual ways. The development of the Internet of Things (IoT) represents an excellent advance that has potential in nearly all areas of human society. Statista reports that the current number of smart mobile customers today is 6.648 billion and that 83.72%

of the world's population now owns a smart mobile. So, through interfacing, the smart mobile, IoT, and the unique sensors technologies that patients require, the enterprise can resolve the foremost issues of the clinical discipline, create masses of jobs, and grow to a trillion-dollar market in the future. The previous decade has seen much research into healthcare services and their technological advancement. Specifically, the IoT has demonstrated software capability to integrate numerous clinical sensors/devices and healthcare experts to provide clinical services in remote regions. This has improved patient safety, lowered healthcare costs, made healthcare services more accessible, and increased operational efficiency in the healthcare business.

Furthermore, future challenges, trends, and issues within the mobile-based health (mHealth) system are discussed. As the wide variety of users grows, cloud-based servers are required to handle and save extensive records, and also enable the use of gadgets through distinctive companions in the mobile healthcare field. So, this chapter discusses various factors of mHealth, and the function and alertness of mHealth is a one-of-a-kind scientific field.

Medical diagnostics account for a sizeable portion of clinic costs. Technology can stream clinical test workouts from a medical institution to the patient's home. Due to increasing costs, healthcare services are becoming out of reach for most of the world's population. Due to the ageing population and the increase in chronic diseases, medical service providers demand quality and affordable healthcare services [1].

IoT is the institution of devices linked to the Internet to carry out these services and to providers that assist our fundamental wishes, economics, fitness, and surroundings. Cloud computing is an archetype wherein dynamical, scalable, and virtualized assets are provided as services through the Internet. Cloud computing alongside the IoT enhances performance competencies and functional resource storage the most. Hence, cloud computing is used to get the right of access to the IoT. The consumerization of the healthcare industry is growing rapidly, which empowers people to stay healthier via connected gadgets, including drugs, wearables, and handheld devices. IoT is an advancing generation that bridges interoperability challenges to trade the way healthcare significantly may be introduced, driving higher results, growing performance, and making healthcare less expensive. Figure 7.1 suggests the basic block diagram of an mHealth device. The patient, who has the smart mobile, has the sensors and app installed and can monitor themselves or through their caretaker. The smart mobile sends those records to the smart hospital, person's home, personal medical doctor, smart pharmacy, and can keep the data on the cloud server to be used and modified for reference on the patient's fitness by the healthcare professional. The patient's prognosis can also be made

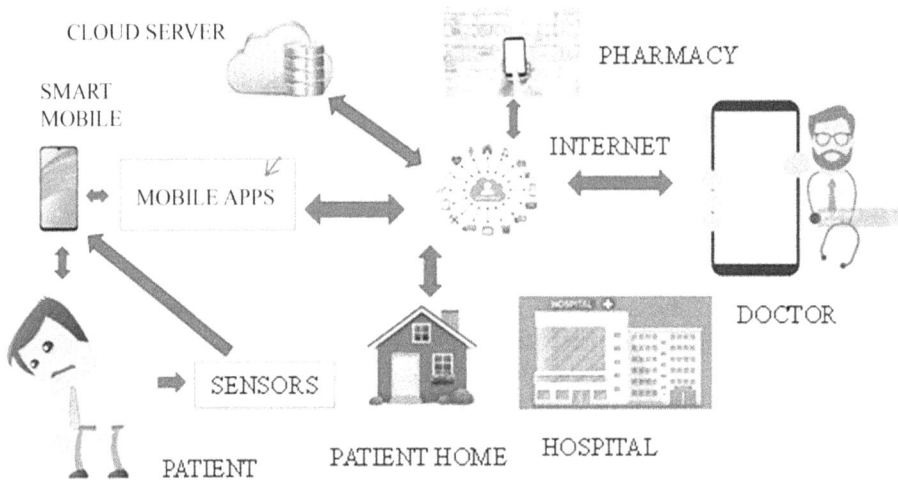

**FIGURE 7.1**  Block diagram of IoT-based mHealth using a cloud server.

using the separate sensors connected to the smart mobile via wireless and wired technologies, such as USB, speaker jack, Wi-Fi, and Bluetooth.

We discuss the human fitness (mHealth) machine in the context of the IoT in this study. First, we discuss the different roles of external sensors attached to the smart mobile and sent to the patient's record. Then we discuss the different sensors present in the smart mobile and their applications in the healthcare sector. After that, the software features to be included in the apps for the smart mobile are discussed and then the benefits of the IoT in mHealth. Finally, the role of the cloud is discussed along with the different applications, challenges and future trends.

mHealth refers to cellular devices that gather patients' fitness data in real time and store it on network servers connected to the Internet. Statistics can be accessed through various customer institutions, for example hospitals, health insurance organizations, and so forth. Doctors use mHealth information to identify, recognize, and treat patients. The availability of wearable scientific devices and body sensor networks promotes mHealth. Integration of mobile fitness devices within the patient's environment allows the detection of fitness irregularities in real time [1]. Recent advancements in micro and nanotechnology, statistical processing, and wireless communication enable smart downsizing, non-invasive biological monitoring, and wearable sensing, processing, and communication.

## 7.2 ROLES OF SENSORS IN mHEALTH

Different sensors have been used for mHealth. Figure 7.2 shows different applications of the sensors used in the healthcare sector. It includes diabetes management, vital signs of the human body, mood monitoring, oncology applications, pulmonary disease monitoring, cardiovascular health,

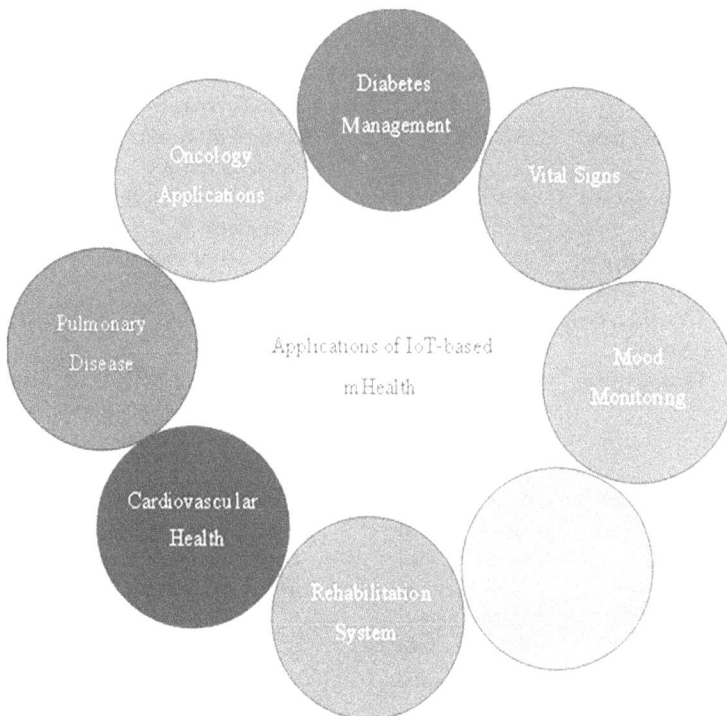

FIGURE 7.2   Applications of IoT-based mHealth.

rehabilitation engineering, and medical management. Based on whether the sensors used are on the smart mobile or interfaced externally with the smart mobile using a wired/wireless protocol, the two classifications are discussed below.

### 7.2.1 Roles of External Sensors Used in mHealth

The traditional sensor for patient healthcare monitoring requires an extensive and costly setup, expert operation, more time, and for patients to go to a distant place. Nowadays, there are options for wearable (external) sensors to monitor the patient's physiological parameters. The wearable sensors come from watches, headbands, bracelets, skin patches, clothing, earphones, and others. The parameters measured by these sensors go to a device such as a smartphone. Then, the information is analyzed with the help of suitable applications and transferred to a remote location for further analysis and suitable action. For example, some doctors, researchers, or patient's family members may be in a remote place. These sensors can measure several body health parameters. The sensors are classified based on the body's physical or chemical parameters. Accordingly, there are force (pressure), optical, audio, acceleration, temperature, GPS, and others. Each sensor can monitor one or more physiological parameters. In addition, they may differ in terms of their manufacturing technique, operating range, and other parameters for that purpose. In Table 7.1, we have discussed different applications, pros and cons of the external sensors used in various

**TABLE 7.1**
**Different External Sensors Applications and Their Pros and Cons**

| Name of Sensor | Application | Physiological Parameter/Disease | Advantages | Limitations |
|---|---|---|---|---|
| Accelerometer | Rehabilitation | Fall detection | Good sensitivity and frequency response | Temperature-sensitive |
| Gyroscope | Rehabilitation | Fall detection | Good sensitivity and frequency response | Temperature-sensitive |
| Magnetometer | Cardiovascular | Functioning of heart | Low power consumption, good range | Less precise, Sensitive to environment |
| Force (resistive, piezoelectric, capacitive and triboelectric) | Cardiovascular, Voice, Bone Structure, Body, Motion | ECG, heart rhythm, strain distribution, sound vibration | Detect several mechanical stimuli of the body | Many of the types are still at the laboratory stage |
| Temperature | General Health | Vital Signs | Easy installation,implementation, and integration, high sensitivity | Measurement sensitivity is low |
| Optical | General Health | Pulse, Oxygen level, Blood pressure Detection: Viral concentration, proteins, ions Glucose monitoring | Immune to EMI, non-invasive measurement, corrosion-resistant, prone to motion artefacts | Ambient light interference, poor penetration in somebody fluids |
| Electrochemical | Different healthcare domains | Cancer, Diabetes | Good accuracy and repeatability | Lower temperature range and shelf life |

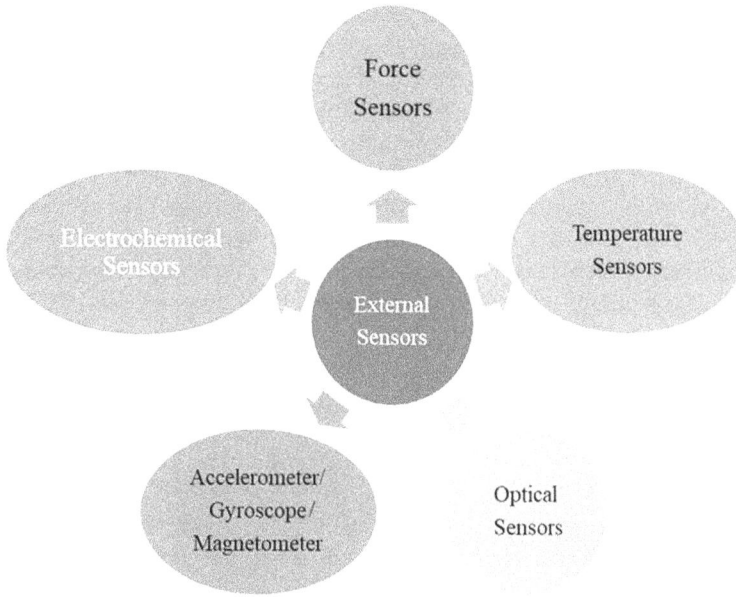

**FIGURE 7.3** External sensors attached with the smart mobile for mHealth.

healthcare applications. Figure 7.3 shows different external sensors used with the smart mobile and are discussed below:

(1) **Force Sensors:** These sensors help measure various biological parameters. They transduce the mechanical signals such as force, pressure, compression, tension, torque, stress, strain, and vibration into electrical signals [2]. Flexible force sensors are mainly used to monitor biological parameters [3]. The sensors are lightweight, require limited space and power, and their signal conditioning is easy to design. The human body generates different mechanical stimuli in body and muscle movements, pulse rate, and blood pressure, and the force sensors are suitable for detecting them. The flexible force sensors are mainly developed from metal, polymer, or carbon-based materials [4], the latter having good biocompatibility and electrical conductivity. Metals also have good conductivity, and polymers have good chemical stability. Flexible force sensors are mainly capacitive, resistive, piezoelectric, and triboelectric [5] and are used for patients undergoing rehabilitation after surgery or an accident to check the body movements of the patients. These sensors are also used for detecting Parkinson's disease. Furthermore, flexible force sensors help monitor the recovery of the hand, detect proper posture, sports training, gait detection, detecting cardiovascular disease, diagnosis of damaged vocal cords, diagnosing facial paralysis, detecting osteoporosis, and others. Based on these applications, the sensors are placed on different parts of the body such as the heart, face, hand, knee, elbow, wrist, foot, throat, and others.

(2) **Accelerometer, Gyroscope, Magnetometer, Barometer:** These sensors are beneficial for fall detection, gait measurement, and mental health analysis. Fall detection is one of the prime parameters useful for aged persons in particular [6]. The accelerometer output relates to the patient's energy expenditure, and the gyroscope output provides the falling risk of a person. In addition, these sensors are useful for accessing diseases like stroke, Parkinson's, cerebral palsy, and others [7]. Accelerometers measure acceleration in three directions, gyroscopes measure angular momentum, magnetometers measure orientation, and barometers measure altitude changes. Furthermore, magnetometers are useful for

cardiology-based applications and analysis of the functioning of the heart. These sensors can be used externally as well as inbuilt sensors of smartphones.

(3) **Optical Sensors:** Optical sensors, such as cameras and other photosensors, are becoming popular alternatives to earlier traditional sensing systems. They can be used in wearable (external) and inbuilt smartphone forms of operation. These sensors are useful for cardiology-based parameters such as heart rate, respiration rate, heart rate variability, blood Pressure, skin-related diseases, ophthalmology-based diseases, glucose level monitoring, and others [8]. Optical sensors' main advantages are their immunity to electromagnetic interference, non-invasive detection up to substantial depth levels, low cost, and others. In addition, the wearable type of optical sensor has the advantage over inbuilt smartphone sensors in terms of using various light sources in different wavelength ranges.

(4) **Electrochemical Sensors:** Currently, amperometric and potentiometric transducers are in use; however, impedance-based transducers have received much interest in the modern age, the reason being that they may be label-unfastened detecting sensors. These sensors are in their early development stage. Besides antibodies, electrochemical devices utilizing DNA hybridization have been created and used for maximum cancer gene mutation detection. An unmarried-stranded DNA series is fixed at the electrode floor. Here the DNA hybridization takes place on this device. Banerjee et al. have shown the use of IoT in most cancer detection [9]. These sensors can also diagnose cancer and diabetes and have other relevant healthcare applications.

(5) **Temperature Sensors:** Temperature is one of the primary physiological parameters of the body. Its irregular value is related directly or indirectly to some disease or problem occurring in the body. For example, the patient must be monitored regularly during some surgical or medical treatment. Flexible temperature sensors as wearable (external) sensors are becoming popular due to several advantages. Flexible resistance temperature detectors (FRTCs), flexible thermistors, flexible thermocouples, and flexible thermochromic are the different types of flexible temperature detectors [10]. FRTCs are highly sensitive, reliable, and flexible [11]; thermistors have high accuracy and repeatability [12] and are mainly used for biological applications. Chen et al. discussed carbon nanomaterial-based flexible temperature sensors [13]. Flexible sensors are also helpful for bionic skin and prosthetics applications.

### 7.2.2   ROLES OF SENSORS OF THE SMART MOBILE IN mHEALTH

Through the mHealth system, it is possible to send the health parameters of the monitored person to the concerned doctors, hospitals, institutes, or organizations. The developments in sensing and communication technologies are helping the healthcare industry achieve its goals. The mHealth-based system is the first step in this direction. mHealth sensors support it using eHealth, Wireless Body Area Networks, Wireless Sensor Networks, and wearable technology for monitoring and diagnosing patients' health parameters [9]. Under these categories, different types of inbuilt smartphone sensors are in use, as shown in Figure 7.4 and their comparisons in Table 7.2, and a few of these are the image (camera) sensors, auditory (microphone) sensors, motion (accelerometer, gyroscope) sensors, temperature sensors, light sensors, pressure sensors, fingerprint sensors, and global positioning system (GPS) sensors. These sensors monitor different physiological parameters of the body. These include cardiovascular, ophthalmic, pulmonary, mental, activity, sleep, hearing, skin, and other health parameters.

(1) **Image Sensor (Camera):** Most modern smartphones have good-quality cameras. These are capable of providing information about various body parameters. These mainly focus on the heart, eye, skin, and other essential parameters. The heart rate (HR) or heart rate variability (HRV) are the primary parameters doctors use to monitor patients as they indicate functional irregularities in the heart. The traditional way to monitor the heart uses multiple-lead electrocardiogram (ECG) systems, which are expensive and not portable. Some portable wearable sensor kits are in the market but require additional accessories.

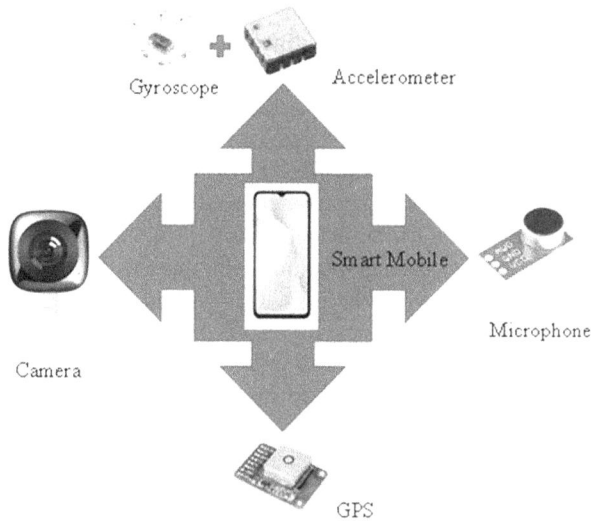

**FIGURE 7.4** Different Components of the Smartphone used as Sensors for the Diagnosis of Different Diseases in mHealth.

**TABLE 7.2**

**Components of Smart Mobile Used for Different Applications for mHealth**

| Name of Components | Application | Physiological Parameter/Disease | Advantages | Limitations |
|---|---|---|---|---|
| Optical (Image) (e.g. camera) | Cardiovascular | HR, HRV, RR, Blood Pressure | Highly sensitive, immune to EMI, lightweight, chemically inert Saves time Saves money Real-time monitoring | Costly, Ambient light interference Accuracy concerns, and Security concern |
| | Skin health | Skin diseases/Skin cancer | | |
| | Ophthalmology | diabetes retinopathy, blindness | | |
| | Glucose monitoring | Glucose level | | |
| Microphone | Pulmonary | Breathing rate, asthma, lung cancer, cough, COPD | Faster processing, Reduced Tinnitus Symptoms, Greater independence | Affected by background noise Security concerns, Adjustment for different levels of sound |
| | Hearing | Hearing Impairment | | |
| | Mental health | Depression, bipolar disorder, schizophrenia | | |
| Accelerometer, Gyroscopes | Rehabilitation | Gait measurement, Fall detection, and others | Highly sensitive, less loading effect, have built-in signal conditioning | Sensitive to temperature, it also requires an external power supply |
| | Mental health | Depression, bipolar disorder, schizophrenia | | |
| GPS | Mental health | Depression, bipolar disorder, schizophrenia, Fitness tracking | Easy patient monitoring | Satellite, Internet dependency, privacy issues |

These issues can be solved with a smartphone camera as an image sensor. It measures the blood volume changes within the finger or face skin with some basic principles of light, and the signal obtained is a photoplethysmograph (PPG) [14]. The smartphone has a light source (near IR, red, or white light) to illuminate the skin. The PPG signal provides the HR or HRV data in the time or frequency domain. The smartphone camera with a suitable application calculates the respiration rate (RR) by detecting chest and abdomen movement.

HR, HRV, and RR results suffer from motion artefacts [15]. Skin cancer caused by abnormal skin tissue growth must be detected early [16]. Among different types of skin cancers, melanoma is the most dangerous one. The main cause of this type of cancer is the over-exposure to harmful ultraviolet rays. This type of skin cancer can be detected by analyzing the size, shape, color, and texture of precancerous lesions. Some skin diseases, such as eczema, moles, and psoriasis, need to be detected. According to researchers, the image sensor in a smartphone camera is utilized to detect, analyze, classify, and predict skin-related diseases. Wadhawan et al. developed a portable library named SkinScan© for melanoma detection. It can be used on handheld devices and is based on certain image processing algorithms using machine learning concepts of Support Vector Machine. The response time mentioned is 15 sec [17]. Do et al. proposed an image processing-based melanoma detection system using a lightweight algorithm suitable to work on a smartphone. According to them, there is a need for a large dataset of skin-related diseases suitable for smartphones [18]. Using smartphones, Kim et al. developed a multispectral imaging technique that discriminates between seborrheic dermatitis and psoriasis on the scalp [19].

Diabetes diseases require timely diagnosis and treatment because they may lead to blindness. The gold standard for its diagnosis is seven field-based stereoscopic-dilated fundus photographs, but it is costly and requires expertly skilled persons to operate. Smartphone camera-based image sensors are becoming useful in diagnosing diabetic retinopathy. Although not as accurate as the seven field-based standards, they provide reasonable accuracy for an initial stage. Myung et al. reported a 3D-printed attachment for smartphones to enable them to capture good fundus photos for ophthalmoscopy [20]. Maamari et al. described a portable smartphone-based camera that captures wide-field fundus images. These images can be transmitted remotely for further analysis [21].

Russo et al. designed an optical attachment for smartphones for ophthalmological monitoring. They captured 200 fields of view of fundus images and used the cross-polarization technique [22]. Russo et al. validated the efficiency of smartphone ophthalmoscopy in detecting glaucoma in patients. They compared smartphone-based ophthalmoscopy with undilated retinal biomicroscopy to grade optical disc's VMware Cloud Disaster Recovery [23]. Giardini et al. discussed a portable eye examination kit (PEEK) which tests visual activity, color, and contrast sensitivity, as well as cataract imaging [24]. Haddock et al. described using a smartphone to take fundus photographs of human and rabbit eyes. It consists of an application and instruments for a smartphone to make it fully functional [25]. Sharma et al. developed a smartphone-based fundus imaging system to measure central and peripheral retina diseases. It is primarily focused on the treatment of the peripheral lesion. According to them, the device is beneficial for the mass screening of patients [26]. Finally, Rono et al. focused on screening children's visual impairment by testing the effectiveness of a smartphone-based sight test system named PEEK eye health. It comprises sight simulation referral cards, a PEEK acuity test, and short message service reminders [27].

(2) **Audio Sensor (Microphone):** Air pollution and tobacco consumption are the main causes of pulmonary diseases like asthma, lung cancer, cough, and chronic obstructive pulmonary disease (COPD). These diseases can be detected with traditional techniques, but they are costlier than the mobile-based sensors system. Early detection of these diseases in a cost-effective manner can be done with the help of a smartphone's microphone. Current research is focused on using a microphone to detect audio signals obtained from the

breathing samples of the patient [28]. Larson et al. provided a mobile-based solution to detect and count individual coughs. The cough detection provided 92% true-positive and 0.5% false-positive rates.

Furthermore, the reconstruction of cough sounds can be done, which is useful for physicians. Their algorithm is based on (a) the generation of the cough model, (b) the extraction of the event, (c) the classification of cough, and (d) cough reconstruction [28]. Larson et al. presented a smartphone application named SpiroSmart for spirometry sensing. It uses a built-in microphone as a sensor for diagnosing different obstructive lung diseases. According to them, this sensing scheme has limitations: the inability to measure inhalation, the requirement of a quiet place for observation, and a decrease in forced vital capacity [29]. Finally, Stafford et al. presented an interactive game called Flappy breath to practice breathing exercises suitable for lung health. This application provides vital information about lung health.

The sensing system is based on acoustic and displacement sensors [30]. Chen et al. proposed a listen-to-nose mobile application based on audio signals obtained from runny/stuffy noses or sneezing. An acoustic recognition model is used to classify the signals [31]. Goel et al. removed the dependency on smartphones to obtain spirometer results. Instead, it utilizes a call-in service and a whistle on any smartphone. The audio signal generated from the spirometer goes through a GSM channel, and the results are calculated on the central server [32]. Hearing impairment is growing in the world, and it affects the quality of life of a person. Therefore, its timely detection and correction is required. Ghanem et al. evaluated using a smartphone application (hear) and audiometer to screen hearing impairment for people older than 65 years of age [33]. Hussein et al. investigated the detection of hearing impairment using smartphone hearing screening (hearScreen™) operated by community health workers [34]. Chen et al. described a smartphone-based method for hearing assessment that uses hearing aids to obtain audio information. The controlling of the hearing aids is done by smartphone using low-power-based wireless communication techniques [35, 36]. Audio sensors were also used to monitor a person's mental health and well-being.

(3) **Accelerometer, Gyroscope, GPS:** Mental health is essential for human health. Under this category, the parameters of importance are a person's emotional state and stress level. The diseases are depression, schizophrenia, bipolar disorder, and others. Cornet and Holden systematically reviewed the ongoing research in smartphone-based passive sensing of human health and well-being. According to them, smartphone sensors are accelerometers and gyroscopes, and apart from them, the location and data usage information are useful for predicting the patient's mental state [37]. Ma et al. proposed a Mood Miner framework to analyze the person's mood information. It uses smartphone data and sensors to detect parameters such as acceleration, sound, light, call log, location, and others [38].

## 7.3 SOFTWARE FEATURE OF THE SMART MOBILE

The most popular health apps all have a few similar key elements, so we've discussed the necessary features for clinical application development:

(1) **Appointment Scheduling:** Most healthcare apps have a feature of appointment schedule. This app is mostly used for scheduling appointments with health practitioners or video consultations. Softermii created a healthcare app called MedRealTime, through which patients get the doctor's appointment details. As a result, the doctors know about their future appointments and the free slots. This app is useful and easy to use for patients.

(2) **Video Conferencing:** The COVID-19 pandemic influenced hospital administration in a significant manner. After the pandemic, there was a decrease of nearly 60% in practitioner

visits. Video consultations are becoming popular in place of in-person consultations. As a result, the long waiting hours of patients at clinics and hospitals get reduced. For this purpose, there is a demand for healthcare apps to include a video conferencing feature. The technology use of WebRTC can implement telemedicine capabilities.

(3) **Online appointments:** Telemedicine packages use different software to book online appointments. After the session, the doctor has to update the patient's wellness report. As a result, most modern telemedicine solutions include a digital health file function. Then, the physician has a separate interface in the app to fill out the fitness shape and upload the facts to the inner gadget.

(4) **E-Prescriptions:** The E-prescription feature helps physicians to generate and send prescriptions to patients. Through this feature, the patient acquires the relevant information and gets the prescribed medicines available in the nearby pharmacy.

(5) **Staff Management:** Telemedicine solutions also help the clinic administration. Additional services such as employee schedules, evaluations, and health data aid in improving staff efficiency. One example is the Locum app, which makes hiring and managing pharmacists easier.

(6) **Messaging:** Notifications are an essential feature. Patients appreciate reminders for medical doctor appointments or prescription administration. However, please don't overdo it with this option. There is a need to transfer only the relevant and important information-carrying notifications. In this way, the app uninstallation rate would go down.

(7) **Dashboard for Users:** A dashboard can provide various statistics depending on your healthcare app's objective. It should be simple to use and informative for patients and physicians. For example, there may be a display of physiotherapy exercises, a person's daily nutritional intake, or appointment reminders. Therefore, a piece of visually appealing and quick information providing the feature of the dashboard is necessary.

(8) **Profiles of Doctors and Patients:** The profile information of both the doctor and patient through the healthcare app will be helpful to both of them. The doctors get a quick summarized view of patient data, and the patient can identify the specialist doctors nearby. The patient can also see the facilities at the doctor's clinic or a nearby hospital.

(9) **Payment Integration:** Online secure payments through the healthcare app will make things easier for the patients. They can fix their meeting with the concerned doctors easily and save time.

(10) **CMS Reporting:** Most consumers are unaware of Content Management System (CMS) features. However, they benefit from the increase in the accessibility of the software. The CMS reporting collects data around the interaction of the app. It provides important insights into updating and developing the app.

(11) **Wearables Integration:** Wearable (or external) sensor-based gadgets are becoming popular due to their advantages. First, the healthcare app must collect data from wearable devices providing heart rate, blood pressure, and blood sugar data. Then, through the app, the doctor analyses the patient information for better treatment.

(12) **Third-Party Integrations:** Application Programming Interfaces (APIs) in healthcare provide patients with regulations over their information. Third-party integration can assist hospitals in recognizing the usefulness of their Electronic Health Records (EHRs). For example, the Box API may be used to save important files. The Doximity API also connects to digital doctors and the Human API for controlling patient data. You might sync data within your healthcare app with a website-impacted individual profile by implementing custom API capabilities in your healthcare app.

(13) **Multi-Language:** Spanish is the predominant language of 13% of the population in the United States. Including multilingual capabilities in users, the software would help it appeal to a broader market. In addition, complex or technical clinical phrases are typically

not something a bi-lingual speaker would acquire in their second language; therefore, providing the option to switch to their native tongue would create a better user experience.

(14) **Usable UX/UI:** Modern customers want simple and engaging mobile apps with a straightforward user interface (UI) that offer a great user experience (UX). Nearly 21% remove the app if they do not find these features. Therefore, the design of mHealth apps should consider the region's demographics.

(15) **Search Feature:** A search function is a major requirement of medical software. Patients can quickly traverse the app and get the required information with a search bar.

(16) **Symptom Checker:** A symptom checker is a prime requirement of a healthcare app. It will help the medical practitioner quickly diagnose the diseases of patients.

## 7.4  IoT ROLE, BENEFITS, AND CUSTOMERS IN MHEALTH

IoT enables records to be sent to a smart mobile, which analyzes and sends the information to the hospital through the Internet. Here, the decisions are taken according to the patient's requirements. Below we've mentioned a number of the roles of the IoT within the mHealth area. IoT has modified people's lives, particularly elderly sufferers, by allowing constant tracking of health situations. This significantly impacts people living on their own and their households.

The expectancy of IoT could be excessive from this interconnected technology, which gives you sizeable benefits in everyday day-to-day lifestyles [1]. These interrelated systems assist plenty in the mHealthcare era and are reliable. It also reduces economic necessities and time, which are the maximum enormous needs of patients. This paper applies mHealth as a small live healthcare gadget using exclusive bio-scientific sensors. These sensors ship each fitness-associated report in numerical waveforms to the caretaker and the clinic server. This generation may want to fulfil the necessities of the impartial affected person who are staying alone, mainly for old age people. If something happens incorrectly or crosses the affected person's edge price after a typical characteristic, an emergency service employing the IoT healthcare service company could be alerted. The service provider keeps all the fitness records of the affected person in the medical institution/caretaker server. As the real-time era is more associated with healthcare gadgets, protection is the most crucial parameter in particular in interconnected and interrelated structures of different technology. Fingerprint scanning should clear up unlawful access and authentication issues in mHealth offerings and allow people in corporate international and army services to be far away from home. The main advantages of the IoT for healthcare agencies are shown in Figure 7.5:

(1) **Patients:** Devices in the form of wearables like health bands and other wireless devices such as blood pressure and coronary heart charge monitoring cuffs, glucometers, and so on, allow the patients access to personalized services. For example, these may be programmed according to workouts, calorie intake, appointments, blood pressure fluctuations, and many such applications.

(2) **Scientists:** The wearables and IoT-based patient monitoring systems help physicians to monitor patients' adherence to treatment programs or the need for prompt clinical activity. Healthcare workers have become more vigilant in patient interactions due to IoT. The collected patient data assist clinicians in deciding the best treatment strategy [36].

(3) **Hospitals:** Aside from tracking patients' health, there are several more applications for IoT devices in hospitals. Sensor-enabled IoT devices monitor the instantaneous position of devices like wheelchairs, nebulizers, defibrillators, oxygen pumps, and other monitoring devices. The deployment of health workers in unique positions may be followed in real time. Infection propagation is a significant issue for patients in hospitals. IoT-based hygiene tracking devices help to prevent patient inflammation.

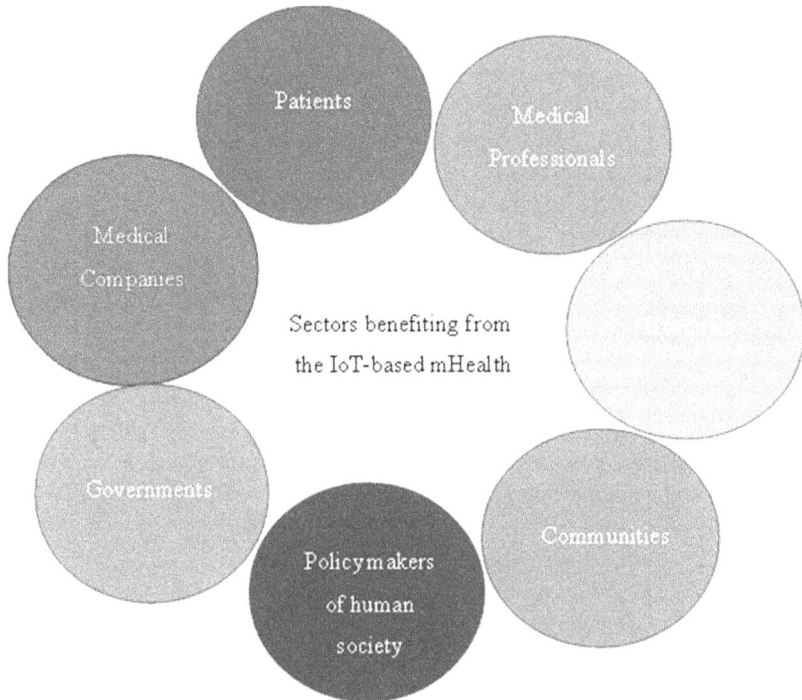

**FIGURE 7.5**  Different benefits of IoT-based mHealth.

(4) **Communities:** Communities can benefit as mHealth may be used for mass diagnoses in remote places, which increases the overall fitness of society.
(5) **Policymakers:** mHealth helps policymakers recognize the spread of any disorder within the population and enable the government to make critical policy decisions to improve the general health of communities.
(6) **Governments:** mHealth also allows the authorities to understand the unfolding of ailments in actual time and take appropriate actions to control pandemics, improve the general fitness of their residents, and accordingly increase the GDP of the country as the performance of the population also increases.
(7) **Medical companies:** Scientific organizations can generate plenty of employment and sales for their countries as there may be a massive capacity in this upcoming market, and in time it will be able to achieve a trillion-dollar marketplace for the medical sector.

The principal advantages of IoT in healthcare encompass the following points:

(1) **Cost Savings:** Doctors may monitor patients in real time using IoT solutions and linked scientific devices. This strategy results in fewer visits to hospital by patients. In addition, it is a green information series and management method. As a result, IoT minimizes costly doctor visits and hospital admissions, making testing and treatment more affordable. As a result, a technology-driven setup lowers healthcare costs by reducing the unnecessary number of visits, using high-quality resources, and optimizing planning.
(2) **Improved Treatment Control:** IoT devices track medication delivery and response to therapy, reducing scientific error. It enables clinicians to make evidence-based, knowledgeable decisions while providing complete openness. Healthcare systems connected with cloud platforms and leveraging massive data give healthcare staff real-time statistics for

decision-making and treatments based on evidence. Medical records accessibility of electronic clinical data enables patients to receive the most appropriate treatment while helping medical personnel to take the best scientific decisions and avoid difficulties. Patients become more engaged in their treatment when linked to the healthcare equipment via IoT, and clinicians improve prediction accuracy since they have access to all relevant patient data. Using IoT devices, hospital administrators may obtain vital information on equipment and employee efficacy and utilize it to advise improvements.

(3) **Early Disease Diagnosis:** Constant patient surveillance and real-time records enable the early detection of illnesses, primarily based on signs and symptoms.

(4) **Proactive Therapy:** Constant health monitoring allows for the provision of proactive clinical treatment.

(5) **Managing Drugs and Equipment:** A primary mission of a healthcare organization is drug and clinical equipment management. Thanks to linked devices, these are handled and administered appropriately with lower costs, improved medication, and medication adherence through mHealth. In addition, IoT solutions save the time of health center personnel by spending less time looking for tablets, tuning supplies and medication, and tuning hygienic procedures in hospitals, reducing clinical infections.

(6) **Error Reduction in Addition to Waste:** Using IoT for statistics requirements and workflow automation reduces waste and systems charges and minimizes errors. Data generated via IoT devices does not only benefit powerful choice-making but makes for better healthcare operations with decreased errors, waste, and device charges.

(7) **Prevention:** Smart sensors analyze fitness situations, lifestyle alternatives and the environment and advise preventative measures to lessen the prevalence of illnesses and acute states.

(8) **Simultaneous Reporting and Tracking:** Clinical emergencies, such as coronary heart failure, asthmatic attacks, diabetes, and others can be handled efficiently with remote tracking and monitoring through connected devices. The information can be transferred to the cloud platform, where the doctor or the concerned person gathers and analyses the information and subsequently takes the necessary steps. IoT technology is also playing an important role in this process.

(9) **End-to-end Connection and Affordability:** In the healthcare domain, the IoT enables artificial intelligence, interoperability, system-to-system interconnection, and others. It also provides data collection and analysis through monitoring and indicators. Moreover, access to the cloud makes it easier to transmit extensive healthcare data in less time.

(10) **Automation:** Manual data collection from many devices and assets and its analysis is challenging, even for healthcare specialists. IoT devices can acquire, document, and analyze real-time data, minimizing the need to preserve raw data. This can all happen in the cloud, with suppliers just seeing the most recent graphical reports. Furthermore, healthcare operations enable firms to get critical healthcare analytics. It reduces the decision time and errors [37].

(11) **Tracking and Indicators:** In a continuous scenario, timely alerting is critical. Medical IoT devices gather important symptoms of any state and send the information to doctors for real-time monitoring, all while notifying individuals about critical aspects via cellular applications and sophisticated sensors. Reports and notifications convey a corporate perspective about a patient's situation, independent of their position or time. It also allows healthcare workers to take correct decisions and deliver timely care. Thus, IoT allows real-time alerting, monitoring, hands-on treatments, improved accuracy, timely decision by doctors, and improved overall patient care shipping outcomes.

(12) **Remote Scientific Assistance:** In an emergency condition, patients can use intelligent mobile apps to contact a health practitioner thousands of kilometers away. With mobile solutions in healthcare, doctors may immediately test patients and diagnose illnesses on the fly. Furthermore, machines must be constructed to dispense medications according to

patients' prescriptions and disease-related data accessible via linked devices. As a result, IoT improves the care of patients in clinics. This, in turn, lowers healthcare costs for people and saves their lives in case of medical emergencies.

(13) **Research:** Another use of IoT healthcare applications is for research purposes. A large amount of patient-related data can be gathered in a very small amount of time. When manually gathering, it would have taken years. This information may then be used for statistical analysis, aiding scientific study. Thus, IoT saves time and money. It allows the development of more effective therapies. It is used in various devices to improve patients' quality of healthcare services. The upgradation of existing equipment using embedded smart healthcare gadget chips is possible due to IoT. This chip improves the support and care that a patient needs. Because IoT devices can gather and analyze vast amounts of data, they offer enormous promise for scientific research.

(14) **Education:** The era additionally helps the present healthcare experts increase their know-how of the biomedical subject with the help of various apps developed in this area and would increase the physicians' performance and productiveness.

## 7.5   ROLE OF THE CLOUD IN mHEALTH

Cloud computing in healthcare refers to establishing remote servers accessible over the Internet for storing, managing, and processing healthcare data. This strategy, unlike the one in which on-site information facilities are provided for website hosting statistics on private computer systems, provides a versatile method to access remote servers where the data is hosted.

Shifting to the cloud has two benefits for both patients and suppliers. On the commercial side, virtualization in cloud computing has shown to be beneficial in lowering operating costs while allowing healthcare organizations to provide excellent and personalized treatment. On the other hand, patients are becoming used to immediate access to healthcare services. Furthermore, healthcare cloud computing boosts patient participation by allowing them access to their healthcare records, which results in better patient outcomes. The expanded accessibility of healthcare brought about by democratizing statistics liberates carriers and patients while breaking down geographical barriers to healthcare access. Cloud computing converges new technologies and presents ones to offer as services with all skills of a computing system to unique styles of users. These services may be accessed from anywhere with the help of an Internet connection. The IoT is a platform that permits the taking of real-time information, enables examination and analysis of these statistics, and shares it with various stakeholders [1]. Cloud and IoT collectively depend on each other. IoT can enjoy the cloud's unlimited abilities and sources to compensate for its technological constraints (for example, garage, processing, and energy). The cloud can gain from the IoT by extending its scope to address real things in the real world [2], such as some hospitals starting to use "smart beds", and the role of cloud computing research might immediately impact some of today's technological problems. However, different approaches are required to handle these issues effectively. Cloud computing research has revealed offerings that can be offered via the cloud. A handful of those services are listed below:

(1) **Software as a Service (SaaS):** With the help of SaaS, the cloud providers host and manage the software and underlying infrastructure, as well as any maintenance, like software upgrades and security patches. Users connect to the utility over the Internet, often via an Internet browser on their smartphone, tablet, or PC.

(2) **Data Storage:** The main advantage of the cloud storage model is eliminating the need to purchase the data storage infrastructure.

(3) **Supply Chain Management:** The cloud presents technology allowing organizations to manage massive amounts of data – from absolutely limitless assets across the complete delivery chain – at speeds and volumes never before seen. Cloud computing offers the

distribution of computing solutions such as servers, databases, storage, software, networking, analytics, and intelligence over the Internet. It enables fast innovation, economies of scale, and flexibility.

(4) **Hosted Services (completely operational IT environment):** A cloud can be non-public or public. A public cloud sells offerings to all and sundry on the Internet. A personal cloud is a registered community or an information middle that materials hosted offerings to a limited number of people, with secure access and permissions settings. Private or public, cloud computing aims to provide smooth, scalable entry to computing sources and IT services.

(5) **Disease Surveillance System:** To provide highly secure, efficient, high-quality, and low-cost care, planned adoption of an interoperable fitness data device infrastructure is required to transition healthcare from a paper-based device to a digital, unified local healthcare device. Research has demonstrated the significance or advantages of a disorder surveillance health statistics device. Access to up-to-date disease surveillance data can help to reduce the number of deaths caused by any illness epidemic. On-time access to diagnostic data can provide workflow, patient care, and disease management advantages. These advantages can also result in potential net value financial savings. Many academic papers and commercial reports have proposed that information and communication technologies might significantly improve fitness and health in developing nations. This chapter investigates the role of mHealth in disease monitoring and how it may be deployed at a regional level to boost local fitness systems in any area throughout the world, focusing on Africa [38].

## 7.6  IoT IN HEALTHCARE: CHALLENGES AND FUTURE TRENDS

Although the IoT has tremendous potential in healthcare, there are significant challenges to overcome before full-scale deployment. The following are the risks and drawbacks of using linked devices in healthcare:

(1) **Security and Privacy:** Security and privacy are significant barriers to people adopting IoT technology for healthcare applications, as healthcare monitoring technologies can be infiltrated or hacked. The disclosure of crucial information about a patient's health and tampering with sensor data may have serious consequences that negate the benefits of IoT. IoT's most significant difficulties in healthcare are data security and privacy. IoT security devices collect and send data in real time [39]. However, most IoT devices lack information standards and security requirements. Furthermore, there is great uncertainty regarding record ownership legislation with electronic devices [1]. These elements make the statistics susceptible to cybercriminals who can get into the system and compromise patients' and physicians' personal health information [37]. In addition, cybercriminals can utilize an afflicted person's digital health record for illegal activities.

(2) **Failure Risk:** Failure or defects in the hardware, as well as power outages, can impair the functionality of sensors and linked devices, putting healthcare processes at risk. Furthermore, delaying a scheduled software update might be riskier than skipping a doctor's check-up. Failure can also come in terms of the Internet connectivity and the slowness of its speed by the service providers that can hamper the IoT carrier of mHealth, and the user cannot send the information in real time.

(3) **Integration of Many Devices and Protocols:** The protocol difference between devices poses integration issues. So, even if a range of devices are linked, the differences in their communication protocols complicate and impede the data-gathering system [1]. This non-uniformity in the protocols of the linked devices slows down the entire operation and limits the scalability of IoT in healthcare. In addition, because there is no agreement on IoT protocols and specifications, devices from various manufacturers do not work effectively together. The lack of homogeneity precludes full-scale IoT integration, limiting its effectiveness.

(4) **Expense:** While IoT promises to save healthcare costs in the long run, the cost of implementing it in hospitals and training employees is relatively high. Were you surprised to discover pricing issues inside the work sections? Most of you are, but the fundamental reality
is that IoT has not made healthcare more affordable to the average person. Healthcare costs
are a concerning indicator for everyone, mainly in developed countries [37]. The issue has
given rise to "Medical Tourism," individuals with critical illnesses admitted to healthcare
facilities in developing nations at a fraction of the cost. The notion of IoT in healthcare is
enticing and intriguing. However, it has not yet resolved the value issues. To successfully
enforce IoT app enhancement and benefit its overall optimization, stakeholders must make
it cost-viable; otherwise, it remains out of reach for everyone except those in the upper
classes [37].

(5) **Data overload and Precision:** Data aggregation is intricate because of different communication protocols and standards. IoT devices report a vast amount of data and provide
critical insights. A vast volume of information creates decision-making problems for physicians. Furthermore, the challenge grows as more devices are networked and more data is
captured [37].

## 7.7   FUTURE ROLE OF IoT AND CLOUD IN mHEALTH

The IoT device is like a tool with sensors interacting with the physical world and transmitting data
via the Internet. The IoT-enabled devices communicate with each other. They conduct important
steps to save the life of patients in a timely manner. The information obtained from IoT devices goes
to the cloud and takes necessary actions. IoT applications in healthcare and medicine cover many
technologies, from future robots and drones to machine learning-based statistics analytics [40].
With the boom in calls for mobile equipment and distant patient tracking technologies at some stage
in the pandemic, governments and traders have also realized that the function of IoT in healthcare
offerings has increased. For example, drones supply tablets to patients whose treatment is completed
at home.

The advancements in miniaturized sensors and gadget learning fuel the possibilities of cellular
fitness care programs, in which those M-IoT gadgets are frame-worn or implanted for continuous
vital sign monitoring. Because bio-alerts derived from the human body include a profusion of data,
the majority of which are unobservable to the human eye or even specialists, better device learning
techniques play a crucial role in mining such facts from the bio-signals. Such information is helpful
for most people going through follow-up fitness situation evaluation and prognosis. In addition, the
importance of signal processing algorithms for noise reduction should also be implemented into
those M-IoT devices to boost the dependability of the measurements provided.

Cloud computing and healthcare are the key components to unlock the new trends in medical
science. The prime advantages of these are better data security, scalability, and accessibility. In
addition, the cloud makes the connection of distant providers with patients and medical experts
possible and removes unequal access to care. The healthcare future seems to be bright with cloud
technology.

## 7.8 CONCLUSIONS

From this vantage point, the IoT is a growing research field in the healthcare domain. It provides an excellent potential for healthcare systems to anticipate fitness problems and a timely diagnosis and treatment of patients at their homes or hospitals. Furthermore, the use of more technology-driven health products enables the fitness structures to offer flexible ways of treatment, and it requires integration with IoT technology for better operation. However, issues such as privacy, cybersecurity, information management, and research gaps in this field need to be considered.

A transformation is going on in the field of clinical research and disease treatment by mobile fitness (m-fitness) devices built around IoT infrastructure. The real-time information gathered by wearable and implant devices goes to the doctors for short analysis, remote tracking, and home rehabilitation. Furthermore, m-fitness significantly saves regular healthcare costs and avoidable hospitalizations.

The study discusses the theoretical knowledge of sensor-driven IoT wearables, big-data platform deployment, cloud-based mHealth, and its influence on impacted person fitness tracking. The Q-type technique was also employed to extend the structure's operationalization size scales. The theoretical hyperlink among IoT wearables, big-records platforms, cloud-based fully mHealth, and its influence on patient health tracking has been implemented with the help of the scientific literature. The study addresses theoretical knowledge by laying the groundwork for future empirical research through dimension scale creation for assembly operationalization.

The mHealth-based healthcare strategy has several advantages. First, they might improve the treatment quality of people of different ages. The IoT-based healthcare system may be considered an integration of ubiquitous computing that interacts with external factors. The information gathered about patient data, and medical personnel's feedback is used to take critical steps for saving of patient's life. An example is continuous glucose level monitoring. Following the collection of data, an IoT healthcare gadget may communicate this crucial information to the cloud so clinicians may act upon it. Based on this, we can conclude that the capability use of IoT in healthcare may boost not only a patient's fitness but also fitness care worker productivity and medical institution operations.

Future research should focus on making standardized IoT protocols that match global and state-of-the-art health systems. Also, the security challenges need to be addressed as the data going through the IoT platform is user sensitive. Governments must also build laws for a data breach in the future as currently there is no law for it, especially in developing countries. The companies providing the solution for mHealth also have to increase the security for digital data transmission so that no one can easily hack the network. The companies will also have to provide a standard platform/protocol for the data transmission, as currently different vendors are using different platforms and protocols for the data transmission from the device to the IoT platform. With the implementation of the above solutions, the path and the direction for mHealth-based diagnostic services looks positive for technology and patients. Additionally, research needs to evaluate blockchain storage to centralized cloud storage regarding IoT-enabled fitness care. Finally, there is a need for scientific guidance on virtual health prescriptions and reimbursement coverage for IoT-based primary and secondary healthcare services. Although we provided a summary of selected literature and not an exhaustive systematic review of the literature, we believe that addressing those areas for future research will go a long way toward allowing a much wider uptake of IoT, which will ultimately save fitness care dollars and improve affected person-centered care.

## REFERENCES

[1] N. K. Narang, "Mentor's musings on the role of disruptive technologies and innovation in making health-care systems more sustainable," *IEEE Internet Things Mag.*, vol. 4, no. 3, pp. 80–89, 2021, doi: 10.1109/MIOT.2021.9548847.

[2] Y. Huang, X. Fan, S.-C. Chen, and N. Zhao, "Emerging technologies of flexible pressure sensors: materials, modeling, devices, and manufacturing," *Adv. Funct. Mater.*, vol. 29, no. 12, p. 1808509, Mar. 2019, doi: 10.1002/adfm.201808509.

[3] M. Cheng et al., "A review of flexible force sensors for human health monitoring," *J. Adv. Res.*, vol. 26. pp. 53–68, 01 Nov. 2020, doi: 10.1016/j.jare.2020.07.001.

[4] P. Jiang, Z. Ji, X. Zhang, Z. Liu, and X. Wang, "Recent advances in direct ink writing of electronic components and functional devices," *Prog. Addit. Manuf.*, vol. 3, no. 1, pp. 65–86, 2018, doi: 10.1007/s40964-017-0035-x.

[5] S. Li et al., "Physical sensors for skin-inspired electronics," *InfoMat*, vol. 2, no. 1, pp. 184–211, Jan. 2020, doi: 10.1002/inf2.12060.

[6] F. Zhao, M. Li, and J. Z. Tsien, "The emerging wearable solutions in mHealth," in *Mobile Health Technologies - Theories and Applications*, InTech, London, 2016.

[7] D. Steins, H. Dawes, P. Esser, and J. Collett, "Wearable accelerometry-based technology capable of assessing functional activities in neurological populations in community settings: a systematic review," *J. Neuroeng. Rehabil.*, vol. 11, no. 1, p. 36, 2014, doi: 10.1186/1743-0003-11-36.

[8] N. L. Kazanskiy, M. A. Butt, and S. N. Khonina, "Recent advances in wearable optical sensor automation powered by battery versus skin-like battery-free devices for personal healthcare—A review," *Nanomaterials*, vol. 12, no. 3, p. 334, Jan. 2022, doi: 10.3390/nano12030334.

[9] A. Banerjee, S. Maity, and C. H. Mastrangelo, "Nanostructures for biosensing, with a brief overview on cancer detection, IoT, and the role of machine learning in smart biosensors," *Sensors*, vol. 21, no. 4, p. 1253, 2021.

[10] W. Zhang, X. Ji, C. Zeng, K. Chen, Y. Yin, and C. Wang, "A new approach for the preparation of durable and reversible color changing polyester fabrics using thermochromic leuco dye-loaded silica nanocapsules," *J. Mater. Chem. C*, vol. 5, no. 32, pp. 8169–8178, 2017, doi: 10.1039/C7TC02077E.

[11] B. Chen, Y. Yang, and Z. L. Wang, "Scavenging wind energy by triboelectric nanogenerators," *Adv. Energy Mater.*, vol. 8, no. 10, p. 1702649, Apr. 2018, doi: 10.1002/aenm.201702649.

[12] R. Liu, L. He, M. Cao, Z. Sun, R. Zhu, and Y. Li, "Flexible temperature sensors," *Front. Chem.*, vol. 9, 22 Sep. 2021, doi: 10.3389/fchem.2021.539678.

[13] Z. Chen et al., "Flexible temperature sensors based on carbon nanomaterials," *J. Mater. Chem. B*, vol. 9, no. 8, pp. 1941–1964, 28 Feb. 2021, doi: 10.1039/d0tb02451a.

[14] S. Majumder, T. Mondal, and M. J. Deen, "Wearable sensors for remote health monitoring," *Sensors (Switzerland)*, 2017, doi: 10.3390/s17010130.

[15] R. Krishnan, B. Natarajan, and S. Warren, "Two-stage approach for detection and reduction of motion artifacts in photoplethysmographic data," *IEEE Trans. Biomed. Eng.*, vol. 57, no. 8, pp. 1867–1876, 2010, doi: 10.1109/TBME.2009.2039568.

[16] R. J. Friedman, D. S. Rigel, and A. W. Kopf, "Early detection of malignant melanoma: the role of physician examination and self-examination of the skin," *CA. Cancer J. Clin.*, vol. 35, no. 3, pp. 130–151, 1985, doi: 10.3322/canjclin.35.3.130.

[17] T. Wadhawan, N. Situ, K. Lancaster, X. Yuan, and G. Zouridakis, "SkinScan©: A portable library for melanoma detection on handheld devices," in *Proceedings - International Symposium on Biomedical Imaging*, 2011, pp. 133–136, doi: 10.1109/ISBI.2011.5872372.

[18] T.-T. Do et al., "Accessible melanoma detection using smartphones and mobile image analysis," *IEEE Transact. Multimedia*, vol. 20, no. 10, pp. 2849–2864, Nov. 2017.

[19] S. Kim et al., "Smartphone-based multispectral imaging and machine-learning based analysis for discrimination between seborrheic dermatitis and psoriasis on the scalp," *Biomed. Opt. Express*, vol. 10, no. 2, p. 879, Feb. 2019, doi: 10.1364/boe.10.000879.

[20] D. Myung, A. Jais, L. He, M. S. Blumenkranz, and R. T. Chang, "3D printed smartphone indirect lens adapter for rapid, high quality retinal imaging," *J. Mob. Technol. Med.*, vol. 3, no. 1, pp. 9–15, Mar. 2014, doi: 10.7309/jmtm.3.1.3.

[21] R. N. Maamari, J. D. Keenan, D. A. Fletcher, and T. P. Margolis, "A mobile phone-based retinal camera for portable wide field imaging," *Br. J. Ophthalmol.*, vol. 98, no. 4, pp. 438–441, 2014, doi: 10.1136/bjophthalmol-2013-303797.

[22] A. Russo, F. Morescalchi, C. Costagliola, L. Delcassi, and F. Semeraro, "A novel device to exploit the smartphone camera for fundus photography," *J. Ophthalmol.*, vol. 2015, 2015, doi: 10.1371/journal.pone.0273633.

[23] A. Russo et al., "Comparison of smartphone ophthalmoscopy with slit-lamp biomicroscopy for grading vertical cup-to-disc ratio," *J. Glaucoma*, vol. 25, no. 9, pp. e777–81, Sep. 2016, doi: 10.1097/IJG.0000000000000499.

[24] "Phone_based_ophthalmoscopy_for_peek." doi: 10.1371/journal.pone.0273633.

[25] L. J. Haddock, D. Y. Kim, and S. Mukai, "Simple, inexpensive technique for high-quality smartphone fundus photography in human and animal eyes," *J. Ophthalmol.*, vol. 2013, 2013, doi: 10.1155/2013/518479.

[26] A. Sharma, S. D. Subramaniam, K. I. Ramachandran, C. Lakshmikanthan, S. Krishna, and S. K. Sundaramoorthy, "Smartphone-based fundus camera device (MII Ret Cam) and technique with ability to image peripheral retina," *Eur. J. Ophthalmol.*, vol. 26, no. 2, pp. 142–144, Sep. 2015, doi: 10.5301/ejo.5000663.

[27] H. K. Rono et al., "Smartphone-based screening for visual impairment in Kenyan school children: a cluster randomised controlled trial," *Lancet Glob. Heal.*, vol. 6, no. 8, pp. e924–e932, Aug. 2018, doi: 10.1016/S2214-109X(18)30244-4.

[28] E. C. Larson, T. Lee, S. Liu, M. Rosenfeld, and S. N. Patel, "Accurate and privacy preserving cough sensing using a low-cost microphone." in *Proceedings of the 13th International Conference on Ubiquitous Computing* (pp. 375–384), Sep. 2011.

[29] E. Larson, M. Goel, G. Boriello, S. Heltshe, M. Rosenfeld, and S. Patel, "SpiroSmart: Using a microphone to measure lung function on a mobile phone," in *UbiComp'12 - Proceeding 2012 ACM Conference Ubiquitous Computing*, Sep. 2012, doi: 10.1145/2370216.2370261.

[30] M. Stafford, F. Lin, and W. Xu, "Flappy breath: A smartphone-based breath exergame," in *Proceedings - 2016 IEEE 1st International Conference on Connected Health: Applications, Systems and Engineering Technologies, CHASE 2016*, 2016, pp. 332–333, doi: 10.1109/CHASE.2016.70.

[31] N. C. Chen, K. C. Wang, and H. H. Chu, "Listen-to-nose: A low-cost system to record nasal symptoms in daily life," in *Proceedings of the 2012 ACM Conference on Ubiquitous Computing*, Sep. 2012, pp. 590–591.

[32] E. C. Larson, M. Goel, M. Redfield, G. Boriello, M. Rosenfeld, and S. N. Patel, "Tracking lung function on any phone," in *Proceedings of the 3rd ACM Symposium on Computing for Development*, Jan. 2013, pp. 1–2.

[33] S. Abu-Ghanem, O. Handzel, L. Ness, M. Ben-Artzi-Blima, K. Fait-Ghelbendorf, and M. Himmelfarb, "Smartphone-based audiometric test for screening hearing loss in the elderly," *Eur. Arch. oto-rhino-laryngology Off. J. Eur. Fed. Oto-Rhino-Laryngological Soc. Affil. with Ger. Soc. Oto-Rhino-Laryngology - Head Neck Surg.*, vol. 273, no. 2, pp. 333–339, Feb. 2016, doi: 10.1007/s00405-015-3533-9.

[34] S. Yousuf Hussein, D. Wet Swanepoel, L. Biagio de Jager, H. C. Myburgh, R. H. Eikelboom, and J. Hugo, "Smartphone hearing screening in mHealth assisted community-based primary care," *J. Telemed. Telecare*, vol. 22, no. 7, pp. 405–412, Oct. 2016, doi: 10.1177/1357633X15610721.

[35] F. Chen, S. Wang, J. Li, H. Tan, W. Jia, and Z. Wang, "Smartphone-based hearing self-assessment system using hearing aids with fast audiometry method," *IEEE Trans. Biomed. Circuits Syst.*, vol. 13, no. 1, pp. 170–179, Feb. 2019, doi: 10.1109/TBCAS.2018.2878341.

[36] E.-A. Paraschiv, E. Tudora, E. Tirziu, and A. Alexandru, "IoT & cloud computing-based remote healthcare monitoring system for an elderly-centered care," in *2021 International Conference on e-Health and Bioengineering (EHB)*, 2021, pp. 1–4, doi: 10.1109/EHB52898.2021.9657585.

[37] R. Marimuthu Narayanan Saravana Kumar, B. Y. Bharath, K. S. Rajendiran Ranjith, and K. S. Kumar, "Use of IoT and mobile technology in virus outbreak tracking and monitoring," in *The Internet of Medical Things: Enabling Technologies and Emerging Applications*, IET Digital Library, USA, 2021.

[38] D. A. M. Lang Loum, "The role of cloud-based mHealth disease surveillance system in regional integration: A case of the ebola crisis in ECOWAS," in *Innovation, Regional Integration, and Development in Africa*, Springer International Publishing, New Delhi, 2019.

[39] A. Haleem, M. Javaid, R. Pratap Singh, and R. Suman, "Medical 4.0 technologies for healthcare: Features, capabilities, and applications," *Internet Things Cyber-Physical Syst.*, vol. 2, pp. 12–30, 2022, doi: 10.1016/j.iotcps.2022.04.001.

[40] N. Sindhwani, R. Anand, R. Vashisth, S. Chauhan, V. Talukdar, and D. Dhabliya, "Thingspeak-based environmental monitoring system using IoT," in *2022 Seventh International Conference on Parallel, Distributed and Grid Computing (PDGC)*, 2022, pp. 675–680, doi: 10.1109/PDGC56933.2022.10053167.

[41] J. Kaur, N. Sindhwani, and R. Anand. "Implementation of IoT in various domains," in *IoT Based Smart Applications* (pp. 165–178). Springer International Publishing, USA, 2022.

# 8 IoT and Cloud Computing
## *Two Promising Pillars for Smart Agriculture and Smart Healthcare*

*A. Sherly Alphonse and S. Abinaya*
Vellore Institute of Technology, Chennai, India

*Ani Brown Mary*
Sarah Tucker College, Tirunelveli, India

*D. Jeyabharathi*
Sri Krishna College of Technology, Coimbatore, India

## CONTENTS

8.1  Introduction ..................................................................................................131
8.2  Benefits of the Internet of Things (IoT) ...................................................132
8.3  Smart Agriculture ........................................................................................133
    8.3.1  Benefits of Smart Agriculture ......................................................133
    8.3.2  Background Literature ....................................................................134
    8.3.3  Methodologies ................................................................................137
    8.3.4  Combined Analysis of IoT End Device Data with Drone Data ............138
8.4  Smart Healthcare .........................................................................................139
    8.4.1  Why is IoT Important in Healthcare? ...........................................139
    8.4.2  How IoT Helps in Healthcare – Process ......................................140
    8.4.3  Applications of IoT in Healthcare ................................................141
    8.4.4  Current Status ................................................................................141
    8.4.5  Challenges of IoT in Healthcare ..................................................142
    8.4.6  Future of IoT in Healthcare ..........................................................142
8.5  Conclusion ...................................................................................................143
References ...............................................................................................................143

## 8.1  INTRODUCTION

The agriculture sector is becoming more important in maintaining a nation's economic status. The massive upsurge in population and considerable income progress in India demands about 2.5 million tons of food grain yearly. However, land degradation, water scarcity, pests and insects are major factors which reduce grain productivity. To meet the projected demand despite the above difficulties, an IoT-enabled autonomous drone-based crop health monitoring and spraying system has been used. Unmanned Aerial Vehicles (UAVs) are becoming more and more popular in a diversity of industries due to robust investment and the relaxation of some governmental regulations. The agricultural UAVs are equipped with infrared cameras, sensors, raspberry pi computers, Wi-Fi, and software to scout out the location and health of crops. They will capture the information for about 30 to 50%

DOI: 10.1201/9781003319238-8

of flight time and then the information will be sent to the cloud for further processing. The drones can also be employed for aerial spraying which reduces the amount of chemicals penetrating into groundwater. The images acquired from the drone are then processed with deep learning algorithms for the classification of diseases. During the classification step, the images captured are compared with the images already collected in the database. Images are classified as having a pest attack, fungal infection, water scarcity, and ageing based upon feature extraction, and in the case of crop failure, the farmer will be able to document losses more efficiently for insurance claims. Recently Industry 4.0 has shown several advantages in the domain of healthcare. The different issues, like pain detection, cardiac and lung failure etc., can be easily detected by incorporating IoT with cloud computing. Mostly aged people and people with severe disability have to be monitored periodically and there is a great significance of IoT in both smart agriculture and healthcare. There are several advancements in the field of healthcare due to industrial IoT and therefore a detailed study is necessary. The IoT and cloud computing techniques connect the health monitoring devices and health-related data with the cloud and this significantly paves way for better patient treatment. The aim of this chapter is to investigate the applicability of the two pillars, namely Internet of Things (IoT) and Cloud Computing, in smart agriculture and in the healthcare domain with suitable authors' findings. In this chapter, Section 2 signifies the benefits of IoT, Section 3 provides an overview of smart agriculture, Section 4 presents the smart healthcare system, and Section 5 concludes the chapter.

## 8.2  BENEFITS OF THE INTERNET OF THINGS (IoT)

The Internet has made human life richer and IoT has the ability to further revolutionize its use. Its influence on a society or a company is generally made felt by the choices formulated while applying it [1]. IoT is the latest trend in controlling and monitoring systems, namely smart cars, smart health, smart transportation, smart grid, smart city, smart environments, and smart wearable devices as represented in Figure 8.1. It has given the public more and smarter options.

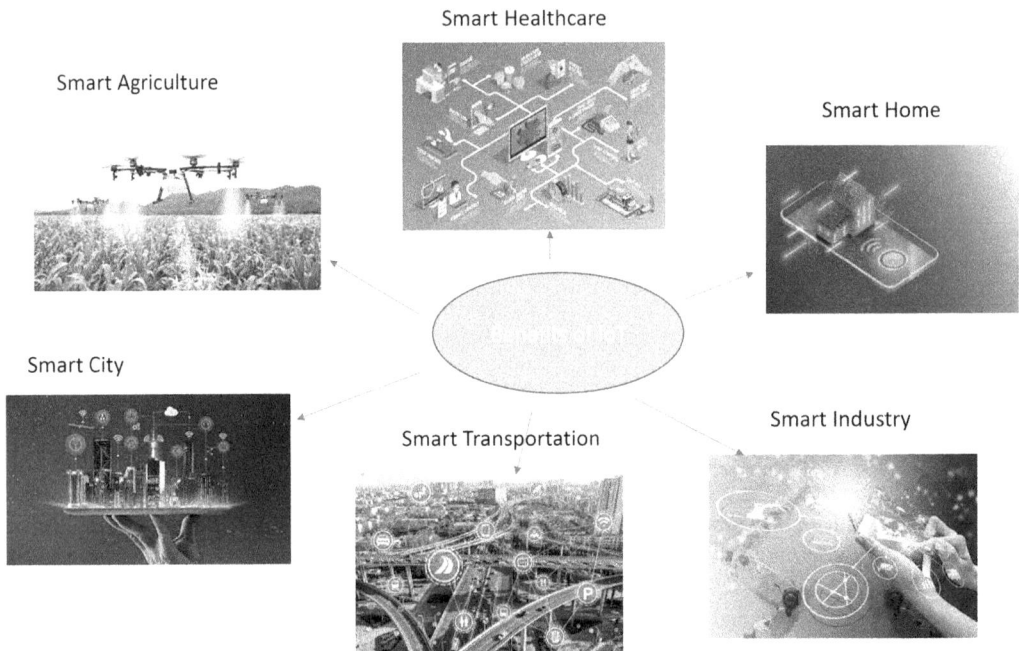

**FIGURE 8.1**  Benefits of IoT.

The potential benefits of the IoT are the monitoring of climate conditions, data analytics, smart greenhouses, precision farming, low-cost results, agricultural drones, and the ease of deployment. The current growth in micro electromechanical systems has provided low-cost IoT solutions for healthcare, with low-cost IoT devices fitted with appropriate radios, memory, sensors, batteries, and processors. It also works for extended periods of times, reducing the cost compared to the usual methods. IoT can be arranged in any environment with comparative ease.

The current rise of IoT is important and it is used in all fields of human machine communication: from urban industry to healthcare, governance to infrastructure administration, from customer services to defense, IoT has shown massive implementation in a period of a just few years [1–4].

## 8.3  SMART AGRICULTURE

The agriculture sector is increasingly important in maintaining a nation's economic status. India has high population pressure on land and other resources to meet its food and development needs. The massive increase in population and substantial income growth demands about 2.5 million tons of food grains annually. However, land degradation, water scarcity, pests and insects are major factors which reduce grain productivity. To meet the projected demand, an IoT-enabled autonomous drone-based crop health monitoring and spraying system has been proposed. UAVs are becoming more and more popular in variety of industries, thanks to increased investment and reduced government regulation. Agricultural UAVs can be equipped with infrared camera, sensors, raspberry pi computers, Wi-Fi, and software to locate and report on the health of crops. They capture the information for between 30 and 50% of flight time and then send the information to the cloud for further processing. The drones can also be employed for aerial spraying which diminishes the amount of chemicals penetrating into ground water. Drone images are then processed with deep learning algorithms and diseases classified by comparison with images already collected in database, the images classified as having a pest attack, fungal infection, water scarcity and ageing based upon the features extracted [2–4]. Once disease is discovered, farmers can apply remedies more precisely, and in the case of crop disaster, losses more easily described for claiming insurance.

### 8.3.1  BENEFITS OF SMART AGRICULTURE

Some of the benefits of smart agriculture as follows:

- Improvement of crop yield by precise crop monitoring and pest management.
- Precise real-time monitoring of crop health using UAVs with the required components.
- Differentiation between fungal infection, pest attack, water scarcity and ageing using image processing and deep learning techniques.
- Triggering a fungicide spray in case of a fungal infection and a pesticide spray in case of a pest attack using drones.
- Spraying the correct quantity of pesticides for soil conservation.
- Eliminating human labor from harmful pesticides by incorporating automation.
- To help farmers document losses due to flood more efficiently for insurance claims.

The pesticide savings is really beneficial in the proposed system using drones as the droplet size is controlled based on the shape and size of the pest in the plant. Therefore, the concentration of the pesticide is less and fewer gallons of pesticides are needed. The cost benefit achieved in the proposed system is greater when compared to the conventional method of tractors and sprayers, and hence less time and labor is required. The different types of attacks in plants are analyzed and pesticide is sprayed only in the needed areas. The size of droplets is changed to save huge amounts of pesticide depending upon the size and shape of the pests. The massive rise in the demand for food due to the upsurge in population internationally is driving the need for smart agriculture with its advanced information and communication innovations like IoT, big data, cloud computing, data analytics, and artificial intelligence. This paves the way for a new revolution in agriculture to the agri-4.0 era. More specifically, this new agri-4.0 era agriculture includes the concept development, design, implementation, and application of modern ways to apply information and communication technology (ICT) in the rural area, with the main objective of precision agriculture [5–8].

These new advancements of ICT in agriculture promote the automated farming, assembly of data from agricultural field, and then assessment of the data. So, farmers have more precise choices to have good-quality crops. In a real-time monitoring system using IoT-enabled drones that improves crop productivity, soil pollution can be reduced by precise pesticide spraying, and farmers can be protected from poisoning and heatstroke. Drones endowed with appropriate cameras, sensors, and integrating modules gather the agricultural field information and process it using a raspberry pi module employed in the drone. The soil moisture and humidity data are collected through Node MCU IoT device. The data from the raspberry pi module of the drone and Node MCU IoT device is further sent to an IoT gateway using a Wi-Fi module for further processing by cloud, computing services, web servers, and finally the information given to user as in Figure 8.2. Different methods in literature employ image processing and deep learning techniques for automatic disease detection on plant leaves. Successively the drone can differentiate a pest and a fungal infection and accordingly spray a pesticide or a fungicide on the plant. This work saves lot of time, cost, and energy compared to spraying pesticides and fungicides individually. The real-time information about the plants available in the cloud and the data analysis helps the farmers to make right decisions on required irrigation to mitigate water scarcity problems.

### 8.3.2 Background Literature

There are various types of works available in the literature. Some of them are discussed in this section. A design of a wheel and pedal functioned sprayer is proposed as a portable device and without needing any fuel to function, it is easy to move and spray the pesticide by touching the wheel and also peddling the apparatus. A multi-nozzle pesticide sprayer pump will achieve spraying at high

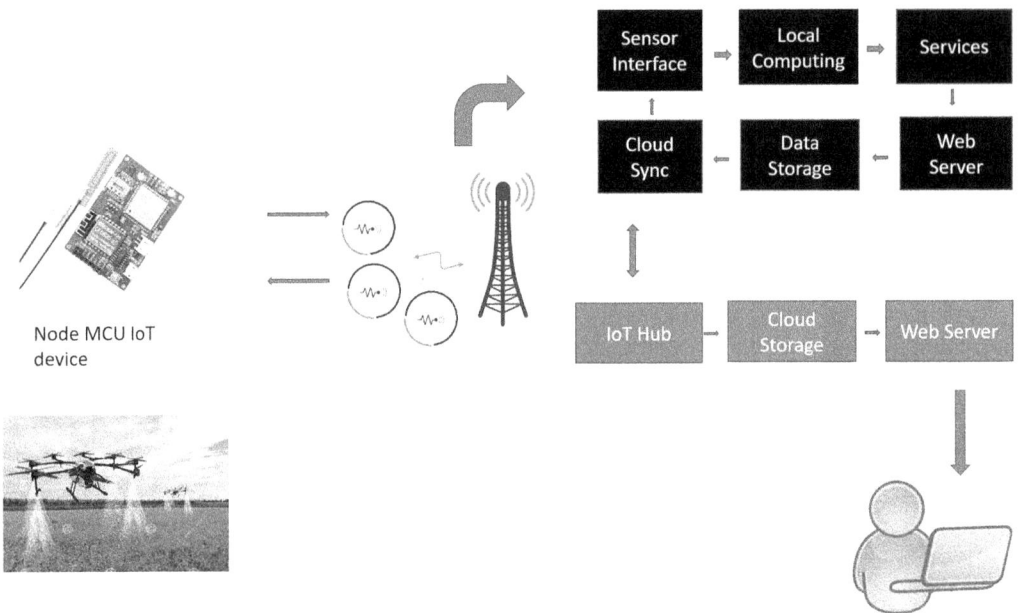

**FIGURE 8.2**  The methodology used in agriculture sector for pesticide spraying using IoT.

rate in least time. Constant flow valves can be applied to nozzles to have unchanging nozzle pressure. A. Thorat et al. [4] used sensors networks for computing the amount of dampness, temperature, and moisture instead of physical check. Numerous sensors are positioned in various sites of farms and measured by a raspberry pi processor. Leaf disease can also be detected using camera interfacing with a raspberry pi. M.N.R. Thorat et al. [5] developed an automatic disease detection system, which is used to detect and identify the diseased part on the leaf images using image processing techniques. Feature extraction, segmentation, and clustering of leaf images are used for efficient disease detection by using IoT. The classification of images can also include genetic algorithm.

V.Gowrishankar and K. Venkatachalam [6] in their paper have proposed an autonomous robot that can sow seed and spray pesticides in order to decrease the load of farmers. The robot, known as agribot, was designed particularly for agricultural needs. It minimizes the work of farmers, increases the efficiency, and reduces the time. It does the basic functions in agriculture like spraying of pesticide and seed sowing. Manual spraying of pesticides is harmful for the health of the farmers. This agribot was controlled through IoT. S.B. Londhe and K. Sujata [7] proposed a method to diagnose the crop disease. This work presents a key to overcome this type of problem and create an automatic system of pesticide spraying that sprays the pesticide only on the particular area of the infected crops. This system uses a sprayer for the spraying of pesticides. The sprayer is controlled by a DC motor so that it moves in all directions at low velocity, based upon the plant height. This system is operated remotely.

T. Watanabe et al. [8] proposed that the function of ecosystems is noteworthy by accepting the composition and allotment of plant resources using spatial-based and temporal-based scales. For a detailed survey, conventional methods are inaccurate and hence not possible in areas that are inaccessible. The sensing methods using the Geographic Information System (GIS), and GPS are much supportive in getting the data in small amount of time that has frequency at various altitudes. This learning used remote sensing and image processing with a GIS-based and integrated model that is geospatial for classifying the plants using DJI Inspired 2 drone photos. The images from the drone were processed using band alignment, mosaicking, and ortho-rectification. Results indicate the zones that are perfect for plant resources using the classification techniques using training data. The analysis of subjective overlap of soil data and categorized zones is done. The result was that the model was geospatial and was an efficient technique for classification of plant resources.

R.F. Olanrewaju et al. [9] proposed a tri-rotor aircraft using control algorithms and Proportional-Integral-Derivative (PID) control with the effect a tilt tri-rotor-based aircraft. The creation of the control algorithms is a model using mathematics which consisted of variables obtained from three motors. It has a tilting technique that is used as a reference. A considerable improvement is obtained and the control algorithm is practical to the system that has pitch overshoot which is greatly minimized and goes to a constant zero. The tilt-based Tri-Rotor Aerial Vehicle is basically an aircraft using three motors that doesn't require pilot involvement in the administration of the flight. There exist circumstances in which the setting is not suitable or may even be hazardous to a human being, such as a war zone region, volatile area, steep terrains, or in areas of armed attacks [4–10]. A design for smart Wi-Fi wireless-based robots for the pesticide spraying process is used in the literature. The microcontroller is monitored using a wireless router. The camera captures video using Android phones. The robot can control the spraying direction and the height of spraying and can also detect if any obstacle is present [11].

Z. Wei et al. [12] proposed an Automatic Toward-target Sprayer that uses dual infrared sensors on two sides of the machine to identify fruit-bearing trees. The control system calculates the rapidity of the tractor using the pulse signal obtained through Hall sensor. On detecting a target, the position and speed used to compute the location and width of spraying is kept in the memory. On detection of any gap, it stops spraying. According to the size of the trees the spraying width can be altered. When the tractor is moving at less than 1.5 m/s, and the tree is inside the detection range (<=70 cm), the system functions accurately.

R. Berenstein and Y. Edan [13] presented a device for a pesticide spraying those deals with different shapes and different sized targets. The device has a single spray nozzle and an adjustable angle of spraying, camera, and distance calculating sensors. This method aims to reduce the amount of pesticide spent by reducing the spotting the location accurately. The spraying diameter is chosen based upon the form and volume of the target. The experiments were done in order to calculate the flow rate of spray and artificial targets. About 45% of pesticide reduction is achieved and spraying time and the amount of spray is reduced.

J. Zhang et al. [14] developed an intelligent master-slave system among agricultural vehicles that enables a semi-autonomous vehicle to go behind a tractor. This paper discusses the recent works and proposed a methodology and mathematical model. They have also conducted some field tests. J. Primicerio et al. [15] proposed an approach that used an aerial vehicle ("VIPtero") which was fully automatic and was assembled. The system was tested with the objective of creating a reliable and dominant tool for managing vineyard sites. The proposed system has six rotors. The system is able to fly automatically and has a multispectral camera for videorecording the images. The accuracy of the camera was measured on ground images. The system VIPtero was experimented in a vineyard in Italy. About 63 images were acquired during 10 minutes of flight. Analysis of the obtained images were done and given for classification. The vigor maps were shaped grounded on the obtained normalized index. The heterogeneity was identified in with ground-based interpretation. The results achieved were accurate.

S. Candiago et al. [16] used UAVs with remote sensing that had many techniques to obtain images in a rapid and flexible way for data in agriculture needs. These types of studies are increasing because of the benefits and returns in agricultural management, in monitoring crops. This work explains some experiences including the analysis of cultivations like vineyards with the help of multispectral data using Tetracam. The Tetracam camera was placed on top of a multi-rotor-based hexacopter. The multispectral information was analyzed with a photogrammetric pipeline to form triband orthoimages of the sites plotted. These orthoimages are employed for extracting Vegetation Indices (VI), the Green Normalized Difference Vegetation Index (GNDVI), Normalized Difference Vegetation Index (NDVI), and the Soil Adjusted Vegetation Index (SAVI). It also examines the vigor for the field. The work illustrates the great possibility of high-resolution UAV data and photogrammetric techniques useful in the technique to collect the images and assess various VI. This signifies that the instruments symbolize a profitable resource in assessing crops for high accuracy-based farming needs and applications.

M. Rieke et al. [17] proposed UAVs that are micro-sized with common receivers using GPS in the civilian domain that will not address situations where high accuracy and positioning is not a requirement. In cases like forming some orthophotos with georeferencing, more positioning techniques are needed. This method fuses Real-Time Kinematic positioning and micro-sized UAVs. It also defines the data synchronization for the workflow and ortho-rectification imagery. The results are illustrated for the precision in farming. This approach has the probability of achieving a high accuracy in a position of 1–3 cm, that is used for georeferencing of aerial images. A. Matese et al. proposed viticulture with cost-methodologies for UAVs that are participating in acquisition platforms with satellite and aircraft having low operative costs. They used the technologies in viticulture. The objective is to compare the NDVI surveys [18].

The entire system architecture based on various design layers are given in Figure 8.2. The end devices in this project are Node MCU IoT with soil moisture, humidity sensor, IoT-enabled drone with raspberry pi module, infrared camera, sensor, and integrating modules. These end devices collect information from the field and send it to the IoT base station through Wi-Fi connectivity. The data is further transported to gateway PC which is mainly a GPU machine to process the data locally. The field data is processed in first level using data analytics to enable required irrigation. From the gateway machine the data is transferred to cloud via the Internet. The information is then preprocessed to clean the data. Thus, cleaned data is further subjected to image processing and deep learning algorithm for disease detection on plant leaves. Plant disease management is a challenging

(a) Fungal infections                                    (b) Pests

**FIGURE 8.3**    (a) Fungal infections, and (b) pests in plants.

task. There are different plant diseases affecting the growth of the plant. The most common problems are fungal infections and pest attack which are mostly seen on the leaves of the plant as shown in Figure 8.3. If they are unnoticed they may affect the entire field thereby the whole productivity will be affected. Hence the main characteristics needed in disease detection are speed and accuracy. The proposed method is automatic, efficient, fast and accurate which is helpful for detecting the leaf disease.

The convolutional neural network is used in the detection of the fungal infection and pest attack affecting on the crops. All information about the disease is sent to the farmer's mobile phone from the cloud. Based on the disease differentiation as a pest or a fungal infection, the actuator in drone will be enabled to spray a pesticide or a fungicide on the plant. This technique saves lot of time, cost, and energy spent spraying pesticides and fungicides individually. The real-time information about the plants available in the cloud and the data analysis helps the farmers to make right decisions on required irrigation to mitigate water scarcity problems.

### 8.3.3    METHODOLOGIES

Agricultural drones are employed to scout the field and monitor the health of the crops. The acquired data is subjected to a series of gateway local data analysis and cloud data analysis to accurately detect pest attack, fungal infection, water scarcity, and ageing problems in the crops. Drones examine the ground and spray the exact amount of liquid, in case of a pest attack by varying the distance from the ground. This will happen more precisely in real environments for even coverage. The result is spraying the correct quantity of pesticides for soil conservation which reduces the amount of chemicals pervading into groundwater and eliminates human labor from harmful pesticide by incorporating automation. Actually, the experts quoted that aerial spraying can be accomplished up to five times quicker with drones compared to conventional systems. Climate conditions play an important and serious role for farming and can worsen the quantity and the quality of the crop production. Here, IoT sensors positioned on and off the agriculture fields accumulate information from the environment and this data helps to choose the best crops which can produce and maintain in the specific climatic conditions. These sensors also help to sense real-time weather conditions, namely moisture, rain, and temperature. There are a lot of sensors that are able to identify and to suit smart farming necessities. These sensors screen the condition of the crops and the weather around them. Suppose any troubling weather conditions are identified, then the corresponding in charges are alerted. During troubling climatic conditions, the productivity is improved and also help farmers to collect more agriculture benefits [18–20].

Precision farming is the most well-known feature of IoT in agriculture. It allows more precise and controlled farming by recognizing smart farming applications, namely vehicle tracking, livestock intensive care, field monitoring, and inventory observations. The aim of smart farming is to investigate the information, created via sensors, and to respond appropriately. This helps farmers to produce information with the help of sensors. This information helps farmers to analyze and to acquire intellectual and rapid decisions. There are several precision farming methods, namely vehicle tracking, irrigation management, and livestock management. Precision farming also helps to investigate soil conditions and to enhance operational efficiency.

Technological advances have revolutionized the intervention of agricultural drones: ground and airborne drones are utilized for spraying pesticides, finding the strength of crops, crop supervising, and field investigation. Using real-time information, drones can revolutionize the agriculture industry: changes in irrigation are identified with thermal or multispectral sensors; when crops begin growing, sensors point towards their strength and compute their plants index. Ultimately smart drones have also concentrated on the ecological contact.

The traditional database arrangement does not have adequate storage space for the information gathered from IoT sensors. The use of the cloud for data storage and as an end-to-end IoT platform plays a significant part in the smart agriculture scheme. These schemes are projected to play a significant position such that improved actions can be executed. Sensors are the primary basis of accumulating information on a big scale and this information is investigated and changed to meaningful data using analytics tools. IoT in the agriculture trade has aided the farmers to preserve the crops and fertility of the land, thus improving the product capacity and value.

### 8.3.4 COMBINED ANALYSIS OF IoT END DEVICE DATA WITH DRONE DATA

Drones equipped with appropriate cameras, sensors, and integrating modules gather the agricultural field information and process it onboard using a raspberry pi module. This work also includes soil moisture and humidity data collected through Node MCU IoT device. The data from the drone's raspberry pi module and the Node MCU IoT device is then sent to IoT gateway using Wi-Fi for combined data analysis through end device layer, gateway layer, and cloud layer, as shown in Figure 8.4. This method employs image processing and deep learning technique for automatic disease detection on plant leaves. Successively, the drone can differentiate a pest and

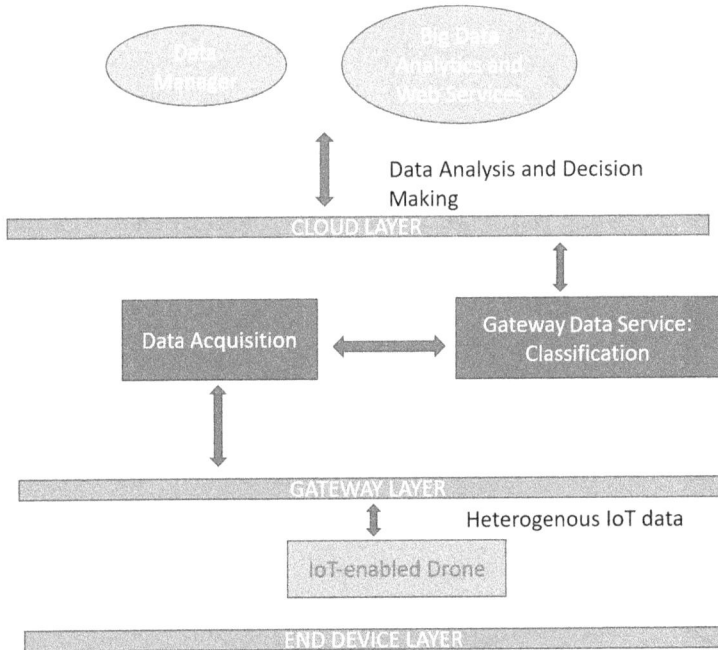

**FIGURE 8.4**  Different layers in applications of IoT in Agriculture Sector.

a fungal infection and accordingly spray a pesticide or a fungicide on the plant. This work saves a lot of time, cost, and energy compared to spraying pesticides and fungicides individually. The real-time information about the plants available in the cloud and the data analysis helps the farmers to make right decisions about required irrigation to mitigate water scarcity problem.

## 8.4  SMART HEALTHCARE

Nowadays in healthcare system, smart devices have enhanced the lot of doctors, patients, and the healthcare industry [2]. IoTs participate in significant responsibility and its assets in human aids life. IoT in medical applications is called the Internet of Health Things (IoHT). Here, wearable devices are able to offer the real-time information associated with health. This technique straightforwardly connects with the appropriate doctor or specialist. IoHT allows disease avoidance and identifies the causes that need to be treated. The most essential consequences are also anticipated for business reasons, namely industrial automation and transportation. This technique will be helpful in supporting and assisting life [21–25]. There are different layers while implementing a system for IoT in healthcare. The perception layer includes the Raspberry Pi2, Radio Frequency Identification (RFID), camera and sensors. The network layer includes the mobile networks 3G/4G, TCP, UDP, IPV4, IPV6. The application layer includes the cloud database and the application for healthcare system as in Figure 8.5.

### 8.4.1  Why is IoT Important in Healthcare?

IoT applications play a vital role in employing linked devices and novel techniques accumulated into agriculture. Smart farming technology mostly comes under IoT. It is used to get rid of farmers, decrease physical work requirements, and increase the production in all possible ways. Health information technology enhances a patient's security by decreasing medication errors, decreasing difficult drug reactions, and refining compliance to training strategies. There is no hesitation to say

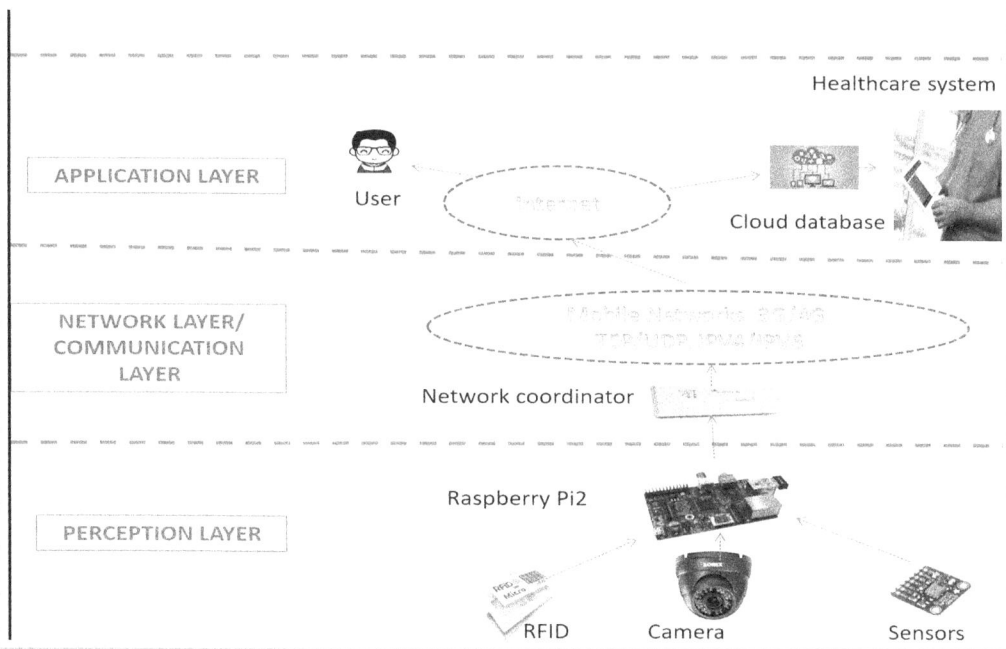

**FIGURE 8.5**  Different layers in Health care using IoT.

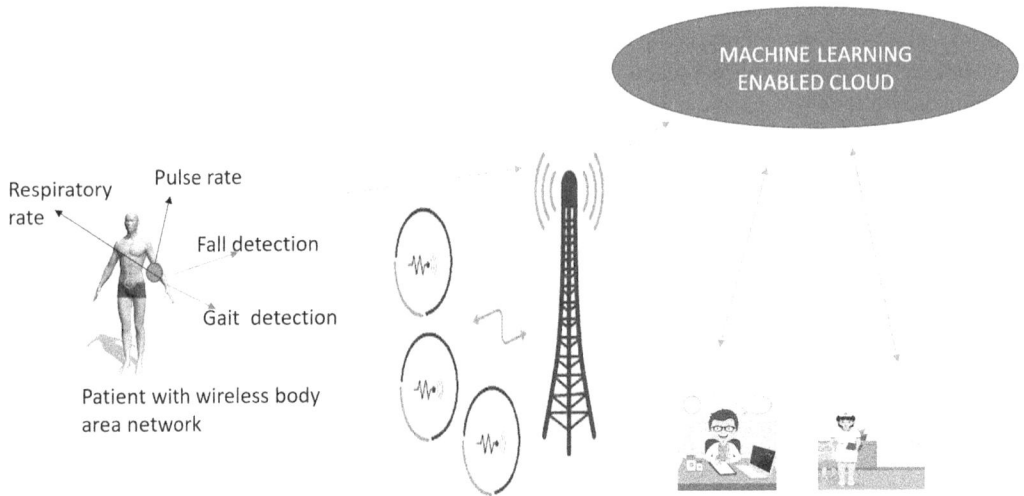

**FIGURE 8.6**   IoT in Healthcare sector.

that health data is significant for improving healthcare excellence and protection. Healthcare organizations technologies have inadequate proof in improving patient safety results [25–30].

Healthcare experts have long publicized the capability of associated and IoT-enabled devices to better engage patients in their individual care rather than leaving it in the hands of surgeons. A lot of wearable devices and fitness trackers intend to facilitate healthy people to enhance their health [31]. Fitness wearables, namely the iWatch, FitBit, Vivosmart, Garmin, and Tom Tom, have enabled numerous people to check and to record their heart rate, work-out activities, sleeping habits, and steps taken per day [26]. Telemedicine permits patients to reside at home for healing rather than having expensive appointment with doctors. A mobile device to converse with doctors can check and transmit glucose levels, blood pressure, temperature, and the outputs of other monitoring devices to doctors. In the healthcare sector, different information, like pulse rate and heart rate, is collected and relayed through a base station to the cloud. The information is then processed and given to the users as in Figure 8.6.

IoT devices serve healthcare production by addressing operational subjects associated to equipment and inventory management for patient tracking [23, 24]. Radio Frequency Identification (RFID) [25–32] is an automated technology which provides communication with the tagged objects through wireless data transmission to recover applicable data. RFID permits automatic recognition of information. RFID equipment and mobile scanners are linked to cloud technology to help organizations control and direct their inventories. This method helps to ensure physicians and patients have a continuous supply of medication and equipment for their continuity of care. Cloud hygiene methods help healthcare organizations considerably decrease hospital-acquired infections [24]. The different applications of IoT in the healthcare domain are given in Figure 8.7.

### 8.4.2   How IoT Helps in Healthcare – Process

IoT provides devices every corner of a farm to determine all kinds of data distantly. It also supplies data to the farmer in real time. IoT devices can accumulate important data, namely livestock health, soil moisture, water tank levels, dam levels, and chemical application. Bringing IoT into medicine facilitates healthier, stronger, and easier patient care. From the starting point of medical devices to smart sensors, IoT speeds up healthcare delivery, allowing doctors to spend less time on travel, identifying illnesses and exchanging a few words with patients. Remote patient monitoring [28–34]

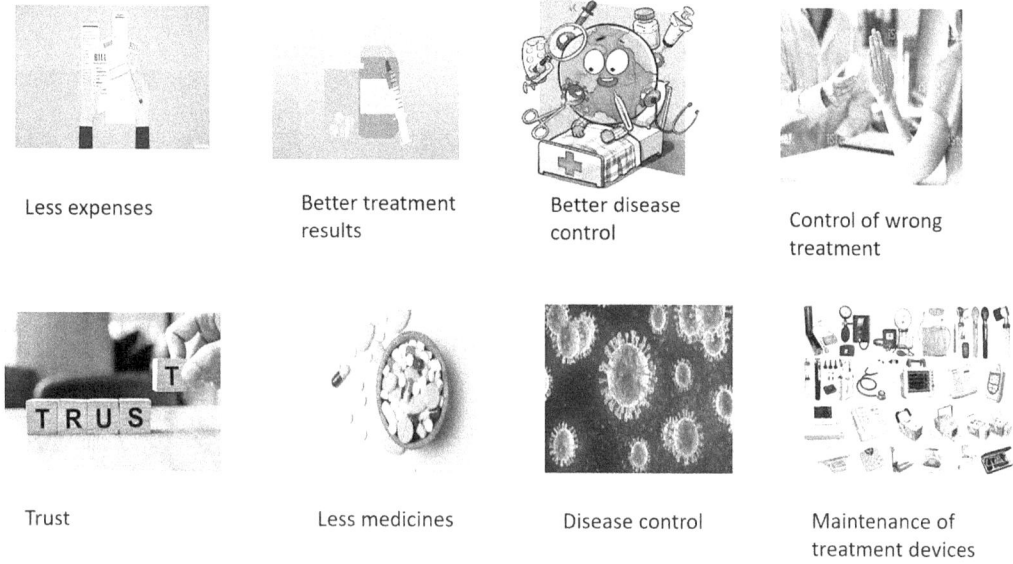

Less expenses

Better treatment results

Better disease control

Control of wrong treatment

Trust

Less medicines

Disease control

Maintenance of treatment devices

**FIGURE 8.7** Applications of IoT in healthcare.

has enhanced patient outcomes for definite chronic problems, namely stroke, heart attack, asthma, and hypertension. Patient Data Management System (PDMS) repeatedly recover information from equipment such as intravenous pumps, patient monitors, and ventilators. To assist healthcare providers, the patient's data is accordingly reviewed and reorganized [15]. Virtual appointments are real-time two-way audio or video communications between the healthcare professional and the patient and are called synchronous telemedicine [34–36]. This technology assists direction from the expert concerning the managing of the patient. These e-consults diminish patient waiting periods for consultant appointments and diagnoses. Telemedicine also gives health data that is accumulated gradually from medical devices and individual mobile devices. This data is utilized to check patients. Patient portals are safe online spaces that give patients access to their personal health data by means of a computer or a cell phone. These patient portals enhance the effects of protective care, disease consciousness, and self-improvement.

### 8.4.3 Applications of IoT in Healthcare

5G is the latest generation of cellular technology with very high speed and responsiveness. Verizon's network has a speed above 1 gigabit per second, that is 100 times quicker than other connections. 5G has low "latency" which reduces the lag time to as small as one millisecond. Thus 5G has great promise in remote healthcare, helping in the connection of a number of devices and thus expanding the Internet of Things (IoT) to connect ambulances and real-time remote surgeries, as in Figure 8.8.

### 8.4.4 Current Status

Some systems have been proposed in the literature for monitoring blood glucose levels [5]. Patients will take the blood glucose readings manually and there are two types of abnormalities: abnormal levels and missed reading. The severity is identified and then either the patient family member or the doctor is notified. Thus, the measurements of blood glucose levels are automated. Heart attacks can be detected using some components and an antenna in [30–35] with ECG sensors used to monitor heart activity, administered by a microcontroller. This data is sent via Bluetooth to the user's

# 5G and its uses in Healthcare

Connected ambulances              Remote surgery              IoT sensor monitoring

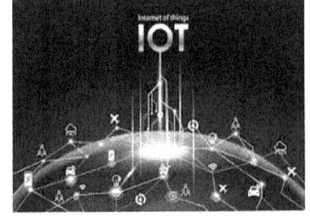

**FIGURE 8.8**   How IoT is Transforming the Healthcare Industry.

mobile, where the data is handled and accessible to a user application. The respiratory rate can also be measured and used to predict a heart attack. SPHERE [29] uses wearable, ecological sensors for monitoring health. Elder people can benefit a lot by monitoring their health while at home. This also permits intervention by doctors in case of critical issues. Environmental sensors are also used in the home. The sensors are used as small and wearable nodes. Bluetooth can be utilized for transferring the data to mobile phones with long-range data transfers to a doctor done using Long-Term Evolution (LTE). The high volumes of data can be stored in the cloud as seen in several preceding works.

## 8.4.5   CHALLENGES OF IoT IN HEALTHCARE

The common challenges of IoT are reliability, interoperability, performance, availability, gig IoT data, privacy, scalability, investment, mobility, compatibility, and precision [2]. Reliability is not presently about transferring trustworthy data but being able to acclimatize to different environmental conditions. This is also challenging in terms of security problems and long-term usability. Major challenges of IoT in healthcare are artificial intelligence, the Innovative Internet of 5G Medical Robotic Things (IIo-5GMRTs), and 5G network technology in the ongoing COVID-19 pandemic. COVID-19 stretched the healthcare sector and obligated healthcare professionals all over the world to come treat infected people with minimal social and physical contact. These techniques assisted the healthcare service sector to deliver on the key mandates and countries to work better in the continuing COVID-19 pandemic situation. The uneasiness of healthcare sector workers motivated the development of several healthcare technologies [35–39].

## 8.4.6   FUTURE OF IoT IN HEALTHCARE

Fast progress in digital technology is redefining society by giving technology to nations and making people relaxed in the use of tech. Digital technology transformation also happens in monitoring the healthcare sector. IoT facilitates devices and empowers physicians to help patients to stay safe and healthy. It also increases speed, innovation, speeding up processes, automation, saving overall costs and civilizing decision making [40–46]. In upcoming works, IoT functionality-based drones will be applied to send drugs to patients for home-based treatment [46–49]. Like in Rwanda, drones are utilized to transfer donor blood. There is a huge rise in the healthcare market due to the growing demand of personalized applications and increasing penetration of medical devices, as in Figure 8.9.

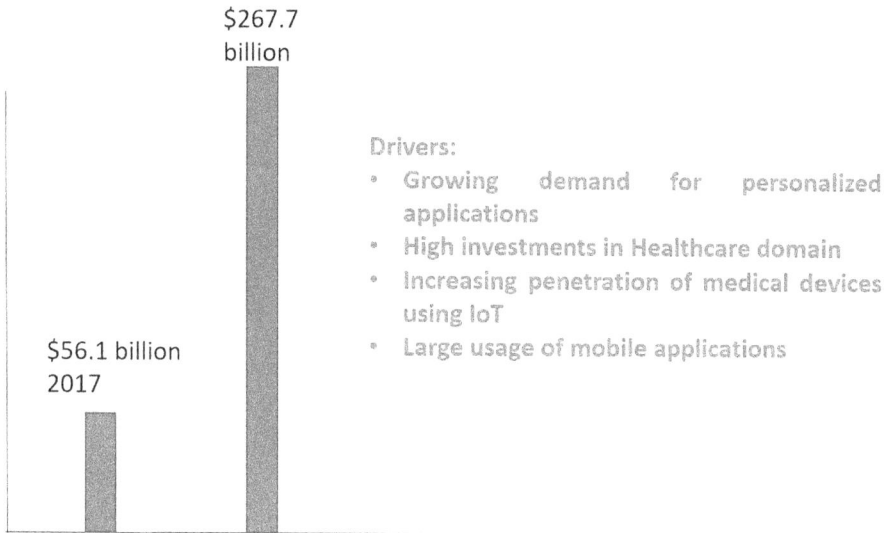

**FIGURE 8.9** Healthcare market due to IoT.

## 8.5 CONCLUSION

IoT has enabled agriculture to employ modern technological results in recent times. This information helps fix the gap between production, quality, and quantity of harvest. Information obtained from numerous sensors stored in a database guarantees swift action and less harm to crops. Hospitals are enabled to perform better through innovative approaches offered by IoT, 5G Medical Robotics for hospital disinfection, and IoHT. Artificial intelligence-based drones for logistics management are also used everywhere [50]. Scientists believed that COVID-19 would drive scientific innovations in 5G medical robotics which would consolidate the Industry 4.0 healthcare computerization. The current techniques employed are cost-optimized healthcare methods for hospital routine works and remote access intensive care using IoT. Medical robotics is also used to reduce the risk of hospital infection in the continuing COVID-19 situations.

## REFERENCES

[1] E. Fleisch, What is the Internet of things? An economic perspective, *Economics, Management, and Financial Markets* 5, no. 2 (2010): 125–157.
[2] A. Ahmed, and G.E. Ahmed, Benefits and challenges of internet of things for telecommunication networks, *Telecommunication Networks-Trends and Developments* (2019): 105–124. doi: 10.5772/intechopen.81891.
[3] J. Kempf, J. Arkko, N. Beheshti, and K. Yedavalli, Thoughts on reliability in the internet of things, in *Interconnecting Smart Objects with the Internet Workshop*, vol. 1, pp. 1–4. Boston, MA: Internet Architecture Board, 2011.
[4] A. Thorat, S. Kumari, and N.D. Valakunde, An IoT based smart solution for leaf disease detection, in *2017 International Conference on Big Data, IoT and Data Science (BID)*, pp. 193–198. IEEE, 2017.
[5] R. Shukla, G. Dubey, P. Malik, N. Sindhwani, R. Anand, A. Dahiya, and V. Yadav, Detecting crop health using machine learning techniques in smart agriculture system, *Journal of Scientific & Industrial Research* 80, no. 08 (2021): 699–706.
[6] V. Gowrishankar, and K. Venkatachalam, IoT based precision agriculture using Agribot, *Global Research and Development Journal for Engineering* 3, no. 5 (2018): 2455–5703.

[7] S.B. Londhe, and K. Sujata, Remotely operated pesticide sprayer robot in agricultural field, *International Journal of Computer Applications* 167, no. 3 (2017): 26–29.

[8] T. Watanabe, A. Raju, Y. Hiraga, and K. Sugimura, Development of geospatial model for preparing distribution of rare plant resources using UAV/drone, *Indian Journal of Pharmaceutical Education and Research* 52 (2018): S146–S150.

[9] R.F. Olanrewaju, R.S.B. Rosli, and B.W. Adebayo, Autonomous control of tilt tri-rotor unmanned aerial vehicle, *Indian Journal of Science and Technology* 9, no. 36 (2016): 1–7.

[10] Y. Shi, Z. Wang, X. Wang, and S. Zhang, Internet of things application to monitoring plant disease and insect pests, in *International Conference on Applied Science and Engineering Innovation (ASEI 2015)*, pp. 31–34, 2015.

[11] P. Jian-Sheng, An intelligent robot system for spraying pesticides, *The Open Electrical and Electronic Engineering Journal* 8, no. 1 (2014): 435–444.

[12] Z. Wei, W. Xiu, D. Wei, S. Shuai, W. Songlin, and F. Pengfei, Design and test of automatic toward-target sprayer used in orchard, in *2015 IEEE International Conference on Cyber Technology in Automation, Control, and Intelligent Systems (CYBER)*, pp. 697–702. IEEE, 2015.

[13] R. Berenstein, and Y. Edan, Automatic adjustable spraying device for site-specific agricultural application, *IEEE Transactions on Automation Science and Engineering* 15, no. 2 (2017): 641–650.

[14] J. Zhang, Z. Cao, C. Geng, and W. Li, Research on precision target spray robot in greenhouse, *Transactions of the Chinese Society of Agricultural Engineering* 25, no. 1 (2009): 70–73.

[15] J. Primicerio, S.F. Di Gennaro, E. Fiorillo, L. Genesio, E. Lugato, A. Matese, and F.P. Vaccari, A flexible unmanned aerial vehicle for precision agriculture, *Precision Agriculture* 13, no. 4 (2012): 517–523.

[16] S. Candiago, F. Remondino, M. De Giglio, M. Dubbini, and M. Gattelli, Evaluating multispectral images and vegetation indices for precision farming applications from UAV images, *Remote Sensing* 7, no. 4 (2015): 4026–4047.

[17] M. Rieke, T. Foerster, J. Geipel, and T. Prinz, High-precision positioning and real-time data processing of UAV systems, *International Archives of the Photogrammetry, Remote Sensing and Spatial Information Sciences* 38, no. 1/C22 (2011): 119–124.

[18] A. Matese, P. Toscano, S.F. Di Gennaro, L. Genesio, F.P. Vaccari, J. Primicerio, C. Belli, A. Zaldei, R. Bianconi, and B. Gioli, Intercomparison of UAV, aircraft and satellite remote sensing platforms for precision viticulture, *Remote Sensing* 7, no. 3 (2015): 2971–2990.

[19] S. Lookinland, and K. Anson, Perpetuation of ageist attitudes among present and future health care personnel: Implications for elder care, *Journal of Advanced Nursing* 21, no. 1 (1995): 47–56.

[20] Y.A. Qadri, A. Nauman, Y.B. Zikria, A.V. Vasilakos, and S.W. Kim, The future of healthcare internet of things: A survey of emerging technologies, *IEEE Communications Surveys and Tutorials* 22, no. 2 (2020): 1121–1167.

[21] V.L. Smith-Daniels, S.B. Schweikhart, and D.E. Smith-Daniels, Capacity management in health care services: Review and future research directions, *Decision Sciences* 19, no. 4 (1988): 889–919.

[22] P. Chawla, A. Juneja, S. Juneja, and R. Anand, Artificial intelligent systems in smart medical healthcare: Current trends, *International Journal of Advanced Science and Technology*, 29, no. 10 (2020): 1476–1484.

[23] J.A. Levy, and R. Strombeck, Health benefits and risks of the Internet, *Journal of Medical Systems* 26, no. 6 (2002): 495–510.

[24] B. Anthony Jr., Use of telemedicine and virtual care for remote treatment in response to COVID-19 pandemic, *Journal of Medical Systems* 44, no. 7 (2020): 1–9.

[25] U.O. Matthew, J.S. Kazaure, O. Amaonwu, U.A. Adamu, I.M. Hassan, A.A. Kazaure, and C.N. Ubochi, Role of Internet of Health Things (IoHTs) and Innovative Internet of 5G Medical Robotic Things (IIo-5GMRTs) in COVID-19 Global Health Risk Management and Logistics Planning, in *Intelligent Data Analysis for COVID-19 Pandemic*, pp. 27–53. Springer, Singapore, 2021.

[26] Y.K. Alotaibi, and F. Federico, The impact of health information technology on patient safety, *Saudi Medical Journal* 38, no. 12 (2017): 1173.

[27] C. Klersy, G. Boriani, A. De Silvestri, G.H. Mairesse, F. Braunschweig, V. Scotti, A. Balduini, M.R. Cowie, F. Leyva, and Health Economics Committee of the European Heart Rhythm Association, Effect of telemonitoring of cardiac implantable electronic devices on healthcare utilization: A meta-analysis of randomized controlled trials in patients with heart failure, *European Journal of Heart Failure* 18, no. 2 (2016): 195–204.

[28] S.C. Inglis, R.A. Clark, F.A. McAlister, J. Ball, C. Lewinter, D. Cullington, S. Stewart, and J.G. Cleland, Structured telephone support or telemonitoring programmes for patients with chronic heart failure, *Cochrane Database of Systematic Reviews* 8 (2010). doi: 10.1002/14651858.CD007228.pub2.

[29] N. Zhu, T. Diethe, M. Camplani, L. Tao, A. Burrows, and N. Twomey, Bridging e-health and the internet of things: The sphere project, *IEEE Intelligent Systems* 30, no. 4 (2015): 39–46.

[30] S.H. Chang, R.D. Chiang, S.J. Wu, and W.T. Chang, A context-aware, interactive M-health system for diabetics, *IT Professional* 18, no. 3 (2016): 14–22.

[31] C.F. Pasluosta, H. Gassner, J. Winkler, J. Klucken, and B.M. Eskofier, An emerging era in the management of Parkinson's disease: Wearable technologies and the internet of things, *IEEE Journal of Biomedical and Health Informatics* 19, no. 6 (2015): 1873–1881.

[32] G. Wolgast, C. Ehrenborg, A. Israelsson, J. Helander, E. Johansson, and H. Manefjord, Wireless body area network for heart attack detection [Education Corner], *IEEE Antennas and Propagation Magazine* 58, no. 5 (2016): 84–92.

[33] M. Elliott, Why is respiratory rate the neglected vital sign? A narrative review, *International Archives of Nursing and Health Care* 2, no. 3 (2016): 050.

[34] B.A. Lieber, B. Taylor, G. Appelboom, K. Prasad, S. Bruce, A. Yang, E. Bruce, B. Christophe, and E.S. Connolly Jr, Meta-analysis of telemonitoring to improve HbA1c levels: Promise for stroke survivors, *Journal of Clinical Neuroscience* 22, no. 5 (2015): 807–811.

[35] K.M. Kew, and C.J. Cates, Home telemonitoring and remote feedback between clinic visits for asthma, *Cochrane Database of Systematic Reviews* 8 (2016). doi: 10.1002/14651858.CD011714.pub2.

[36] S. Omboni, and A. Guarda, Impact of home blood pressure telemonitoring and blood pressure control: A meta-analysis of randomized controlled studies, *American Journal of Hypertension* 24, no. 9 (2011): 989–998.

[37] A. Cheung, F.H. Van Velden, V. Lagerburg, and N. Minderman, The organizational and clinical impact of integrating bedside equipment to an information system: A systematic literature review of patient data management systems (PDMS), *International Journal of Medical Informatics* 84, no. 3 (2015): 155–165.

[38] K. Tan, and N.M. Lai, Telemedicine for the support of parents of high-risk newborn infants, *Cochrane Database of Systematic Reviews* 6 (2012). doi: 10.1002/14651858.CD006818.pub2.

[39] K.M. Kew, and C.J. Cates, Remote versus face-to-face check-ups for asthma, *Cochrane Database of Systematic Reviews* 4 (2016). doi: 10.1002/14651858.CD011715.pub2.

[40] T.L. Gregersen, A. Green, E. Frausing, T. Ringbaek, E. Brøndum, and C.S. Ulrik, Do telemedical interventions improve quality of life in patients with COPD? A systematic review, *International Journal of Chronic Obstructive Pulmonary Disease* 11 (2016): 809.

[41] A. Salmoiraghi, and S. Hussain, A systematic review of the use of telepsychiatry in acute settings, *Journal of Psychiatric Practice®* 21, no. 5 (2015): 389–393.

[42] P.J. Hofstetter, J. Kokesh, A.S. Ferguson, and L.J. Hood, The impact of telehealth on wait time for ENT specialty care, *Telemedicine and e-Health* 16, no. 5 (2010): 551–556.

[43] J.N. Olayiwola, D. Anderson, N. Jepeal, R. Aseltine, C. Pickett, J. Yan, and I. Zlateva, Electronic consultations to improve the primary care-specialty care interface for cardiology in the medically underserved: A cluster-randomized controlled trial, *The Annals of Family Medicine* 14, no. 2 (2016): 133–140.

[44] H. Daniel, L.S. Sulmasy, and Health and Public Policy Committee of the American College of Physicians, Policy recommendations to guide the use of telemedicine in primary care settings: An American College of Physicians position paper, *Annals of Internal Medicine* 163, no. 10 (2015): 787–789.

[45] R.H. Weber, and R. Weber, *Internet of Things*, vol. 12. Heidelberg: Springer, 2010.

[46] K. Dewangan, and M. Mishra, Internet of things for healthcare: A Review, *International Journal of Advanced in Management, Technology and Engineering Sciences* 8, no. 3 (2018): 526–534.

[47] E. Vasilomanolakis, J. Daubert, M. Luthra, V. Gazis, A. Wiesmaier, and P. Kikiras, On the security and privacy of internet of things architectures and systems, in *2015 International Workshop on Secure Internet of Things (SIoT)*, pp. 49–57. IEEE, 2015.

[48] J. Duffy, and A. Colon, Best fitness trackers for 2015, *PC Magazine*, 2016.

[49] I. Erguler, A potential weakness in RFID-based Internet-of-Things systems, *Pervasive and Mobile Computing* 20 (2015): 115–126.

[50] N. Sindhwani, R. Anand, R. Vashisth, S. Chauhan, V. Talukdar, and D. Dhabliya, Thingspeak-based environmental monitoring system using IoT, in *2022 Seventh International Conference on Parallel, Distributed and Grid Computing (PDGC)*, pp. 675–680, 2022. doi: 10.1109/PDGC56933.2022.10053167.

# 9 Splitter with Cryptographic Model for Cloud Data Transmission Security

*Ashima Arya*
CSIT Department, KIET Ghaziabad, India

*Mitu Sehgal, Aarzoo, and Sangeeta*
PIET, Samalkha, Panipat, India

*Junaid Rashid*
Kongju National University, South Korea

## CONTENTS

9.1 Introduction ............................................................................................147
    9.1.1 Cloud Computing .......................................................................147
    9.1.2 Need for Proposed System ........................................................149
    9.1.3 Various Attacks ..........................................................................149
9.2 Literature Review ..................................................................................150
9.3 Objective of Proposed Work .................................................................150
9.4 Proposed Model ....................................................................................151
9.5 Results ...................................................................................................152
    9.5.1 File Splitter ................................................................................153
    9.5.2 GUI for Client Interface ............................................................153
    9.5.3 GUI for Server Interface ...........................................................154
    9.5.4 Secure Files ...............................................................................154
9.6 Fog Implementation ..............................................................................155
9.7 Conclusion ............................................................................................156
9.8 Future Scope ........................................................................................158
References .....................................................................................................158

## 9.1 INTRODUCTION

### 9.1.1 CLOUD COMPUTING

Users can use hardware, software, and data from cloud servers thanks to a new IT paradigm known as cloud computing. Users of cloud computing can access wide range of services [1]. The development of cloud computing has been a long-term, slow-but-steady process that began with the first computer. This is a new paradigm which has emerged in the wake of grid computing, which revolutionized both data storage with computation. In this modern period, [2] it has gained popularity as new infrastructure with little investment in hardware platforms, human training, software tool licensing. A group of IT experts is dedicated to providing secure cloud services on a round-the-clock

DOI: 10.1201/9781003319238-9

basis. All those services are offered over the Internet by utilizing current networking protocols and formats which are either subscription-based or pay-per-use. This can be utilized as "infrastructure as a service (IaaS Cloud)," "software as a service (SaaS Cloud)," or "platform as a service (PaaS Cloud)". In recent years, the concept of "data storage as a service" (also known as "DaaS Cloud") has also arisen to give customers access to storage [3].

Parallel to this development, Technologies like big data and Hadoop are used for data storing, accessing and processing for cloud computing platform. These technologies are being utilized widely and more frequently to produce services and applications across a variety of industries, including health, web, and energy. Additionally, it covers a wide range of topics, such as business, science, and public as well as private administration [4]. In other words, big data and cloud computing are essential for both present and future research frontiers. Similar to any IT infrastructure, the cloud may be targeted by hackers due to flaws and vulnerabilities in underlying technologies. The security and data storage aspects have a significant impact on how well a cloud operates. Users are very worried about the safety and privacy of data stored on cloud. This is a main factor that limited cloud computing's ability to be used anywhere in the world. The book discusses the security of data storage and intrusion detection in context of cloud computing. In context of distributed network, intrusion detection is the most frequently asked about problem in the cloud scenario. The following security issues are brought up by cloud computing:

- The two most pressing issues with cloud computing today are access control and security [5].
- Access control: A method in which only authorized user can store and process data from a cloud server. There are numerous issues that arise while accessing data are security of data, extensive access times, loss of data, overhead, redundant data, etc. [6].
- There is one evident disadvantage to cloud computing: since cloud servers are managed by a third party, your data is housed by an external entity and is thus not totally secure against insider and outsider attacks and invasions [7].
- The second major issue is how to implement security level for different cloud services. With great range of knowledge related to the security field and heterogeneity of cloud service descriptions are among the issues that have emerged [8].

Nevertheless, despite all the advantages of cloud computing, there are numerous security issues with regard to virtualization, hardware, data, networks, and service providers which operate as a major impediment to cloud adoption in the IT sector [9].

In a cloud computing environment, a datacenter is made up of numerous servers, cooling equipment, and power delivery equipment. Complex frameworks require a tremendous amount of computational power to run. The datacenter has evolved into the focal point for a notable rise in power consumption, heat dissipation, and server temperature as a result of the increased need for computing capacity. Because user workloads now require more computing, the energy consumption of cloud datacenters has skyrocketed. Thus, it has become crucial to address the issue of energy conservation [10].

- ***The infected application:*** *The service provider needs administrative access to server in order to operate and maintain it.*
- ***Account or service traffic hijacking:*** *If login credentials are taken, an account may be hijacked.*
- ***An unreliable application program interface:*** *This interface will have power over third parties. It also verifies user's identification.*
- ***Denial of Service:*** *This is what happens when a shared service is requested by a large number of individuals. In this case, the hacker exploits the circumstance.*
- ***Insiders with malicious inclination:*** *This occurs when someone gains access to our login information [11].*

Due to the presence of diverse applications, workloads such as content delivery, MapReduce, network web applications, etc., have confrontational allocation requirement in terms of ICT resource capacities. Due to limited resources and rising consumer demands, this task of resource allocation becomes more difficult [12].

## 9.1.2 Need for Proposed System

A secure cryptographic prototype has been established for the movement of data across different companies in a secure manner. A splitter was used to split data into a cloud layer and fog layer for security reasons. Since the data is dispersed over two tiers, there is no chance of data loss. The study effort gave the IP address and port numbers, the recommended model offers security at session layer. Comparison of conventional work with suggested work has also been done. Cloud computing has produced significant changes involving the centralization and relocation of software, platforms, and infrastructure onto cloud by sharing [13] IT resources through the Internet. Industry 4.0, or increased digitization in manufacturing facilities which is directly related to Industrial Internet of Things (IIoT), connects various types of machineries and forecasts new uses: business models in addition to more efficient production. Cloud computing is limited in Industry 4.0 for a number of reasons: First of all, Industry 4.0 produces large volumes of data that must be processed, stored, and then analyzed. It is frequently technically difficult or very expensive to move data to and from the cloud due to constrained bandwidth or unstable Internet connection in production site. Second, it's impossible to guarantee real-time processing and interactions with cloud services [14]. Thirdly, since various industrial sensors and controllers are constrained by their own resources, Industry 4.0 scenarios need far more processing power. Fog computing [15] is development of cloud computing that processes data locally rather than sending it to the cloud in order to speed up processing and lessen the load on Internet. Fog computing is a decentralized strategy which provides growing architecture for computation, storage, processing, control, and networking and distributes its services along the "cloud-to-things continuum". A current security model was contrasted with the earlier approach [16], and it was discovered that the advised work experienced packet loss less frequently. It has been established that conventional security measures are worthless. The suggested method secures data by isolating and encrypting with cutting-edge cryptographic techniques. This system is less susceptible to congestion and packet loss. In order to provide network security on a variety of levels, this research examines both active as well as passive threats. The suggested method splits packets into several pieces to safeguard them [17].

## 9.1.3 Various Attacks

Although there have been various methods and approaches for secure data transfer over the years, it is true that new types of assaults are becoming extra powerful plus efficient in getting past security measures. This assault could involve denial of service (DoS) or distributed denial of service (DDoS) [13, 18].

From client-server context, a distributed denial of service (DDoS) assault would bring down the entire system; however, in a cloud setting, it is less effective but still tries to disrupt the system's normal operations by:

a. Overloading of the flow table;
b. Saturation of the control plane;
c. A byzantine assault;
d. Malware;
e. A cross-site scripting attack;
f. Eavesdropping;
g. A side-channel attack [12];
h. Attack using SQL injection [19].

## 9.2   LITERATURE REVIEW

Numerous studies have examined the security variables that play a significant part in a transmission model. Here is a list of some of these:

**Prachi S Deshpande addresses the issues of accessing the security in cloud computing** [2]. This study is concerned about how to use intrusion detection in cloud computing for security and privacy of data. Due to the distributed nature of operation, intrusion detection is the most frequently asked about problem in the cloud scenario.

**In 2021, Asif Ali Laghari [10] predicted that over the last few years, data processing shifted from cloud computing to the fog computing**. Basically, fog computing is enlargement of cloud computing that progressions data locally rather than transmitting it to the cloud in order to speed up processing and lower the load on the Internet. The author discusses the technology, architecture, and applications of fog computing. The author specifically discusses the most recent advancements in platforms, security, privacy, and cloud at the edge networking.

**Nabil Abubaker [20] thought of the Privacy-Preserving Fog Computing Paradigm**. To achieve end user or IoT device secrecy using the described strategies, the design purpose of fog computing must be altered. This is one of the main challenges. End-location user's identity is not kept secret, which could result in an unintentional open closet to fog node as second problem. The author discusses these problems in context of actual circumstances and suggests fog computing paradigm that protects privacy as a solution.

**Yunguo Guan [13] focused on data security and privacy in fog computing**. Due to great distance among cloud and end users, drawbacks of cloud computing (such high latency) eventually becomes apparent. Fog computing is a solution that would take advantage of network's edge to extend the cloud. To process communication of data between cloud and end users, fog computing specifically provides an intermediate layer called fog. As a result, fog computing is typically seen as a development of cloud computing. The author explored design concerns for data privacy and security in fog computing.

**Jiyuan Zhou presented a hierarchical secure cloud storage mechanism depending on fog computing in 2017 [21]**. Such plans are crucial for protecting cloud storage in a fog environment. However, studies show that putting in place such a security system is quite challenging.

**Shafat Khan [22] considered issues and risks of embracing cloud in business environment**. Cloud computing has developed over the past few years from a potential business idea to one of the IT industry's fastest growing sectors. Despite the popularity of the cloud and its many benefits, such as financial advantage, a quick elastic resource pool, on-demand benefit, corporate clients are still cautious to move their operations to cloud, and paradigm also presents challenges for both customers and suppliers. The purpose of this study is to clarify setup challenges of cloud computing.

**A systematic study of load-balancing approaches in the fog computing environment by Mandeep Kaur [23]**. Fog operates in a real-time setting, thus offers from connected device must be handled instantly. Virtual machines (VMs) at the fog layer get overburdened due to the rise in user requests. A load-balancing method can evenly spread the load among all the VMs. To overcome the load-balancing issue, a fog computing architecture is proposed in the article. It also addresses existing problems and difficulties that can be remedied in upcoming research projects.

## 9.3   OBJECTIVE OF PROPOSED WORK

The objective of this research is to provide defenses beside intrusions at application layer, where a user interacts with network directly. The most frequently used protocols at application layer are FTP, TELNET, and HTTP [24, 25]. The primary aim of research is to develop an application layer system that is quicker and includes the following:

1. To gain knowledge of numerous attack types, like active as well as passive attacks.
2. To provide network security at different tiers by utilizing cryptographic methods.
3. Creating and implementing a secure mechanism for data protection at application layer from active as well as passive threats [26].
4. Using a splitter to separate data while providing data security on both layers.
5. To compare the suggested model to the current security paradigm and show how suggested model is superior [27].

## 9.4 PROPOSED MODEL

The suggested technique has provided an additional secure way to safeguard data from active as well as passive types of attacks at application layer. In this research, the proposed model and the existing security model were contrasted. When comparing packet drops between the suggested work and the traditional work, it was found that proposed work has less chance of packet loss. It is found that conventional security measures were useless. Data is separated and encrypted using complex cryptographic processes as part of the proposed method to protect it. This system is likely to experience less congestion and packet loss [28].

To provide network security at several layers, this study looked into active and passive attacks. The proposed method divides packets into multiple fragments in order to provide packet security. A new security system is needed due to the shortcomings of traditional security measures. Here, an IP filter has been used to prevent packets from being sent from a server to a client without authentication. The improved AES ENCRYPTION module operates if the packet is genuine. Here, proposed data transmission process flow has been shown in Figure 9.1.

**Steps that follow while sending a hand include:**

1. USE SIMPLE TEXT (256 bits)
2. Set counter to 1 and apply the round key
3. If counter is fewer than N-1, then (here N is number of iterations)
4. Perform the SubByte
5. Execute a shift row
6. Mix columns
7. Counter=counter+1; 8. if not
8. Handle SubBytes
9. Reversal rows
10. Use the round key

Cipher text would be produced 12. (256 bits)

**At the receiver end, the steps include:**

1. CIPHER THE TEXT (256 bits)
2. Set counter to 1 and apply the round key
3. If counter is fewer than N-1, then
   a. Process the reverse shift row;
   b. Inverse Mix Columns;
   c. Inverse SubByte;
   d. Counter=counter+1.
4. In any case
   a. Rows with an inverse shift;
   b. Apply round key;
   c. Reverse the SubByte;
5. The output would be in plain text (256 bits).

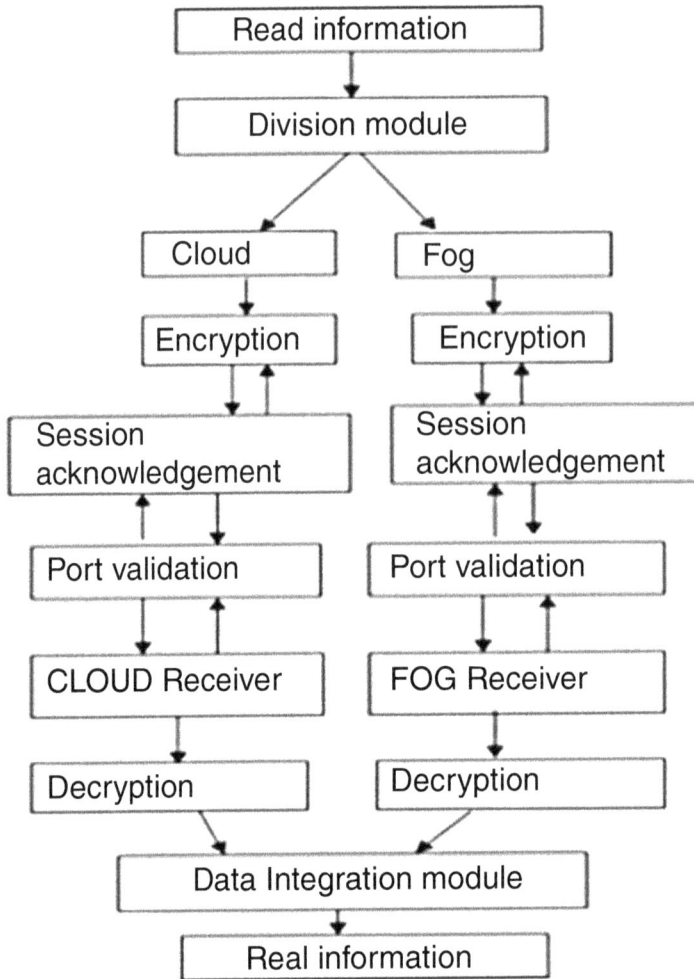

**FIGURE 9.1**    Process Flow of Proposed Work.

The purpose of this study is to offer defense against attack at app level, where a person can directly interact with a network. At application layer, protocols including HTTP, TELNET, FTP are utilized. The most important objective of research is to make an effective application layer system. Numerous types of assaults, including passive and active ones, as well as the employment of cryptographic techniques [29] to provide various levels of network security have been the topic of research. Additionally, safe data security mechanisms [30] against active as well as passive application layer threats are intended to be designed and implemented as part of the work being proposed. The proposed method uses a splitter to divide packets into two layers and offer security of data. In conclusion, the study is done to demonstrate the superiority of the suggested model over the current security paradigm [31].

## 9.5   RESULTS

Here it has been explained how the suggested task was carried out and what the outcome was.

### 9.5.1 FILE SPLITTER

Using FILE splitter, the data is divided into two separate files, one for cloud and the other for fog. The file has two sections, one for cloud and one for fog, and is identified by its name and security code. This improves the transmission's dependability and security. [32]. A file splitter is shown in figure 9.2.

### 9.5.2 GUI FOR CLIENT INTERFACE

The server would receive data from this file transmitter interface. Here, you input your user name, password, IP address, port number, file path, security token, and AES code. Figure 9.3 shows the application at sender side.

# File splitter

Enter File name

Enter code

Split file for cloud and fog

**FIGURE 9.2**  File Splitter.

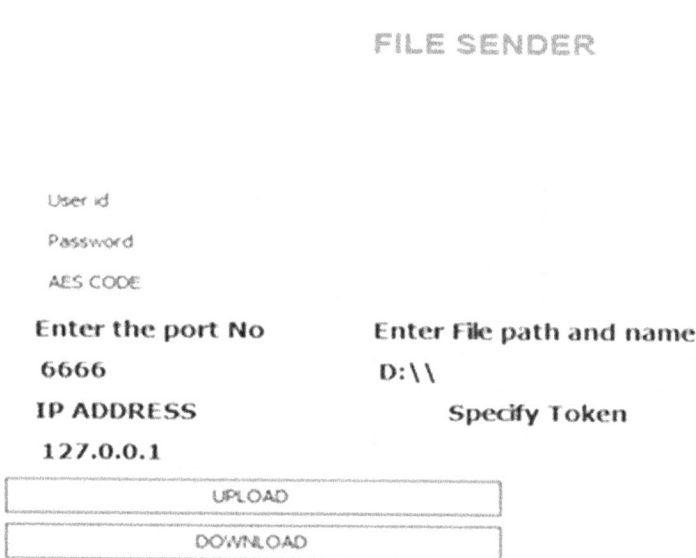

## FILE SENDER

User id

Password

AES CODE

| Enter the port No | Enter File path and name |
|---|---|
| 6666 | D:\\ |
| IP ADDRESS | Specify Token |
| 127.0.0.1 | |

UPLOAD

DOWNLOAD

**FIGURE 9.3**  Application at Sender Side.

### 9.5.3 GUI FOR SERVER INTERFACE

Through this file transmitter interface, data is sent over to servers. After sending from the sender end, the following file would be delivered to the recipient, taking into account port number, AES CODE, file path, and security token. Figure 9.4 and figure 9.5 show the receiver side application and file for transfer respectively.

At other end, following file would be received (as shown in figure 9.6). The file content would be same as what was sent by the sender.

### 9.5.4 SECURE FILES

Here, the contents of the file were displayed during transmission. It is unintelligible since the text is encrypted (as shown in figure 9.7). The knowledge would not be understood if it were compromised.

Data from cloud would be delivered to end user using the transmitter module. The data would be encrypted using an authentication code. The sender side's port would be same as the end user's port. Here, the IP address of the end user would be supplied.

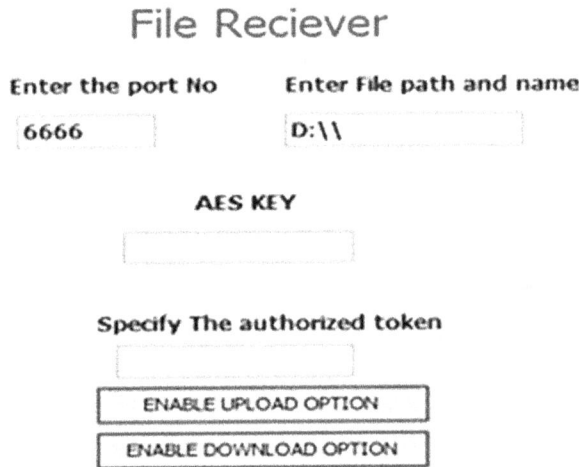

FIGURE 9.4   Receiver Side Application.

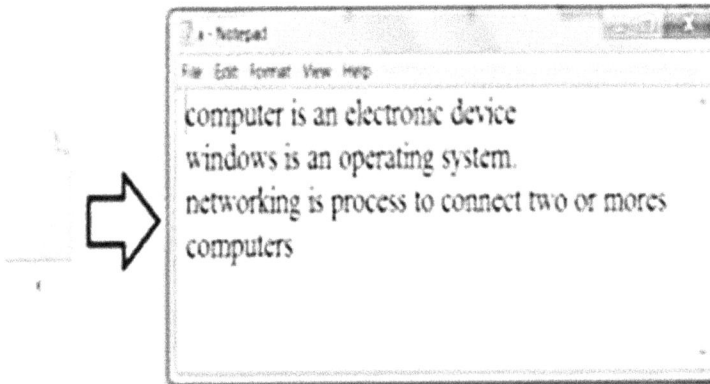

FIGURE 9.5   File for Transfer.

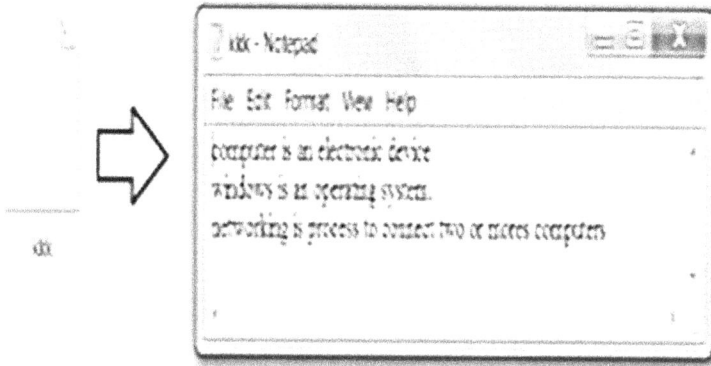

**FIGURE 9.6**   File Received.

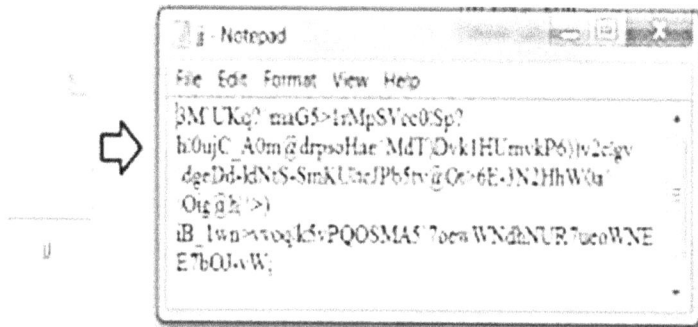

**FIGURE 9.7**   Encrypted File.

The figure 9.8 depicts the layout of data transmitter module for cloud. There are three input boxes. A port number, server IP address, and an authentication code would be input into the first, second, and third input boxes, respectively, to encrypt the data that would be transmitted.

## 9.6   FOG IMPLEMENTATION

Data from the fog would be transmitted to the user by the transmitter module. The data would be encrypted using an authentication code. The sender side's port would be same as the end user's port. Here, the IP address of the end user would be supplied. The image shown in figure 9.9 depicts the layout of the data transmitter module for fog.

**Three parts make up the end user module.**

1. **Ready to receive from fog:** Data collection from the fog side is made possible by the first component, which opens the fog port. There should be access from both sides of the port. Given that both parties would utilize the same shared authentication code, data transmitted from the fog side might be decoded.
2. **Prepared to take cloud data:** This action opens cloud port, allowing data to be gathered and sent from cloud side. Both sides of the port should be able to access it. Data transmitted from cloud side might be decrypted because the common authentication code would be the same.
3. **Merge and decode:** Using the authentication code, this stage combines and decodes the incoming data.

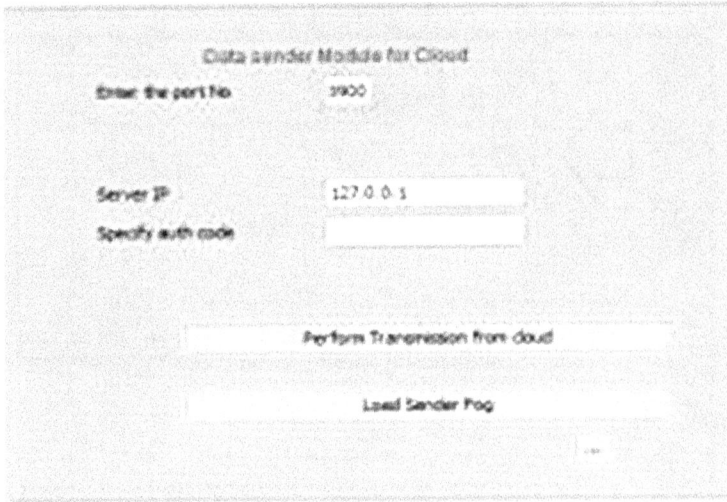

**FIGURE 9.8**    Design of Data Sender Module for Cloud.

**FIGURE 9.9**    Design of Data Sender Module for Fog.

The figure 9.10 shows the end user module. Table 9.1 indicates the comparison of the security between the existing work and the proposed work while Table 9.2 shows the comparison of packet dropping analysis.

Figure 9.11 shows the analysis of packet dropping with the help of a graph.

## 9.7   CONCLUSION

In comparison to previous implementations, the suggested methodology provides more secure means of protecting data from both active and passive types of attacks at application layer. During the investigation, the proposed security model was contrasted with the existing security model. When comparing the likelihood of packet loss between traditional and proposed works, it was noticed that there is minimal chance of packet loss with the suggested work. Traditional security systems have been noted. By segregating and encrypting data using complex cryptographic processes, this system

## End User

Enter the port number    3900

Authentication

Ready to receive from FOG

Ready to receive from CLOUD

Merge and Decode

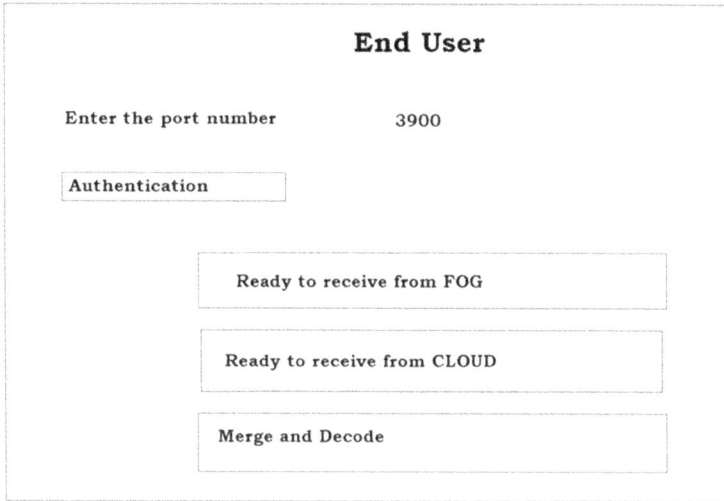

**FIGURE 9.10** End User Module.

## TABLE 9.1
## Comparison Factors That Affect the Data Security of Traditional and Proposed Work

| Parameters | Traditional Work | Proposed Work |
|---|---|---|
| Integration_to Fog [33] | Doesn't employ fog. | The proposed project will utilize fog. |
| Level of Security [34] | Employment is less safe. | The intended task is safer because split data is transferred from two different sites. |
| Reliability [35] | Traditional job is less trustworthy. | The proposed job is more trustworthy. |
| Packet dropping [36] | There is a greater chance of a packet dropping. | Fewer packets are likely to be dropped. |
| Congestion [37] | There is a greater likelihood of congestion. | There is less likelihood of congestion. |
| Transmission path [38] | To transfer data, only one path would be picked. | There would be several paths for data to travel. |
| Port | Predefined. | User defined. |

## TABLE 9.2
## Comparison of Packet Dropping Analysis

|  | Traditional | Proposed |
|---|---|---|
| 100 | 6 | 4 |
| 200 | 10 | 4 |
| 300 | 11 | 5 |
| 400 | 13 | 5 |
| 500 | 15 | 8 |
| 600 | 19 | 9 |
| 700 | 31 | 13 |
| 800 | 41 | 24 |

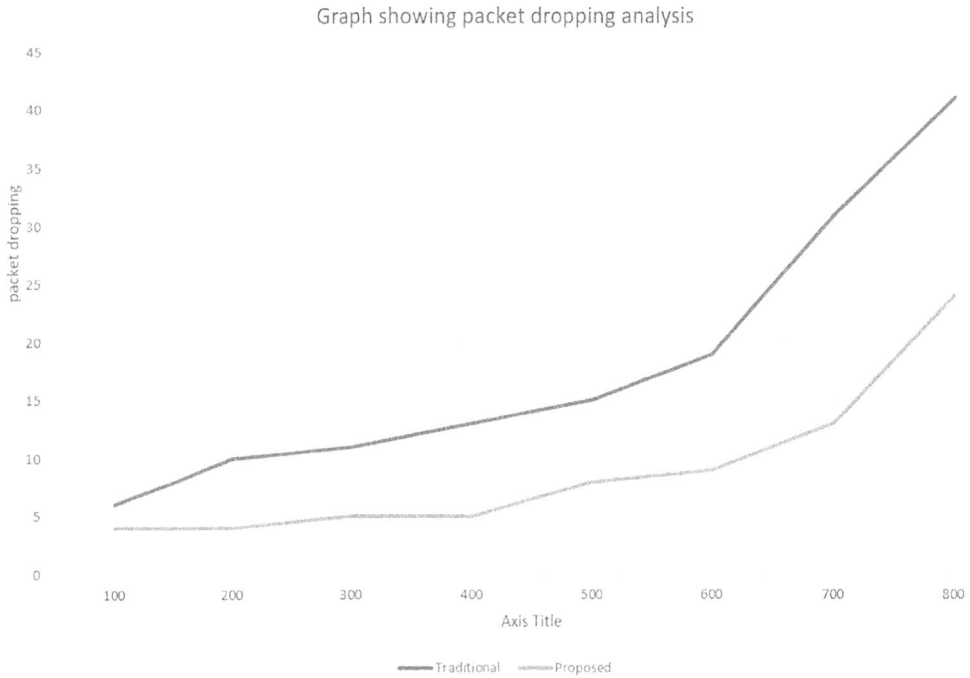

**FIGURE 9.11**    Graph Showing Packet Dropping Analysis.

safeguards data. With this setup, the possibility of packet loss and congestion is reduced. In order to maintain network security at different stages during the inquiry, potential threats have been investigated. In particular, the suggested method segments packets into various pieces to provide packet security. A new security system is now necessary due to the shortcomings of old security measures.

## 9.8  FUTURE SCOPE

The security of data transmission would benefit from the adoption of cryptographic techniques. In the case of the traditional work, data protection has only been available at the application layer. The possibility of total data loss could be reduced by one with the application of separating data into two layers. Both IP address with port number have been utilized in this study work to give the system security. In the suggested work, packet security [39, 40] is provided. This mechanism would make the security system resistant to various attackers.

## REFERENCES

1.  Arya, A., & Sidhu, J. "A Survey on Big Data Storage Issues in Cloud Computing Environment." *Journal of Advanced Database Management & Systems*. 4.3 (2017): 1–5.
2.  Deshpande, S. P., Sharma, C. S., Sateesh, K, & Peddoju, K. S. *Security and Data Storage Aspect in Cloud Computing*, Part of the Book Series: Studies in Big Data, vol. 52. SBD, 2019.
3.  Namasudra, S. "Cloud Computing: A New Era." *Journal of Fundamental and Applied Sciences* 10.2 (2018). https://doi.org/10.4314/JFAS.V10I2.9.
4.  Sampson, D., & Chowdhury, M. "The Growing Security Concerns of Cloud Computing." In *2021 IEEE International Conference on Electro Information Technology (EIT)*, 050–055, 2021.

5. Zamfiroiu, A., Petre, J., & Boncea, R. "Cloud Computing Vulnerabilities Analysis." In *Proceedings of the 2019 4th International Conference on Cloud Computing and Internet of Things*, 48–53, 2019.

6. Maroc, S., & Zhang, B. J. "Context-Aware Security Evaluation Ontology for Cloud Services." In *2019 IEEE 4th Advanced Information Technology, Electronic and Automation Control Conference (IAEAC) 1*, 1012–1018, 2019.

7. Kazim, M., & Zhu, S. *A Survey on Top Security Threats in Cloud Computing, Science and Information (SAI)*. New York: Organization Ltd., 2015.

8. Kaur, A., Singh, V. P., & Gill, S. S. "The Future of Cloud Computing: Opportunities, Challenges and Research Trends." In *2018 2nd International Conference on I-SMAC (IoT in Social, Mobile, Analytics and Cloud) (I-SMAC) I-SMAC (IoT in Social, Mobile, Analytics and Cloud) (I-SMAC)*, 2018.

9 Abid, A., et al. "Challenges and Issues of Resource Allocation Techniques in Cloud Computing." *KSII Transactions on Internet and Information Systems (TIIS)* 14.7 (2020): 2815–2839.

10. Laghari, A. A., et al. "Review and State of Art of Fog Computing." *Archives of Computational Methods in Engineering* 28.5 (2021): 3631–3643.

11. Abdulqadder, H. I., et.al "Defeating Vulnerable Attacks through Secure Software-Defined Networks." *IEEE Access* 6 (2018): 8292–8301.

12. Pooranian, Z., et al. "Defeating Side Channels Based on Data-Deduplication in Cloud Storage." In *IEEE Conference on Computer Communications Workshops (INFOCOM WKSHPS)*, 444–449, 2018.

13. Guan, Y., & Shao, J. "Data Security and Privacy in Fog Computing." *IEEE Network* 32.5 (2018): 106–111.

14. Khakimov, A., Muthanna, A., & Muthanna, M. S. A. "Study of Fog Computing Structure." In *2018 IEEE Conference of Russian Young Researchers in Electrical and Electronic Engineering (EIConRus)*, 51–54, 2018.

15. Yi, S., Hao, Z., Qin, Z., & Li, Q. "Fog Computing: Platform and Applications." In *2015 Third IEEE Workshop on Hot Topics in Web Systems and Technologies (HotWeb)*, 73–78, 2015.

16. Zhang, Y., Jia, D., Jia, S., Liu, L., & Lin, J. "Splitter: An Efficient Scheme to Determine the Geolocation of Cloud Data Publicly." In 2020 *29th International Conference on Computer Communications and Networks (ICCCN)*, 1–11. IEEE.

17. Sinha, K., Priya, A., & Paul, P. "K-RSA: Secure Data Storage Technique for Multimedia in Cloud Data Server." *Journal of Intelligent & Fuzzy Systems* 39.3 (2020): 3297–3314.

18. Shao, Cuili, et al. "IoT Data Visualization for Business Intelligence in Corporate Finance." *Information Processing & Management* 59.1 (2022): 102736.

19. Uppal, Mudita, et al. "Cloud-Based Fault Prediction Using IoT in Office Automation for Improvisation of Health of Employees." *Journal of Healthcare Engineering* 2021 (2021). https://doi.org/10.1155/2021/8106467.

20. Abubaker, N., Dervishi, L., & Ayday, E. "Privacy-Preserving Fog Computing Paradigm." In *The 3rd IEEE Workshop on Security and PrivaAcy in Cloud (SPC 2017)*, 2017.

21. Shukla, D. K., Dwivedi, V. K., & Trivedi, M. C. "Encryption Algorithm in Cloud Computing." *Materials Today: Proceedings* 37 (2021): 1869–1875.

22. Khan, S. "Cloud Computing: Issues and Risks of Embracing the Cloud in a Business Environment." *International Journal of Education and Management Engineering* 9.4 (2019): 44.

23. Kaur, M., & Aron, R. "A Systematic Study of Load Balancing Approaches in the Fog Computing Environment." *The Journal of Supercomputing* 77.8 (2021): 9202–9247.

24. Kour, Kanwalpreet, et al. "Controlling Agronomic Variables of Saffron Crop Using IoT for Sustainable Agriculture." *Sustainability* 14.9 (2022): 5607.

25. Sharma, Monika, et al. "IoT Application for Healthcare." In *Healthcare Solutions Using Machine Learning and Informatics*, pp. 187–204. USA: Auerbach Publications, 2022.

26. Singamaneni, Kranthi Kumar, et al. "A Novel QKD Approach to Enhance IIOT Privacy and Computational Knacks." *Sensors* 22.18 (2022): 6741.

27. Shivankit, A., et al. "Deep Learning Approach for Traffic Sign Recognition on Embedded Systems." In *Handbook of Machine Learning for Computational Optimization*, pp. 113–135. USA: CRC Press, 2021.

28. Uppal, Mudita, et al. "Cloud-Based Fault Prediction for Real-Time Monitoring of Sensor Data in Hospital Environment Using Machine Learning." *Sustainability* 14.18 (2022): 11667.

29. Dhall, Aakash, et al. "Machine Learning Algorithms for Industry Using Image Sensing." *Healthcare Solutions Using Machine Learning and Informatics*, pp. 75–97. USA: Auerbach Publications, 2022.

30. Sharma, Sheetal, et al. "SWOT: A Hybrid Hardware-Based Approach for Robust Fault-Tolerant Framework in a Smart Day Care." *Security and Communication Networks* 2022 (2022). https://doi.org/10.1155/2022/2939469.

31. Monga, Chetna, et al. "Sustainable Network by Enhancing Attribute-Based Selection Mechanism Using Lagrange Interpolation." *Sustainability* 14.10 (2022): 6082.

32. Joseph, M., & Mohan, G. "A Novel Algorithm for Secured Data Sharing in Cloud using GWOA-DNA Cryptography." *International Journal of Computer Networks and Applications (IJCNA)* 9.1: 114–124.

33. Dave, J., & Gayathri, M. "Hybrid Encryption Algorithm for Storing Unimodal Biometric Templates in Cloud." In *Inventive Communication and Computational Technologies*, pp. 251–266. Singapore: Springer, 2022.

34. Rezapour, R., Asghari, P., Javadi, H. H. S., & Ghanbari, S. "Security in Fog Computing: A Systematic Review on Issues, Challenges and Solutions." *Computer Science Review* 41 (2021): 100421.

35. Long, K., Dong, J., Fan, S., Geng, Y., Cao, Y., & Zhao, H. "Application of Data Encryption in Chinese Named Entity Recognition." arXiv preprint arXiv:2208.14627, 2022.

36. Suda, M., Dißauer, G., & Prawits, F. "True Random Number Generation with Beam Splitters under Combined Input Scenarios using Defined Quantum States to Increase the Security of Cryptographic Devices." In *ICQNM 2019*, 17, 2019.

37. Kumar, R., & Joshi, K. "Enhancing Network Security for Image Steganography by Splitting Graphical Matrix." *International Journal of Information Security Science* 9.1 (2020): 13–23.

38. Chaeikar, S. S., Jolfaei, A., & Mohammad, N. "AI-Enabled Cryptographic Key Management Model for Secure Communications in the Internet of Vehicles." *IEEE Transactions on Intelligent Transportation Systems* 24.4 (2022): 4589–4598.

39. Sindhwani, N., Anand, R., Vashisth, R., Chauhan, S., Talukdar, V., & Dhabliya, D. "Thingspeak-Based Environmental Monitoring System Using IoT," in *2022 Seventh International Conference on Parallel, Distributed and Grid Computing (PDGC)*, 675–680, 2022. https://doi.org/10.1109/PDGC56933.2022.10053167.

40. Kaur, J., Sindhwani, N., & Anand, R. "Implementation of IoT in Various Domains." In *IoT Based Smart Applications*, pp. 165–178. USA: Springer International Publishing, 2022.

# 10 IoT in Connected Electric Vehicles for Smart Cities

*Manish Bhardwaj and Shweta Singh*
KIET Group of Institutions, Delhi-NCR, Ghaziabad, India

*Yu-Chen Hu*
Providence University, Taiwan

## CONTENTS

10.1 Introduction ...................................................................................................162
10.2 Research Gap and Motivation ........................................................................164
10.3 Components of a Smart City ...........................................................................164
    10.3.1 Smart Concept in Agriculture ............................................................165
    10.3.2 Services Available in a Smart City .....................................................165
    10.3.3 Power System of a Smart City ............................................................165
    10.3.4 Innovation in Health ...........................................................................165
    10.3.5 Automated Home .................................................................................166
    10.3.6 New Generation Industries .................................................................166
    10.3.7 Smart Structure ...................................................................................166
    10.3.8 Automated Transport ..........................................................................166
10.4 Usage of IoT in Smart Cities .........................................................................167
    10.4.1 Architecture of a Smart City with IoT ...............................................167
    10.4.2 Smart City Challenges with IoT .........................................................167
        10.4.2.1 Privacy and Security ..........................................................167
        10.4.2.2 Sensors for Smart Cities ....................................................169
        10.4.2.3 Concept of Networking ......................................................169
        10.4.2.4 Data Analytics ....................................................................169
    10.4.3 Technologies Used for Sensing Data ..................................................170
    10.4.4 Technologies Used for Networking .....................................................170
        10.4.4.1 Network Topologies ...........................................................170
        10.4.4.2 Architectures ......................................................................171
    10.4.5 Privacy and Security with IoT in a Smart City ..................................172
10.5 SWOT Method for Data Analysis ..................................................................172
    10.5.1 Strengths ..............................................................................................173
    10.5.2 Flaws ....................................................................................................173
    10.5.3 Prospects ..............................................................................................173
    10.5.4 Coercions .............................................................................................173
10.6 Conclusion ......................................................................................................174
10.7 Future Recommendations ...............................................................................174
References ...............................................................................................................174

DOI: 10.1201/9781003319238-10

## 10.1  INTRODUCTION

A network comprising of networked physical objects is generally referred to as Internet of Things (IoT), such as cars, buildings, and other objects, that are combined with electronic hardware, software, sensors, actuators, and network connectivity. It not only minimizes the need for human interaction, but also makes it possible for objects to be sensed or controlled over an already-existing network, enabling a closer connection between the physical world and computer-based systems [1].

An IoT usually comprises a varied series of devices, from heart monitor implants to farm animal biochip transponders to electric clams in coastal waters, and everything in between, including cars with built-in sensors and DNA analysis devices for monitoring the environment, food, and pathogens to field operation devices used by firefighters in search and rescue operations, and so on and so forth.

It is a development vision to combine information communication with the IoT in a safe manner so that a city's assets may be effectively managed [2]. People in modern smart cities have higher expectations than just high-quality and dependable transportation. The next step in the automotive revolution is projected to be the inclusion of wireless communication capabilities in the vehicles themselves [3]. This article examines the latest IoT strategies for car connectivity, as well as the implementation issues and end-use applications of these technologies in smart cities [4].

The term "connected car" refers to automobiles equipped with wireless connectivity that can communicate both internally and outside [5]. As connected vehicles constitute the foundation of the Internet of Vehicles (IoV), extensive research and various industrial projects have laid the groundwork for this new era of connected vehicles. Vehicle wireless connectivity is being pushed by two major drivers [6]. The first is the pressing need to enhance road transit efficiency and safety. Mobile data usage by road users, on the other hand, is steadily rising.

There is still a lot of uncertainty about the viability of many off-the-shelf and new technologies required for the development and implementation of fully connected vehicles. To support the interactions of V2S, V2V, V2R, and V2I, as depicted in Figure 10.1, we have addressed a comprehensive view of wireless technologies and potential problems that can be overcome in order to provide vehicle-to-x communication [7].

Since recent days, the automotive industry has seen significant transformations as all companies strive to produce vehicles more efficiently and intelligently. As a result, obsolete engines have been phased out in favor of newer models that are less harmful to the environment [8]. It has been done this way so that electric vehicles, which cause far less pollution, have been introduced with many of the same features as conventional automobiles.

When compared to vehicles powered by diesel or gasoline, electric vehicles are virtually as fast, travel as far, and are as efficient as those powered by those fossil fuels [9]. Electric vehicles also feature a battery source that makes them ideal for long-distance trips because of their range and efficiency. Users can choose the type of battery and even check the battery's charge with the location of a charging station, making it easier for them to keep track of their devices.

A new monitoring device, which replaces the previous Global Positioning System (GPS) technology, has been installed into necessary sections of electric vehicles during this monitoring stage [10]. An intelligent device known as a wireless sensor performs the same function as GPS. Even with electric automobiles, sensor technology will be superior to GPS gadgets.

As a result, the incorporation of wireless sensors in critical areas of vehicles can track characteristics such as battery efficiency, travel distance, charging station alerts, and more.

When wireless sensors are employed, car maintenance becomes easier for everyone. A user-friendly environment may be created even with the introduction of electric vehicles, which will lead to the creation of smart cities and smart villages in all locations [11]. In addition, the sensors installed in these electric vehicles will allow them to keep track of all traffic conditions, making them extremely useful during times of heavy traffic. Even in the case of a modest heart attack, this technology can be used to prevent countless accidents [12].

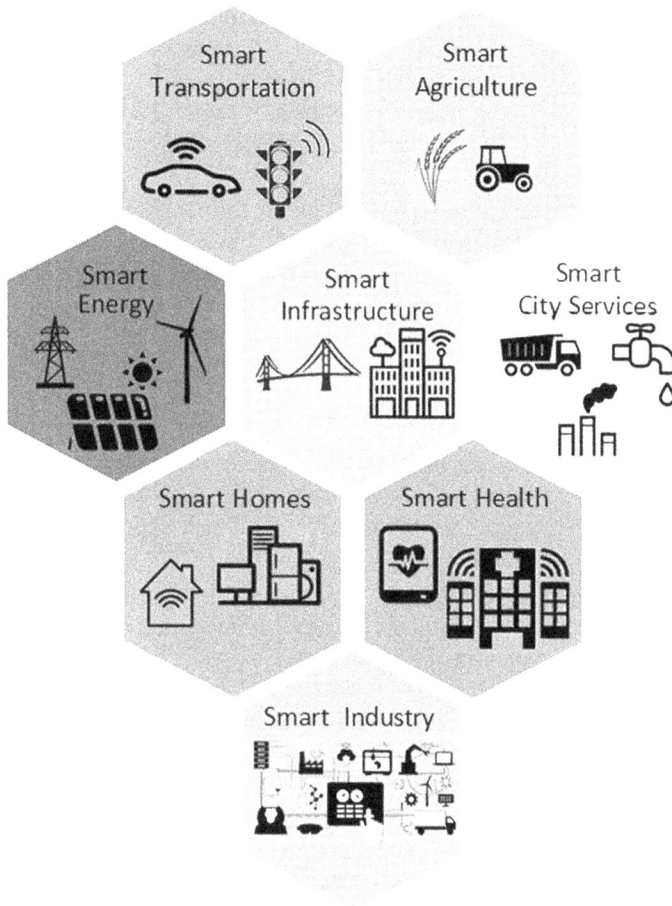

**FIGURE 10.1**  Different Areas of Smart Cities.

However, this new technology cannot be implemented in standard automobiles because of the greater implementation costs and the inability of the systems to withstand the higher temperatures [13]. Electric vehicles will have a significantly longer lifespan than older models with internal combustion engines, which means that they will be more accessible to the general public.

As a result, it is challenging to define what a "Smart City" is. Several factors can be used to determine if a city is "smart," including the deployment of unique e-governance plans, the development of social learning initiatives and community involvement efforts, and the more common use of ICT for innovation [14].

A smart city is one that makes use of various information and communication technologies to raise the standard of living for its residents, including a wide range of other applications of this technology in government, travel, housing, commerce, sustainability, social learning, and community involvement [15]. Smart cities may function more cohesively and actively by removing the traditional barriers between social and governmental systems. The ultimate objective of a smart city is this.

Smart cities have a number of advantages over traditional city ecosystems in terms of value.

1. Climate goal achievement: With the help of cutting-edge technologies, smart cities are paving the path for governments to achieve their climate targets. The idea of a "smart city" concentrates on three key areas: transportation system, management, and the administration

of the proposed model. These three areas are important because they help reduce cities' carbon footprint and facilitate the boom for new applications and development of technologies that assures better livelihood [16].

2. Worth of Fund: Smart city projects for 2025 will market in USD 1 Trillion, which gives a substantial financial incentive for both governments and commercial firms to participate in smart city development.

3. Impact on society: With this in mind, smart city projects seek to enhance people's daily lives and build an inclusive society where everyone has access to opportunities and their opinions are valued [17]. By facilitating interactions between citizens and the urban environment and improving daily living, information and communication technologies are crucial to the delivery of public services in the context of smart cities.

## 10.2   RESEARCH GAP AND MOTIVATION

Not all aspects – like capacity, distance, temperature, and cost – are monitored completely in existing approaches [18–25]. This research work identifies this gap. Despite the fact that the References include an authoritative security process, the primary parameters that influence the growth of the electrical process in the designing part are misplaced.

A battery management system with high temperature changes in the circuit that detects overcharging and excess current flow through the use of an IoT method has been presented [26]. Data unit capacity for circuit integration is significantly less. The number plate of a car can also be used to identify the vehicle in a smart parking system that gives a fundamental understanding of how to adjust the procedure [27]. It has also been designed by employing a direction-enabled Android platform to avoid excessive traffic conditions [28].

A high-end prototype has been constructed for both of the aforementioned procedures, but even more study has been done to design the same employing a low-cost CPU by combining wireless sensors.

To fully grasp IoT and the significance of the sensing device, it is important to run all of the electric vehicle's parameters through a calculator [29]. Motivated by the usage of electric vehicles to reduce environmental threats by integrating IoT, the authors have implemented their proposed method in real time to eliminate all gaps in parametric evaluation.

## 10.3   COMPONENTS OF A SMART CITY

Figure 10.1 depicts the various components that make up a "smart city." The gathering of data, transmission/reception, storage, and analysis of that data are the four main components of most smart city applications [30]. Application-specific data collecting has been a major motivator of sensor development in a variety of fields. To complete the second phase, data must be sent from collecting devices to cloud storage and analysis.

In order to accomplish this goal, a variety of methods have been employed, including citywide Wi-Fi networks, the deployment of 4G and 5G technologies, and a variety of local networks capable of transmitting data both locally and globally [31].

The third step is cloud storage comprising various strategies that aid in grouping and organizing the data to make the fourth stage, data analysis, easier to complete. To improve decision-making, data analysis involves searching for patterns and making assumptions based on the information acquired [32]. In some circumstances, basic analysis techniques like decision-making and aggregation may be sufficient.

Statistical approaches, machine learning, and deep learning algorithms may all be used in real time to analyze large amounts of data for more sophisticated decisions thanks to the cloud's availability [33].

### 10.3.1 SMART CONCEPT IN AGRICULTURE

The 2030 UN Sustainable Development Goals give food security a major emphasis [34]. Due to a growing global population and a deteriorating climate that is causing unpredictable weather in food-producing regions, governments around the world are scrambling to safeguard the production of food in a more sustainable manner which uses water resources more wisely [35].

The incorporation of sensors into plants and fields to gather data on a range of features that may be used to improve decisions and fend against diseases, pests, and other issues is a crucial component of smart agriculture [36]. As part of the smart agriculture paradigm, sensors are put in plants to offer exact measurements and enable the implementation of particular care methods. [37] Future food security will depend on precision agriculture, making it a key component towards production of food that is sustainable. There are various essential AI applications related to agriculture incorporating with IoT, such as crop monitoring, disease detection, data-driven crop care, and decision-making [38].

### 10.3.2 SERVICES AVAILABLE IN A SMART CITY

Services that support a city's inhabitants, such as water supply, waste management, environmental control, and monitoring, are all included in "smart city services." It is possible to monitor the condition of the city's water supply and detect leaks using sensors for water quality.

As previously mentioned, waste management is a common component of smart city initiatives. This includes everything from chutes and bins in Barcelona equipped with sensors and cloud connectivity to alert local authorities when they should be emptied, as well as using artificial intelligence (AI) to figure out which route is the most cost-effective [39]. Sensors can also be used to keep tabs on pollution levels in cities [40] and direct drivers to the nearest open parking spot to save money on fuel [41].

### 10.3.3 POWER SYSTEM OF A SMART CITY

Generally, in electrical networks, fossil fuel or hydroelectric-based power plants are utilized as the principal generators of energy. The power production method used with these schemes requires production of power significantly beyond users' needs to ensure continuity of supply since there is no evidence of response from the consumer end.

Detecting and correcting errors in these systems is a lengthy process that takes a lot of time. Furthermore, as renewable energy technologies have become more affordable, today's consumers are able to generate their own power in addition to receiving it from the main utilities [42, 43].

Information and communication technology (ICT) is used by smart networks to increase the visibility of both old and new grids, enable distributed energy generation at both the utility and customer levels, and give the grid the ability to repair itself [44]. With the use of prediction models built from consumption data obtained by smart grids and the network's self-healing [45] to maintain a continuous supply, it is now able to regulate the generation of power more effectively.

### 10.3.4 INNOVATION IN HEALTH

Using information and communications technology (ICT) in healthcare is known as "Smart Health." Researchers and healthcare professionals alike have focused their attention on this issue as the population grows and healthcare expenditures rise [46]. Overburdened current healthcare systems are unable to meet the rising demand from the public.

Telemedicine and AI are two examples of smart health's goal of ensuring that healthcare is accessible to as many people as feasible [47–49]. Because of the extensive usage of various health trackers and mobile phones to accumulate real-time data related to a person's medical condition (ECG readings; temperature; body oxygen saturation; and other biosensors), it is now possible to use cloud computing to analyze this data and make better healthcare decisions. As a result, healthcare costs and burdens are reduced [50].

### 10.3.5 Automated Home

The "Smart Home" is a key element of smart cities since it is the hub of daily life for city residents. In order to create a smart home, sensors are placed throughout a residence to collect data about both the residence itself and its residents. It is possible that these sensors might include environmental sensors, motion trackers, and power/energy consumption monitors [51].

### 10.3.6 New Generation Industries

It's no secret that businesses around the world are constantly striving to become more efficient and productive while slashing costs. The Industry 4.0 paradigm envisions a connected factory in which all of the production's intermediary functions are seamlessly integrated and functioning in unison. The IoT [52] makes this possible.

The usage of IoT and cyber-physical systems (CPS) that integrate employees and machines has benefitted manufacturing and production processes. This has led to quicker and better innovation, greater resource utilization, enhanced product quality, and increased worker safety in factories [53]. The requirement for cyber-physical systems to be flexible in configuration, fast to connect, and be able to manage a wide range of heterogeneous devices and equipment are just a few of the challenges that IoT applications for smart industry confront.

The creation and application of AI and the IoT have accelerated the development of Industry 4.0 services. Data from these sources can be used to improve various services with automation, implementing intelligence in business operations, etc., thanks to sensors that are integrated into machines and other production processes. Frameworks for combining AI and IoT for Smart Industry have been proposed by researchers [54–56]. Four crucial AI applications in the sector are fault detection, monitoring concerning machine health, production management, and predictive maintenance.

### 10.3.7 Smart Structure

City governments are expected to build new roads and bridges for the use of their population and to maintain them so that they can be used continuously since a city's infrastructure is essential to the quality of life of its residents. Smart materials and accelerometer-based sensors for monitoring the structural health of buildings and bridges [57] help cities make sure their infrastructure is in good condition and useable. To achieve predictive maintenance of metropolitan infrastructures, information received via sensors is utilized [58].

### 10.3.8 Automated Transport

Many urban areas are plagued by traffic, pollution, and issues with the timing and cost of public transportation. The contact between automobiles and people has reported increasingly with the rapid development and acceptance of various information and communication technologies [59]. Technologies, such as Vehicle to Infrastructure (V2I), Vehicle to Vehicle (V2V), Vehicle to Person (V2P) are utilized for intelligent transportation networks [60].

Many methods of measuring vehicle behavior and traffic patterns use GPS data because every driver has a cell phone [61]. These real-time data are already used for trip planning on public transit as well as route planning in apps like Waze and Google Maps. Additionally, drivers might be directed to the closest open parking space via parking systems with sensors.

## 10.4 USAGE OF IOT IN SMART CITIES

Thanks to the broad digitalization made possible by this enabling technology, smart cities – which have IoT at their core – are now a reality. Many projects have IoT [62] at their core with the system gathering and analyzing data to support decision- and policy-making.

By the year 2025, it is predicted that over 75 billion gadgets will be linked to the Internet [63], paving the way for even more innovative software solutions. Using IoT in the context of smart cities, sensors may gather and communicate data about the state of the city to a central cloud where it can be analyzed for patterns and used to make decisions.

### 10.4.1 ARCHITECTURE OF A SMART CITY WITH IoT

Through cloud services, the IoT combines data sensing, transmission/reception, processing, and storage. A typical IoT design is shown in Figure 10.2 and has five layers, each of which uses data from the one before [64]. The three different IoT designs are also shown in the diagram.

The "sensing layer" is referred to as a layer comprising of actuators and sensors, such as RFID tags, readers, etc. The sensing layer is also known as the "perception layer." In each application, sensors can collect information on relevant physical quantities [65]. Data is transferred from the sensing layer to the "middleware layer" via the "networking layer" using Wi-Fi, cellular Internet, Zigbee, Bluetooth, or other wireless network technologies [66].

As an illustration, the middleware layer acts as a standard interface between the sensors and the "application layer", which uses this data to offer services to customers. Planning and implementing policies that help manage the entire system take place at this layer, which is connected to the application layer [67].

The business layer manages the entire IoT system including security, privacy, applications etc. [67].

### 10.4.2 SMART CITY CHALLENGES WITH IoT

The IoT will digitalize everything. This digitalization process will lead to an increase in sensors across all facets of urban operations in smart cities [68]. With such a broad range of applications, there are significant challenges associated with the formation and successive positioning of IoT systems in smart cities [69].

Designers face a variety of challenges when deploying IoT systems in smart city applications. We'll examine some of the problems that academics have been debating in relation to the IoT in smart cities in this paper [70]. Implementing smart city IoT system are usually challenging in terms of privacy and security of the sensors utilized, big data analytics, and networking.

#### 10.4.2.1 Privacy and Security

Smart cities are more concerned with safety and privacy. This means that if there is a malfunction in the city's key infrastructure, inhabitants will be inconvenienced, and lives and property will be at risk [71]. As a result, in smart cities, safety is a major concern.

Smart cities are increasingly vulnerable to malevolent cyberattacks in today's world when such crimes and warfare have become standard tactics in international politics. This scenario necessitates the use of network encryption. Citizens' trust and involvement are essential for the success of smart city initiatives [72].

There is a risk of citizens' daily actions being exposed to outsiders due to the proliferation of sensors in smart cities. On the other hand, the IoT network allows organizations and corporations to collect data from citizens and utilize it without their permission for targeted advertising and other activities [73]. In order to find a solution, it will be necessary to implement data gathering techniques that protect the context of the measured task while also anonymizing it.

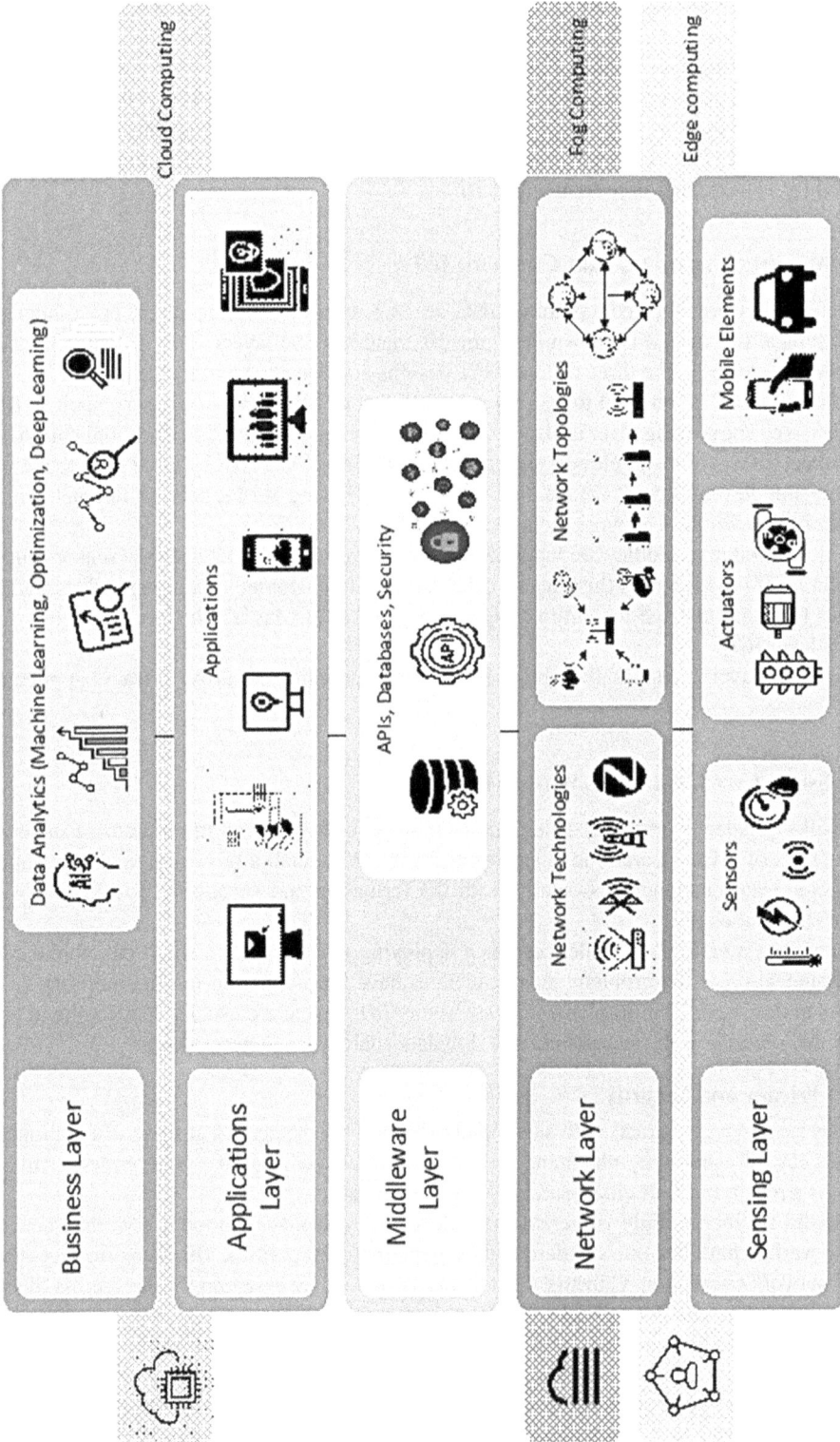

**FIGURE 10.2** Different Layered Architecture of IoT.

### 10.4.2.2 Sensors for Smart Cities

Smart sensors are the hardware elements used in smart cities to gather data. These gadgets are produced by numerous different companies, each of which follows a different data format, measuring standard, and networking protocol [74].

For the implementation of a smart city, all of these devices must be able to communicate, schedule tasks among themselves, and collect data in order to draw conclusions. As a solution to this problem, open protocols and data formats should be developed and used by manufacturers in order to facilitate the implementation of IoT systems.

Building "standard" access points for IoT systems that can speak with devices utilizing a range of protocols and decode the data they receive is another approach. Juneja et al. [75] assert that some manufacturers should actually make their products compatible with alternate protocols.

The reliability and robustness of smart sensors are not their only challenges. The terms "reliability" and "robustness," respectively, allude to the IoT system's dependability and accuracy. Future smart cities will be supported by the IoT, which means that it must enhance the user experience [76]. Hence, demand of service requests made by program users are answered promptly and correctly.

Every resident of the smart city should have access to high-quality services. Decentralized systems should be used to provide essential services like transportation and power. As a result, the system will be more resilient and reliable. Self-healing in smart grids is one such example.

Many of the sensors in smart cities will run on batteries, but many of the networking protocols currently in use were created for devices that have access to constant power. They will also need to collect, measure, convey, and store data [77]. As a result, new memory and storage technologies must be created, along with new low-power gadgets that can maximize battery life.

In order to store this enormous volume of data, compression algorithms and database design will need to be developed when smart cities and IoT go up. To ensure long-term functionality, new battery technologies must be developed, possibly with energy harvesting techniques included [78].

### 10.4.2.3 Concept of Networking

Sensors and other IoT devices must be able to interact with one another and the cloud in order for the IoT to function. Keeping these devices connected when new smart city applications appear is a significant challenge [79]. The methods used to provide network services currently are found ineffective when incorporated with a smart city. To meet the acceptance level of the services in such an environment, gadgets must achieve higher ranges for throughout and mobility.

To address this issue, a number of different strategies for identifying access points, local networks, and the like have been put forth. Many present protocols do not adequately satisfy the needs of the IoT, which necessitates the development of efficient and dynamic routing protocols capable of functioning with both stationary and mobile devices [80].

### 10.4.2.4 Data Analytics

By the year 2018, 13.6 Zetta Bytes of data had been produced by IoT-based devices, and it is probable that the figures would increase to 79.4 Zetta Bytes by 2025. Services provided in smart cities are further required to enhance the data in a continuous manner, for which development of new data analytics techniques is requested.

Due to the sheer number of characteristics which are tracked by smart cities, adaptable algorithms must be used to operate with both structured and unstructured data, and more effective data fusion methods must be created in order to combine data in meaningful ways and extract patterns and inferences [81]. Since deep learning can utilize this enormous amount of data to deliver improved results for a number of applications, it has attracted interest in this area. Such algorithms must be generated which are more scalable, can be utilized for the intended purpose, and include generality.

Usually, a convolutional neural network (CNN) model which has been trained for activity detection on a particular dataset does not guarantee to perform as per expectations on another, as demonstrated by the authors, whereas in [82] it was found that a deep learning network results drastically poor for a tomato having color different on which it was trained. Another issue is that ongoing data collection could lead to changes in the characteristics of data with time. Practices with incremental learning have been reported as useful for this situation.

Under smart city concept, analytics must be extensively implemented, and mainly in the province of smart health, they must be simple to understand. For the purpose of applying flood monitoring, a hybrid deep learning classifier and semantic web technologies-based solution has been demonstrated in [83]. With encouraging outcomes, a deep learning-based architecture for healthcare system [84] concerning COVID-19 at the edge is shown in [85]. More work needs to be put into integrating explain ability approaches like distillation, visualization, and intrinsic methodology in order for smart city applications based on machine and deep learning to grow.

### 10.4.3 TECHNOLOGIES USED FOR SENSING DATA

Smart city technology relies heavily on sensors. In order to generate new smart city technologies, sensors give the information and data necessary to do so. Because smart city projects and their varied components are so diverse, a wide variety of sensors are employed [86]. Environmental, motion, electric, biosensor, identity and presence sensors, hydraulic and chemical sensors are some examples of IoT sensors that can be split into different categories as shown in Figure 10.3.

Sensors are an essential part of IoT systems for smart cities since they enable the development of new services and interaction between the system and city residents. One thing to remember is that, as we've seen, many sensors have a variety of applications. Additionally, each application will demand the use of different sensors to detect different physical variables [87].

This includes sensors that monitor anything from temperature and humidity to airflow and movements. Working with various sensors, each with a different output data type, is a hurdle that must be overcome when using a variety of sensors.

### 10.4.4 TECHNOLOGIES USED FOR NETWORKING

When it comes to smart city applications, wireless technology is essential since physical connections would be prohibitively expensive (anywhere they can be used) and would not be able to meet the mobility requirements that are normal. Computers, smartphones, and other electronic gadgets all over the world are now interconnected via the Internet, enabling the instant movement of data between them. IoT, however, may not necessitate an Internet connection as many applications lack edge devices capable of connecting to the Internet.

Using a multi-hop communication protocol, a local network of sensors can exchange data and send it to a centralized hub, gateway, or a node. If the gateway is permanently installed and associated through Internet, it may be able to send any data that it discovers to the cloud for processing or use.

A system like this was provided as an example in [88], and several hubs are commercially available. In the smart home, for instance, manufacturers create devices incorporating trademarked or incompatible protocols that can be used with a hub; it is probable that devices in an application may utilize various protocols such that the central node communicate with every other node.

#### 10.4.4.1 Network Topologies

Network topologies in an IoT-based architecture are generally star, mesh, or point-to-point [69]. Devices are connected in a sequence of sequential connections using a point-to-point architecture. In a point-to-point network, packets must pass through each node on their route in order to exchange data between two nodes. There are numerous data hops as a result.

The network will collapse if one of the intermediary nodes fails. One of the reasons networks with point-to-point connections aren't used very frequently in IoT-based architectures is due to this. Since each unit in a star topology is connected to a central gateway, data cannot be transmitted

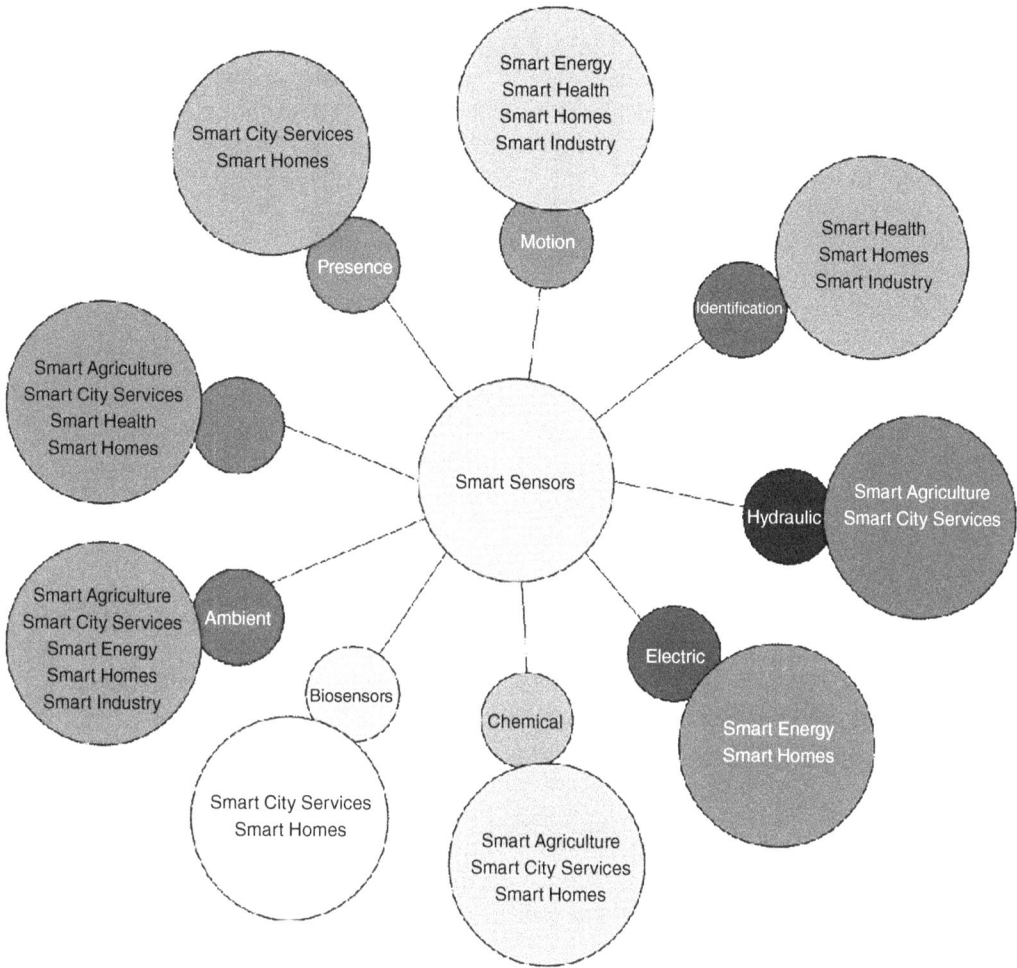

**FIGURE 10.3** IoT Smart Cities Sensing Technologies.

between units in a star topology. For devices to communicate and share data, the central node is necessary [89].

Star topology networks, a natural data aggregation strategy for the IoT, feature a central node structure. However big networks with numerous devices, such as an application related to a smart city, can result in higher latency rates and bottlenecks with information transmission. Star topology is employed in environmental sensing and disaster prevention [90].

The third type of network topology is IoT mesh networks. Mesh networks provide cross-network communication between all linked devices. Mesh architecture allows nodes to connect with one another, extending the network's range and resilience by enabling data to go across the network more frequently in the case that a node malfunctions and is unable to deliver a packet. These topologies are utilized in architecture with smart grids [91] together with smart homes. Other topologies, such as the tree, are not mentioned.

### 10.4.4.2 Architectures

The architecture of a certain application's network is referred to as "network architecture." Since we've already shown that IoT devices don't always have to be linked over the Internet to share data, it is possible to create a distributed connection structure that allows for only one network node to be able to communicate with the cloud.

Smart cities based on IoT use three different forms of network architecture, as discussed in [92]. HANs, WANs and FANs/NANs (field/neighborhood/area networks) are all examples of local area networks (LANs). Various short-range networks are utilized frequently to transmit data to the centralized node before it is sent to the cloud via home area networks (HANs).

To connect within the network, low-power communication technologies like Zigbee, Bluetooth, and Wi-Fi are employed [93]. Home automation systems (HANs) collect data on power consumption and operating time from a variety of appliances as part of a smart grid and send it to a smart meter. [94] A second sort of network architecture is called field area networks (FANs), commonly referred to as neighborhood area networks (NANs).

Using field area networks (FANs), utility providers and customers can communicate (such as in a smart grid). Compared to HANs, FANs have a wider communication range. Wide area networks (WANs) are the best choice for networks that need to connect across large distances.

Many of these networks, which use technologies like fiber optics and low-power protocols developed expressly for WANs, are not quite as dense as those found in HANs [75]. Wide area network (WAN) technology is frequently used in smart grids to link substations or transmit data concerning substation and customer.

### 10.4.5 PRIVACY AND SECURITY WITH IoT IN A SMART CITY

Local and wide area networks in smart cities are used to transfer sensing data and control information. As a result, a variety of smart city components are essential to a city's operation and are closely linked to the mutual and isolated life of its inhabitants. Smart city security and confidentiality are of utmost relevance and have received significant attention [95].

Security of the various designs of the IoT have been discussed and difficulties and solutions presented. To bring our examination of IoT security in smart cities to a close, we address the difficulties that are unique to this setting. Smart cities collect data from residents and city sensors, process it, and then use it to improve the lives of those who live there. With the use of these sensors, it is possible to get an idea of the condition of various city components, such as transportation and electricity systems, as well as the condition of buildings and movement of people [96].

In the cloud, these data are analyzed and mined. But there are a number of concerns about the integrity, protection, and secrecy of this process as a result of how these data are transported and used. As a matter of fact, a cyberattack on Ukraine's power grid in 2015 that left 225,000 people without electricity exposed the globe to the dangers presented by cyberattackers.

It is likely that information obtained by smart city applications will be utilized for a range of bad deeds, such as tracking a person's identity, position, and habits using GPS sensors, which are found in the majority of smartphones and automobiles. A building's power usage and ambient sensor data can also identify its residents.

Bad actors could exploit this information to commit crimes that put people's lives and property at risk. IoT for smart cities' security and privacy challenges necessitate the development of novel security and privacy measures that are not commonly used in the IoT for the smart cities context.

Security and privacy issues are unique to each of the three IoT architecture levels. The layers included in an IoT architecture are such as, network, perception, and network layers, as well as some system-wide issues. Below, these issues are covered in more detail.

An instant of the security and confidentiality challenges highlighted by IoT for smart cities as well as the various solutions to those issues.

## 10.5 SWOT METHOD FOR DATA ANALYSIS

After conducting a thorough analysis, we come to a conclusion by conducting a SWOT analysis on an IoT-based design for smart cities. This analysis looks at the advantages and disadvantages of IoT in addition to the opportunities and risks of IoT-based application aspects concerning smart cities in terms of future work [97].

### 10.5.1 STRENGTHS

The advantages of IoT smart cities include enhanced quality of life for the city's residents, reduced operating expenses, and the ability to maintain a city's sustainability.

The IoT makes it possible to watch key city services like transportation, electricity, gas, and water hoarding, in addition to misconduct monitoring, in real time. This real-time information can be used by cities, businesses, and other stakeholders to enhance citizen services, increase productivity, and cut expenses through more cost-effective operations.

Using IoT data, the city can now better assess the many services it offers its residents, as well as their interrelationships, and use that information to make better decisions and make life easier for the community. Distributed systems and flexible architectures that allow for fluidity through the mobility of sensing units make it possible to modify and expand existing systems with little additional cost.

The fact that these systems can self-heal thanks to a distributed architecture increases their reliability and makes them more resistant to failures.

### 10.5.2 FLAWS

As indicated, the current deployment scenario includes a number of technologies linked to networks, hardware platforms, and software frameworks that don't necessarily work well together. These same technology flaws also affect IoT in smart cities. The development of standards for communication, network discovery, and device administration has been a collaborative effort amongst numerous standards bodies, including the Internet Engineering Task Force (IETF), ETSI, IEEE, and other organizations.

Interoperability hasn't been fully resolved because there are so many "standards." Without a significant reworking of system components, this can be a significant barrier to IoT expansion. IoT systems face additional difficulties due to the absence of data laws and policies. There is a concern that information strategies are not advanced adequately to normalize the handling of data in IoT devices, according to past discussions. People's concerns about how their personal information is utilized in a connected society are growing.

### 10.5.3 PROSPECTS

Researchers and businesses alike can benefit from IoT in smart cities by addressing existing problems and developing new city services. It is possible to employ big data techniques to generate new apps and services based on the information gathered by sensors in IoT systems. It's a tremendous chance for researchers in the field of data analytics to design new algorithms for service delivery using this heterogeneous data set.

The development of computationally affordable encryption algorithms, effective data storage systems, and networking technology can simplify and reduce the cost of IoT implementation. With the advancement of new sensor technologies, researchers working on the IoT for smart cities have a second opportunity. IoT services will be used more frequently and extensively as a result of the development of better, more effective, low-cost sensors.

### 10.5.4 COERCIONS

The IoT for smart cities poses a number of security risks owing to its interconnected nature, including challenges with user trust, privacy concerns caused by network attacks, and possible data theft [97]. Since IoT applications allow for such a high degree of personalization between users and devices, privacy breaches, data piracy, and other security issues are a major worry for both service providers and their customers.

Numerous cyberattacks on smart city architectures have been discussed in [98]. IoT stakeholders have more privacy and security issues as a result of the fact that many IoT implementations would not be able to use conventional security measures and procedures, such as routing, networking, and authentication. This can put off customers from participating in smart city programs.

## 10.6  CONCLUSION

In this chapter, the IoT in smart cities is thoroughly examined. We introduce the IoT as a significant aspect concerning smart city amenities, examine numerous smart city plans, and deliberate the challenges confronted while implementing smart city applications. This chapter provides a comprehensive discussion of smart cities and all its various subfields. Next, we'll examine the networking and sensing technologies utilized in these applications and discuss how AI might be applied to smart cities.

Based on the multiple uses that each component has, this section gives an overview of recent research on IoT-based smart cities. On the basis of the technologies and architectures presented, we have taken into account the deployment type. Security and confidentiality subjects for IoT-based smart cities are discussed, and a SWOT analysis is offered. This survey aims to provide a comprehensive starting point for academics interested in IoT in smart cities.

## 10.7  FUTURE RECOMMENDATIONS

There are a number of recommendations that can be made when adopting IoT for smart city initiatives based on the discussion in this article. The security and privacy of IoT in smart cities is a significant field for research focusing on encoding methods, validation protocols, data anonymization methods, and additional approaches to circumvent unauthorized admission into an IoT network.

The majority of IoT data transport standards now in use are not compatible. It is crucial that sensor nodes in the network can interact with one another utilizing a variety of protocols while using little power.

Another issue that needs to be addressed in the near future is the creation of hardware with low power consumption and energy-efficient storage techniques. Decentralized solutions have been suggested as a way to increase the application's stability. Techniques like federated learning can be applied to the implementation of decentralized deep learning systems.

Additionally, there is a tonne of future development in the subject of AI. The use of heterogeneous data sources is being facilitated by data fusion and intelligent feature selection algorithms, while superfluous or "uninteresting" information is being removed from the AI development process. Deployments will proceed more swiftly and function better as a result. To develop machine learning and deep learning algorithms further understandable for the numerous practices in a smart city, modern and creative techniques must be used.

## REFERENCES

1  Worldometers. World Population Forecast—Worldometers. 2019. Available online: https://www.worldometers.info/worldpopulation/world-population-projections/ (accessed on 9 March 2021)
2  Ahvenniemi, H.; Huovila, A.; Pinto-Seppä, I.; Airaksinen, M. What are the differences between sustainable and smart cities? *Cities* 2017, 60, 234–245.
3  United Nations. About the sustainable development goals—United nations sustainable development. Available online: https://sdgs.un.org/goals (accessed on 9 March 2021)
4  Cardullo, P.; Kitchin, R. Being a 'citizen' in the smart city: Up and down the scaffold of smart citizen participation in Dublin, Ireland. *GeoJournal* 2019, 84, 1–13.
5  Desdemoustier, J.; Crutzen, N.; Giffinger, R. Municipalities' understanding of the Smart City concept: An exploratory analysis in Belgium. *Technol. Forecast. Soc. Chang.* 2019, 142, 129–141.

6  Khan, M.S.; Woo, M.; Nam, K.; Chathoth, P.K. Smart city and smart tourism: A case of Dubai. *Sustainability* 2017, 9, 2279.

7  Wu, S.M.; Chen, T.C.; Wu, Y.J.; Lytras, M. Smart cities in Taiwan: A perspective on big data applications. *Sustainability* 2018, 10, 106.

8  Ejaz, W.; Anpalagan, A. Internet of things for smart cities: Overview and key challenges. *Internet Things Smart Cities* 2019, 1–15.

9  Janssen, M.; Luthra, S.; Mangla, S.; Rana, N.P.; Dwivedi, Y.K. Challenges for adopting and implementing IoT in smart cities: An integrated MICMAC-ISM approach. *Internet Res.* 2019, 29, 1589–1616.

10  Sánchez-Corcuera, R.; Nuñez-Marcos, A.; Sesma-Solance, J.; Bilbao-Jayo, A.; Mulero, R.; Zulaika, U.; Azkune, G.; Almeida, A. Smart cities survey: Technologies, application domains and challenges for the cities of the future. *Int. J. Distrib. Sens. Netw.* 2019, 15.

11  Silva, B.N.; Khan, M.; Han, K. Towards sustainable smart cities: A review of trends, architectures, components, and open challenges in smart cities. *Sustain. Cities Soc.* 2018, 38, 697–713.

12  Atat, R.; Liu, L.; Wu, J.; Li, G.; Ye, C.; Yang, Y. Big data meet cyber-physical systems: A panoramic survey. *IEEE Access* 2018, 6, 73603–73636.

13  Hollands, R.G. Will the real smart city please stand up? Intelligent, progressive or entrepreneurial? *City* 2008, 12, 303–320.

14  Anthopoulos, L.G.; Reddick, C.G. Understanding electronic government research and smart city: A framework and empirical evidence. *Inf. Polity* 2016, 21, 99–117.

15  Khan, Z.; Anjum, A.; Soomro, K.; Tahir, M.A. Towards cloud based big data analytics for smart future cities. *J. Cloud Comput.* 2015, 4, 1–11.

16  Koubaa, A.; Aldawood, A.; Saeed, B.; Hadid, A.; Ahmed, M.; Saad, A.; Alkhouja, H.; Ammar, A.; Alkanhal, M. Smart palm: An IoT framework for red palm weevil early detection. *Agronomy* 2020, 10, 987.

17  O'Grady, M.; Langton, D.; O'Hare, G. Edge computing: A tractable model for smart agriculture? *Artif. Intell. Agric.* 2019, 3, 42–51.

18  Kaur, J.; Sindhwani, N.; Anand, R. "Implementation of IoT in Various Domains." In *IoT Based Smart Applications*; Springer International Publishing, USA, 2022; pp. 165–178.

19  Pardini, K.; Rodrigues, J.J.; Kozlov, S.A.; Kumar, N.; Furtado, V. IoT-based solid waste management solutions: A survey. *J. Sens. Actuator Netw.* 2019, 8, 5.

20  Dutta, J.; Chowdhury, C.; Roy, S.; Middya, A.I.; Gazi, F. Towards smart city: Sensing air quality in city based on opportunistic crowd-sensing. In *Proceedings of the 18th International Conference on Distributed Computing and Networking*, 2017.

21  Al-Turjman, F.; Malekloo, A. Smart parking in IoT-enabled cities: A survey. *Sustain. Cities Soc.* 2019, 49, 101608.

22  Shirazi, E.; Jadid, S. Autonomous self-healing in smart distribution grids using multi agent systems. *IEEE Trans. Ind. Informatics* 2018, 3203, 1–11.

23  Andreão, R.V.; Athayde, M.; Boudy, J.; Aguilar, P.; de Araujo, I.; Andrade, R. Raspcare: A Telemedicine Platform for the Treatment and Monitoring of Patients with Chronic Diseases. In *Assistive Technologies in Smart Cities*; IntechOpen, London, UK, 2018.

24  Keane, P.A.; Topol, E.J. With an eye to AI and autonomous diagnosis. *NPJ Digit. Med.* 2018, 1, 10–12. [PubMed]

25  Trencher, G.; Karvonen, A. Stretching "smart": Advancing health and well-being through the smart city agenda. *Local Environ.* 2019, 24, 610–627.

26  Haverkort, B.R.; Zimmermann, A. Smart industry: How ICT will change the game! *IEEE Internet Comput.* 2017, 21, 8–10.

27  Tao, F.; Cheng, J.; Qi, Q. IIHub: An industrial internet-of-things hub toward smart manufacturing based on cyber-physical system. *IEEE Trans. Ind. Inform.* 2018, 14, 2271–2280.

28  Trakadas, P.; Simoens, P.; Gkonis, P.; Sarakis, L.; Angelopoulos, A.; Ramallo-González, A.P.; Skarmeta, A.; Trochoutsos, C.; Calvo, D.; Pariente, T.; et al. An artificial intelligence-based collaboration approach in industrial IoT manufacturing: Key concepts, architectural extensions and potential applications. *Sensors* 2020, 20, 5480.

29  Wan, J.; Yang, J.; Wang, Z.; Hua, Q. Artificial intelligence for cloud-assisted smart factory. *IEEE Access* 2018, 6, 55419–55430.

30  Huang, Y.; Dang, Z.; Choi, Y.; Andrade, J.; Bar-Ilan, A. High-precision smart system on accelerometers and inclinometers for Structural Health Monitoring: Development and applications. In *Proceedings of the 2018 12th France-Japan and 10th Europe-Asia Congress on Mechatronics*, Tsu, Japan, 10–12 September 2018; pp. 52–57.

31  Farag, S.G. Application of Smart Structural System for Smart Sustainable Cities. In *Proceedings of the 2019 4th MEC International Conference on Big Data and Smart City (ICBDSC)*, Muscat, Oman, 15–16 January 2019; pp. 1–5.

32  Wang, Y.; Ram, S.; Currim, F.; Dantas, E.; Sabóia, L.A. A big data approach for smart transportation management on bus network. In *Proceedings of the IEEE 2nd International Smart Cities Conference: Improving the Citizens Quality of Life, ISC2 2016—Proceedings*, Trento, Italy, 12–15 September 2016; pp. 1–6.

33  Lele, A. Internet of things (IoT). *Smart Innov. Syst. Technol.* 2019, 132, 187–195.

34  Mell, P.; Grance, T. *The NIST-National Institute of Standars and Technology- Definition of Cloud Computing; NIST Special Publication 800-145*; NIST, Gaithersburg, MD, USA, 2011.

35  Bar-MagenNumhauser, J. *Fog Computing Introduction to a New Cloud Evolution*; University of Alcalá, Alcalá de Henares, Spain, 2012.

36  Aazam, M.; Zeadally, S.; Harras, K.A. Fog computing architecture, evaluation, and future research directions. *IEEE Commun. Mag.* 2018, 56, 46–52.

37  El-Sayed, H.; Sankar, S.; Prasad, M.; Puthal, D.; Gupta, A.; Mohanty, M.; Lin, C.T. Edge of things: The big picture on the integration of edge, IoT and the cloud in a distributed computing environment. *IEEE Access* 2017, 6, 1706–1717.

38  Yousefpour, A.; Fung, C.; Nguyen, T.; Kadiyala, K.; Jalali, F.; Niakanlahiji, A.; Kong, J.; Jue, J.P. All one needs to know about fog computing and related edge computing paradigms: A complete survey. *J. Syst. Archit.* 2019, 98, 289–330.

39  Gupta, A.; Srivastava, A.; Anand, R.; Chawla, P. Smart vehicle parking monitoring system using RFID. *International Journal of Innovative Technology and Exploring Engineering* 2019, 8(9S), 225–229.

40  Bhardwaj, M. Research on IoT Governance, Security and Privacy Issues of Internet of Things. In *Privacy Vulnerabilities and Data Challenges in the IoT*; Tylor and Francis, CRC Press, pp. 115–120, 23 Nov. 2020. https://doi.org/10.1201/9780429322969.

41  Bhardwaj, M.; Jindal, Shivani Internet of Things with Protocols and Applications. In *Internet of Things Businesses in Disruptive Economy*; NOVA Science Publications, New York, ISBN 978-1-53618-958-2, 2021.

42  Rahnemoonfar, M.; Sheppard, C. Deep count: Fruit counting based on deep simulated learning. *Sensors* 2017, 17, 905. [PubMed]

43  Thakker, D.; Mishra, B.K.; Abdullatif, A.; Mazumdar, S. Explainable artificial intelligence for developing smart cities solutions. *Smart Cities* 2020, 3, 1353–1382.

44  Rahman, A.; Hossain, M.S.; Alrajeh, N.A.; Guizani, N. B5G and explainable deep learning assisted healthcare vertical at the edge: COVID-19 perspective. *IEEE Netw.* 2020, 34, 98–105.

45  Marques, G.; Garcia, N.; Pombo, N. A Survey on IoT: Architectures, Elements, Applications, QoS, Platforms and Security Concepts. In *Advances in Mobile Cloud Computing and Big Data in the 5G Era*; Springer, Berlin/Heidelberg, Germany, 2017; pp. 115–130.

46  Zhang, K.; Ni, J.; Yang, K.; Liang, X.; Ren, J.; Shen, X.S. Security and privacy in smart city applications: Challenges and solutions. *IEEE Commun. Mag.* 2017, 55, 122–129.

47  Mehmood, Y.; Ahmad, F.; Yaqoob, I.; Adnane, A.; Imran, M.; Guizani, S. Internet-of-Things-based smart cities: Recent advances and challenges. *IEEE Commun. Mag.* 2017, 55, 16–24.

48  Rong, W.; Xiong, Z.; Cooper, D.; Li, C.; Sheng, H. Smart city architecture: A technology guide for implementation and design challenges. *China Commun.* 2014, 11, 56–69.

49  Ahmed, E.; Yaqoob, I.; Gani, A.; Imran, M.; Guizani, M. Internet-of-things-based smart environments: State of the art, taxonomy, and open research challenges. *IEEE Wirel. Commun.* 2016, 23, 10–16.

50  Chen, S.; Xu, H.; Liu, D.; Hu, B.;Wang, H. A vision of IoT: Applications, challenges, and opportunities with China Perspective. *IEEE Internet Things J.* 2014, 1, 349–359.

51  Ahlgren, B.; Hidell, M.; Ngai, E.C. Internet of Things for smart cities: Interoperability and open data. *IEEE Internet Comput.* 2016, 20, 52–56.

52  Lee, I.; Lee, K. The Internet of Things (IoT): Applications, investments, and challenges for enterprises. *Bus. Horizons* 2015, 58, 431–440.

53  Morais, C.M.; Sadok, D.; Kelner, J. An IoT sensor and scenario survey for data researchers. *J. Braz. Comput. Soc.* 2019, 25, 1–17.

54  Sharma, N.; Solanki, V.K.; Davim, J.P. Basics of the Internet of Things (IoT) and Its Future. In *Handbook of IoT and Big Data*; CRC Press, Boca Raton, FL, USA, 2019; p. 165.

55  Hancke, G.P.; Hancke, G.P., Jr. The role of advanced sensing in smart cities. *Sensors* 2013, 13, 393–425. [PubMed]

56  Penza, M.; Suriano, D.; Villani, M.G.; Spinelle, L.; Gerboles, M. Towards air quality indices in smart cities by calibrated low-cost sensors applied to networks. In *Proceedings of the SENSORS, 2014 IEEE*, Valencia, Spain, 2–5 November 2014; pp. 2012–2017.

57  Folianto, F.; Low, Y.S.; Yeow, W.L. Smartbin: Smart waste management system. In *Proceedings of the 2015 IEEE Tenth International Conference on Intelligent Sensors, Sensor Networks and Information Processing (ISSNIP)*, Singapore, 7–9 April 2015; pp. 1–2.

58  Bedi, G.; Venayagamoorthy, G.K.; Singh, R.; Brooks, R.R.;Wang, K.C. Review of Internet of Things (IoT) in electric power and energy systems. *IEEE Internet Things J.* 2018, 5, 847–870.

59  Song, E.Y.; Fitzpatrick, G.J.; Lee, K.B. Smart sensors and standard-based interoperability in smart grids. *IEEE Sens. J.* 2017, 17, 7723–7730.

60  Abdellatif, A.A.; Mohamed, A.; Chiasserini, C.F.; Tlili, M.; Erbad, A. Edge computing for smart health: Context-aware approaches, opportunities, and challenges. *IEEE Netw.* 2019, 33, 196–203.

61  Fan, K.; Zhu, S.; Zhang, K.; Li, H.; Yang, Y. A lightweight authentication scheme for cloud-based RFID healthcare systems. *IEEE Netw.* 2019, 33, 44–49.

62  Prati, A.; Shan, C.; Wang, K.I. Sensors, vision and networks: From video surveillance to activity recognition and health monitoring. *J. Ambient. Intell. Smart Environ.* 2019, 11, 5–22.

63  Ding, D.; Cooper, R.A.; Pasquina, P.F.; Fici-Pasquina, L. Sensor technology for smart homes. *Maturitas* 2011, 69, 131–136. [PubMed]

64  Srivastava, A.; Gupta, A.; Anand, R. Optimized smart system for transportation using RFID technology. *Mathematics in Engineering, Science and Aerospace (MESA)* 2021, 12(4), 953–965.

65  Guerrero-Ibáñez, J.; Zeadally, S.; Contreras-Castillo, J. Sensor technologies for intelligent transportation systems. *Sensors* 2018, 18, 1212.

66  Gharaibeh, A.; Salahuddin, M.A.; Hussini, S.J.; Khreishah, A.; Khalil, I.; Guizani, M.; Al-Fuqaha, A. Smart cities: A survey on data management, security, and enabling technologies. *IEEE Commun. Surv. Tutor.* 2017, 19, 2456–2501.

67  Talari, S.; Shafie-Khah, M.; Siano, P.; Loia, V.; Tommasetti, A.; Catalão, J.P. A review of smart cities based on the internet of things concept. *Energies* 2017, 10, 421.

68  Park, D.M.; Kim, S.K.; Seo, Y.S. S-mote: SMART home framework for common household appliances in IoT Network. *J. Inf. Process. Syst.* 2019, 15, 449–456.

69  Yaqoob, I.; Ahmed, E.; Hashem, I.A.T.; Ahmed, A.I.A.; Gani, A.; Imran, M.; Guizani, M. Internet of things architecture: Recent advances, taxonomy, requirements, and open challenges. *IEEE Wirel. Commun.* 2017, 24, 10–16.

70  Sakhardande, P.; Hanagal, S.; Kulkarni, S. Design of disaster management system using IoT based interconnected network withsmart city monitoring. In *Proceedings of the 2016 International Conference on Internet of Things and Applications, IOTA 2016*, Pune, India, 22–24 January 2016; pp. 185–190.

71  Kang, L.; Poslad, S.; Wang, W.; Li, X.; Zhang, Y.; Wang, C. A public transport bus as a flexible mobile smart environment sensing platform for IoT. In *Proceedings of the 12th International Conference on Intelligent Environments, IE 2016*, London, UK, 14–16 September 2016; pp. 1–8.

72  Adiono, T.; Fathany, M.Y.; Putra, R.V.W.; Afifah, K.; Santriaji, M.H.; Lawu, B.L.; Fuada, S. Live demonstration: MINDS – Meshed and internet networked devices system for smart home: Track selection: Embedded systems. In *Proceedings of the 2016 IEEE Asia Pacific Conference on Circuits and Systems, APCCAS 2016*, Jeju, Korea, 25–28 October 2016; pp. 736–737.

73  Ghosh, A.; Chakraborty, N. Design of smart grid in an University Campus using ZigBee mesh networks. In *Proceedings of the 1st IEEE International Conference on Power Electronics, Intelligent Control and Energy Systems, ICPEICES 2016*, Delhi, India, 4–6 July 2016; pp. 1–6.

74   Yan, Y.; Qian, Y.; Sharif, H. A secure data aggregation and dispatch scheme for home area networks in smart grid. In *Proceedings of the GLOBECOM—IEEE Global Telecommunications Conference*, Houston, TX, USA, 5–9 December 2011; pp. 1–6.

75   Juneja, S., Juneja, A., Bali, V., & Upadhyay, H. Cyber Security: An Approach to Secure IoT from Cyber Attacks Using Deep Learning. In *Industry 4.0, AI, and Data Science*; CRC Press, USA, 2021; pp. 135–146.

76   Kuzlu, M.; Pipattanasomporn, M. Assessment of communication technologies and network requirements for different smart grid applications. In *Proceedings of the 2013 IEEE PES Innovative Smart Grid Technologies Conference, ISGT 2013*, Washington, DC, USA, 24–27 February 2013; pp. 1–6.

77   Al-Sarawi, S.; Anbar, M.; Alieyan, K.; Alzubaidi, M. Internet of Things (IoT) communication protocols: Review. In *Proceedings of the ICIT 2017—8th International Conference on Information Technology*, Amman, Jordan, 17–18 May 2017; pp. 685–690.

78   Mekki, K.; Bajic, E.; Chaxel, F.; Meyer, F. A comparative study of LPWAN technologies for large-scale IoT deployment. *ICT Express* 2019, 5, 1–7.

79   Samuel, S.S.I. A review of connectivity challenges in IoT-smart home. In *Proceedings of the 2016 3rd MEC International Conference on Big Data and Smart City, ICBDSC 2016*, Muscat, Oman, 15–16 March 2016, pp. 364–367.

80   Kuppusamy, P.; Muthuraj, S.; Gopinath, S. Survey and challenges of Li-Fi with comparison of Wi-Fi. In *Proceedings of the 2016 International Conference onWireless Communications, Signal Processing and Networking (WiSPNET)*, Chennai, India, 23–25 March 2016; pp. 896–899.

81   Heile, B.S.A.; Liu, B.; Zhang, M.; Perkins, C.F. Wi-SUN FAN Overview. 2017. Available online: https://tools.ietf.org/id/draftheile-lpwan-wisun-overview-00.html (accessed on 9 March 2021).

82   Hammi, B.; Khatoun, R.; Zeadally, S.; Fayad, A.; Khoukhi, L. IoT technologies for smart cities. *IET Netw.* 2018, 7, 1–13.

83   Juneja, A., Bali, V., Juneja, S., Jain, V., & Tyagi, P. (Eds.). *Enabling Healthcare 4.0 for Pandemics: A Roadmap Using AI, Machine Learning, IoT and Cognitive Technologies*; John Wiley & Sons, USA, 2021.

84   de Souza, J.T.; de Francisco, A.C.; Piekarski, C.M.; Prado, G.F.D. Data mining and machine learning to promote smart cities: A systematic review from 2000 to 2018. *Sustainability* 2019, 11, 1077.

85   Rayan, Z.; Alfonse, M.; Salem, A.B.M. Machine learning approaches in smart health. *Procedia Comput. Sci.* 2018, 154, 361–368.

86   Varghese, R.; Sharma, S. Affordable smart farming using IoT and machine learning. In *Proceedings of the 2018 Second International Conference on Intelligent Computing and Control Systems (ICICCS)*, Madurai, India. 14–15 June 2018; pp. 645–650.

87   AlZu'bi, S.; Hawashin, B.; Mujahed, M.; Jararweh, Y.; Gupta, B.B. An efficient employment of internet of multimedia things in smart and future agriculture. *Multimed. Tools Appl.* 2019, 78, 29581–29605.

88   Pratyush Reddy, K.S.; Roopa, Y.M.; Kovvada Rajeev, L.N.; Nandan, N.S. IoT based smart agriculture using machine learning. In *Proceedings of the 2nd International Conference on Inventive Research in Computing Applications, ICIRCA 2020*, Coimbatore, India, 15–17 July 2020; pp. 130–134.

89   Goap, A.; Sharma, D.; Shukla, A.K.; Rama Krishna, C. An IoT based smart irrigation management system using Machine learning and open source technologies. *Comput. Electron. Agric.* 2018, 155, 41–49.

90   Rodríguez, S.; Gualotuña, T.; Grilo, C. A system for the monitoring and predicting of data in precision agriculture in a rose greenhouse based on wireless sensor networks. *Procedia Comput. Sci.* 2017, 121, 306–313.

91   Kitpo, N.; Kugai, Y.; Inoue, M.; Yokemura, T.; Satomura, S. Internet of things for greenhouse monitoring system using deep learning and bot notification services. In *Proceedings of the 2019 IEEE International Conference on Consumer Electronics, ICCE 2019*, Las Vegas, NV, USA, 11–13 January 2019.

92   Saha, A.K.; Saha, J.; Ray, R.; Sircar, S.; Dutta, S.; Chattopadhyay, S.P.; Saha, H.N. IOT-based drone for improvement of crop quality in agricultural field. In *Proceedings of the 2018 IEEE 8th Annual Computing and CommunicationWorkshop and Conference (CCWC)*, Las Vegas, NV, USA, 8–10 January 2018; pp. 612–615.

93   Juneja, A., Juneja, S., Bali, V., & Mahajan, S. (2021). Multi-criterion decision making for wireless communication technologies adoption in IoT. *Int. J. Syst. Dyn. Appl.*, 10(1), 1–15. https://doi.org/10.4018/ijsda.2021010101

94 Araby, A.A.; AbdElhameed, M.M.; Magdy, N.M.; Abdelaal, N.; Abd Allah, Y.T.; Darweesh, M.S.; Fahim, M.A.; Mostafa, H. Smart iot monitoring system for agriculture with predictive analysis. In *Proceedings of the 2019 8th International Conference on Modern Circuits and Systems Technologies (MOCAST)*, Thessaloniki, Greece, 13–15 May 2019; pp. 1–4.

95 Nandhini, S.A.; Radha, R.H.S. Web enabled plant disease detection system for agricultural applications using WMSN. *Wirel. Pers. Commun.* 2018, 102, 725–740.

96 Juneja, A.; Juneja, S.; Bali, V.; Jain, V.; Upadhyay, H. Artificial Intelligence and Cybersecurity: Current Trends and Future Prospects. In *The Smart Cyber Ecosystem for Sustainable Development*, vol. 27; Wiley, USA, 2021; pp. 431–441.

97 Anand, R.; Sindhwani, N.; Juneja, S. Cognitive Internet of Things, Its Applications, and Its Challenges: A Survey. In *Harnessing the Internet of Things (IoT) for a Hyper-Connected Smart World*; Apple Academic Press, USA, 2022; pp. 91–113.

98 Alibasic, A.; Al Junaibi, R.; Aung, Z.; Woon, W.L.; Omar, M.A. Cybersecurity for smart cities: A brief review. In *Data Analytics for Renewable Energy Integration: 4th ECML PKDD Workshop, DARE 2016*, Riva del Garda, Italy, 23 September 2016. Revised Selected Papers 4, 2017; pp. 22–30. Springer International Publishing.

# 11 Artificial Intelligence and Machine Learning for Smart Farming Using Cloud Computing

*Sapna Juneja, Abhinav Juneja, and Arti Sharma*
KIET Group of Institutions, Ghaziabad, India

*Vishal Jain*
Sharda School of Engineering and Technology, Sharda University, Greater Noida, India

*Amena Mahmoud*
Kafrelsheikh University, Egypt

## CONTENTS

11.1 Introduction ................................................................................................. 181
11.2 Machine Learning Approaches ................................................................... 182
    11.2.1 Literature Review ............................................................................ 182
11.3 ML and IoT-based Framework for Smart Farming ..................................... 184
    11.3.1 IoT- and AI-based Sensor System for Effective Farming .................. 184
11.4 Usability of Smart Farming: How AI and IoT are Benefitting Agriculture ........ 186
11.5 Benefits of AI in Environment Supported Agriculture ................................ 186
11.6 Demerits of AI in Agriculture and Environment ........................................ 187
11.7 Challenges of AI and IoT Approach in Agriculture .................................... 187
11.8 Conclusion .................................................................................................. 187
References ............................................................................................................. 188

## 11.1 INTRODUCTION

Agriculture or farming is the most ancient and remarkable occupation in the whole world. But people involved in this field are facing difficulty controlling threats because of diseases and bugs in the crops which have been caused by climate change, sharecropping, and extensive usage of pesticides [1]. For all these reasons the agriculture industry is now switching to artificial intelligence (AI). AI technologies help produce better quality crops, control bugs, keep track of soil to make it more productive, collect the information required by farmers, help deal with problems, and other cultivation-related tasks [2]. AI-powered solutions will not only enable farmers to improve efficiency, but they will also improve quantity, quality, and ensure faster go-to-market for crops. In today's agriculture market, AI is suitable for different farming methods. Machine learning (ML) is the fastest growing technology, and it can be used in smart farming [3]. The usage of AI and ML in

DOI: 10.1201/9781003319238-11

AI IN AGRICULTURE MARKET, BY REGION (USD BILLION)

4.0

1.0

| 2017 | 2018 | 2019 | 2020-e | 2022 | 2024 | 2026-p |

■ Americas    ■ Europe    ▨ APAC    ░ RoW

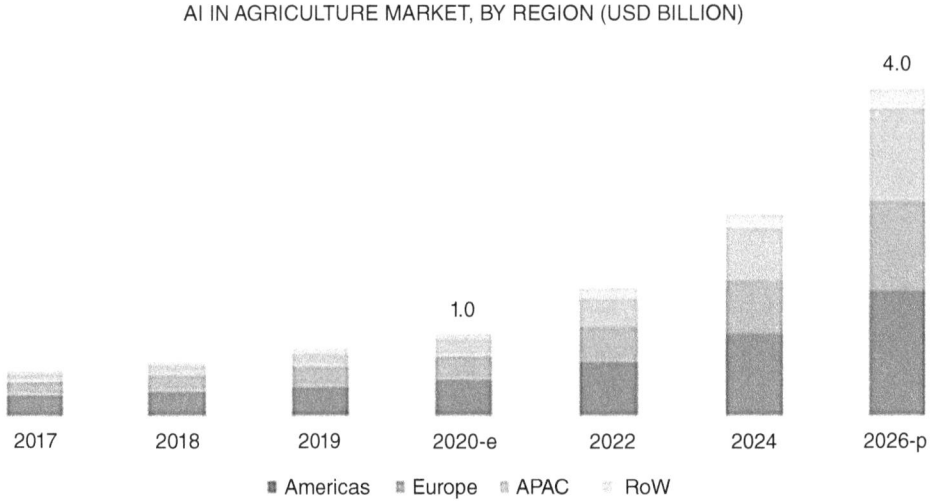

**FIGURE 11.1**   Growth rate of Artificial Intelligence in the Agriculture Market.

farming helps to generate healthier crops and seeds. Agriculture plays a major role in our country's economy. As the population is increasing, there is a persistent pressure on the farming industry to increase the production of the crops to satisfy everyone's needs [4]. Smart agriculture or smart farming is the usage of new technologies for the improvement of the agricultural process in terms of quality and quantity as well [5]. It basically involves the usage of various IoT-based devices in farming to make it more effective and productive. Figure 11.1 shows the rate of growth of artificial intelligence in the market of agriculture.

## 11.2   MACHINE LEARNING APPROACHES

In smart farming which uses AI, the techniques are generated from the learning process [6]. These technologies are required to learn from previous experience and to perform the next function. Performance metrics have been used to compute the performance of the machine learning algorithm and these algorithms improve over a period of time by gaining more and more experience [7]. There are two types of machine learning approach that can be used in smart farming:

1. Supervised Machine Learning: In this approach of smart farming, the input data and the output data is defined and the main aim is to generate a rule which maps that particular input to the output. The final obtained model is known as a trained model which is further used to predict the yield obtained in future [8].
2. Unsupervised Machine Learning: In this type of machine learning farming, the trained model and the trained set remains the same, and unidentified data is used. The main aim of this approach is to find out the hidden pattern of data [9].

### 11.2.1   LITERATURE REVIEW

A number of researchers have worked upon the usage and implementation of AI in the agricultural field and tried to make agriculture "Smart". Table 11.1 shows the most commonly used AI-based IoT technologies used for smart agriculture.

With the usage of AI and ML, farmers can use knowledgeable data and various tools that helps in better decision making, better outcomes, reduction of wastage of food and hence minimize any negative impacts upon the environment [16].

**TABLE 11.1**

**Previous Research Work of AI in Agriculture**

| Author | Title and Year | Summary of the Paper |
|---|---|---|
| P.P. Ray [10] | Internet of things for smart agriculture: Technologies, practices and future direction, 2017 | The researcher defined various IoT devices that can be used in smart farming also mentioned the challenges for implementing these devices. Various case studies were used by them to elaborate the IoT techniques that the farmers are using in today's scenario. |
| Nobrega et al. [11] | An IoT-based solution for intelligent farming, 2019 | The paper elaborated the current development in smart farming using IoT devices and M2M communication. The paper also defined the possibility of scalability and feasibility of these IoT devices if used at a large scale. |
| Julien Roux et al. [12] | A New Bi-Frequency Soil Smart Sensing Moisture and Salinity for Connected Sustainable Agriculture, 2019 | In this paper, the researchers focused on designing of a sensor that was cylindrical in shape, that consumes less power, and can connect with mobile phones without any wired connection. |
| Rajeshwari et al. [13] | A smart agricultural model by integrating IoT, mobile and cloud-based big data analytics, 2017 | In this paper, the researchers evaluated various characteristics in farming like soil quality, humidity level, temperature of the place, etc., using various sensors and all the sensors are controlled by a common controller. Using these sensors, real-time data can be obtained which can further be useful in identifying the patterns according to their suitability of time and weather. |
| Andreas Kamilaris et al. [14] | A review on the practice of big data analysis in agriculture, 2017 | The paper mainly focuses on the usage of big data analysis in the field of agriculture. The reviewers stated how bid data can be efficiently used to make the farming techniques smarter using the available software and hardware. |
| Mekala et al. [15] | A Survey: Smart agriculture IoT with cloud computing | The researchers in this paper tried to define an IoT-based model that works in wireless environment, made up of several sensors used to sense various farming conditions. |

Some of the most important benefits of AI and ML in Agriculture Industry are as follows:

A). Digitalized Cultivation. Digital cultivation of crops increases the accuracy in the production of the crop by using farming decisions generated by AI and ML based upon previous experiences [17]. Further, farmers can use various sensors in their fields that can use AI applications in order to identify the quality of crops and diagnose any crop-related diseases.

1. Quality Evaluation of the crops: Using various ML techniques evaluation of the yields can be done to check whether required features are present in the crop or not, up to how much extent the damage in the crop is acceptable and to identify some other features that can affect the whole crop [18].

2. Identification of variety/type of crops: There are many crops which are quite similar to each other in terms of their shape, size, color, and structure, etc., so it becomes difficult for an individual to identify them. Here, many machine learning approaches can be used to diagnose their patterns and to differentiate each of them [19].

3. Detection of diseases in the crops: In order to check whether the crop is having some disease or not, farmers can use machine learning algorithms, e.g., image processing. Using these algorithms, it becomes very easy to detect the disease of the crop if any like any kind of bacteria or fungus, etc. Apart from this, various AI and ML Algorithms can be used to decide which treatment can be taken if the plant is suffering from some particular disease [20].

B). Livestock Farming. ML techniques can be used in livestock farming to keep a check on the health of livestock, dairy farming, to look after the kettle, and for selective reproduction [21].

    1. Health conservation of livestock: AI and ML helps the farmers to keep an eye on the health of the livestock, used to predict the pattern of breeding of animals, early diagnosis of health disorders, etc. [22].

    2. Data analysis done by ML is being used by farmers to get more precise, high-quality dairy products and to accurately manage their production.

    3. ML algorithms can be used with the help of robots which further support the herding of animals. Drones are also used when trained with ML techniques to herd the animals [23].

    4. ML algorithm uses previous genetic information of the animals to predict the best pregnancy time of the livestock which further helps in obtaining best quality breed of the livestock with characteristics such as quality and quantity of milk production, healthy animal, etc. [24].

C). Water and Soil Management. ML helps in analyzing various farming conditions like the condition of the soil, chemical composition required, moisture content present in the soil, etc. by using the data which helps in better yield production [25].

D). Cultivation of Plants. By analyzing the genes of the crops and plants, an ML approach helps in detection of features that promote the best quality crop production and which plant can be cultivated at what time with which favorable conditions [26].

E). Usage of chat bots by farmers. Farmers can use the ML-assisted chabot environment to communicate with government officials and sellers. Farmers can get the information about the farming problem which they face and hence the proper monitoring of the crops can be done even if the farm is at a remote location [27].

F). AI-supported pesticide usage device. With the help of both AI applications and IoT sensing devices, it becomes easy for the farmers to save the crop from pests. Now farmers can treat all the plants separately using patch treatment. With this treatment, it becomes feasible to apply less chemical treatment as treatment will be done only on the infected plant [28].

G). Usage of drones to keep track of crops [29]. Drones can be used for keeping an eye on the fields to check any risk on crops. With the help of sensors installed in the field, drones can collect information about the crops even at the smallest level. After gathering the data, the ML approach is used to provide the complete information about the yield [30].

These overall applications of AI and IoT in smart farming are represented in Figure 11.2.

## 11.3 ML AND IOT-BASED FRAMEWORK FOR SMART FARMING

Figure 11.3 defines the architecture of smart farming that uses the concept of ML [31]. Here the sensors and smart monitoring devices, i.e., the devices that are monitoring the environment and connected to the Internet all the time, have been used to control the farming activities [32]. These sensing devices are detecting the information and sending this gathered data for analysis and planning via the cloud-based infrastructure [33]. This analysis and planning uses ML algorithms to decide what preventive measures should be taken based upon the previous data and send that information to the smart control system which usually comprises of robots that controls the farm or the system according to the information provided by planning team [34].

### 11.3.1 IoT- AND AI-BASED SENSOR SYSTEM FOR EFFECTIVE FARMING

The IoT- and AI-based system can be represented here in Figure 11.4 that can be used to implement the techniques of smart farming thus making farming more effective and productive.

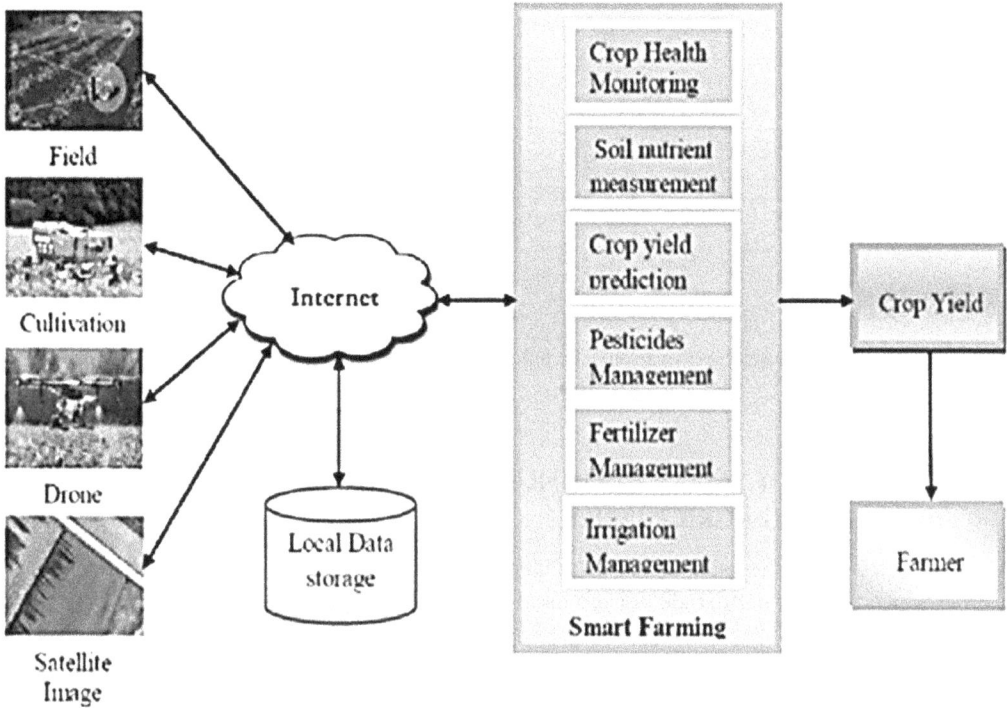

Smart Farming Architecture

**FIGURE 11.2**    Smart Farming Architecture.

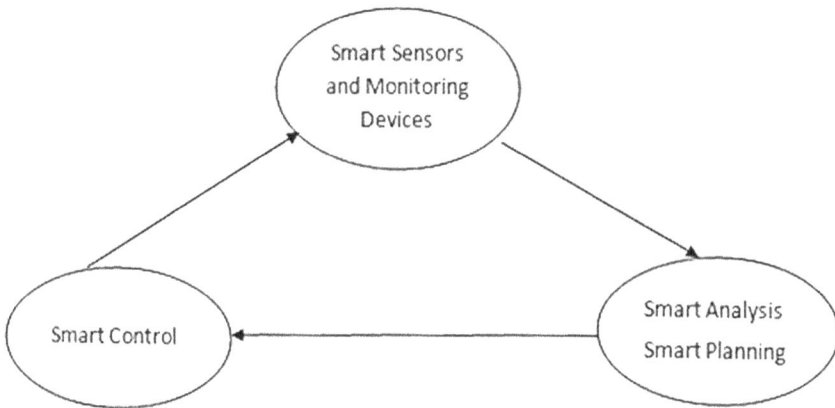

**FIGURE 11.3**    Cloud-based ML-based Architecture for Smart Farming.

The sensors are installed in the fields to gather the physiological data and cameras to collect the environmental data. With the help of historical data, differences in the outcome of the growth information and best optimal environment for the plant growth are analyzed and then the best environmental conditions suitable for the growth and development of the crops are generated [35]. The AI-based device is used to control and regulate the environment in which plant will grow [32]. The purpose of blockchain is to make sure that the data stored in all the sensors has not been updated or changed to ensure high quality [36].

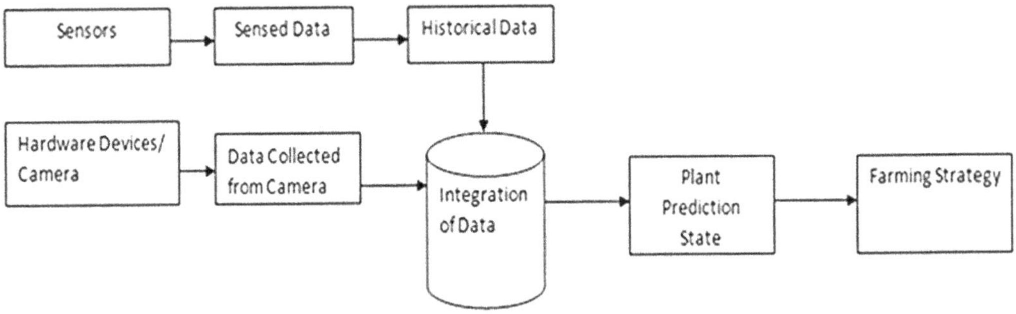

**FIGURE 11.4**   IoT/AI-based Smart Farming Technique.

## 11.4   USABILITY OF SMART FARMING: HOW AI AND IoT ARE BENEFITTING AGRICULTURE

AI along with IoT has the capability to alter the agricultural field in many aspects. There are some ways in which AI/IoT can upgrade the agriculture as:

- A huge amount of data is gathered by sensors using a smart farming approach and hence this data can be used in the improvement of business strategies, performance of employees, and increasing the efficiency of various devices involved [37].
- Farmers can do better crop management as it becomes easy to predict the quality and quantity of the upcoming yield by sensing the status of the soil, weather, health of the plant, etc. [37].
- With the better crop management, farmers can manage the overall cost of farming in better way and hence can reduce the overall wastage of yields, if any [38].
- By using the automated devices, sensors, etc., the overall production can be increased thus increasing the overall business [39].
- Better quality crops can be obtained with the least use of pesticides and other chemicals [40].

Taken together, all these aspects can finally lead to more profits.

## 11.5   BENEFITS OF AI IN ENVIRONMENT SUPPORTED AGRICULTURE

If properly used, AI can be useful for the betterment of the environment. Using this approach, farmers can grow crops even in non-rural areas [41]. AI-based algorithms help with crop planting in very small areas even with the minimum usage of water and electricity. The countries where the main population is in metropolitan cities are getting benefits from this approach [42]. The environmental benefit of using this technique of crop plantation lessens deforestation in rural areas. Another merit of AI for the environment is that it helps in detecting carbon sinks from the area of the jungle that sucks up extra $CO_2$ from the environment and thus helps in maintaining oxygen level. One more feature provided by AI for the environment is that is capable of locating the area in the fields that contains weeds and thus stops the farmers overspreading the chemicals leading to protecting the environment from extra chemicals. Further AI can help in fighting and controlling global warming [43]. AI technology monitors the energy consumption of various devices and gadgets used by the people and thus based upon that consumption, predicts the heat produced that helps in officials making policies in order to handle such huge amount of heat [44].

## 11.6   DEMERITS OF AI IN AGRICULTURE AND ENVIRONMENT

1. As AI requires a large amount of data for executing its processes, it releases a lot of energy during its execution which is not environment friendly at all. The stored data and the servers where data has been processed also emit heat thus increasing global warming. So, in order to use this AI technology, first some environment friendly solution must be used so as to decrease the negative impact of this technology upon the environment [45].

2. Although this technology brings a lot of change in the farming industry, still it is not easy for the farmers to adapt this technology as this can lead to the increase in the overall farming cost which is sometimes not feasible for every farmer to pay. Further implementation of this method also requires proper maintenance of devices, resources, and data and this also causes an increase in the farming budget. Hence it is not always feasible for in small-scale farming with lower budgets to implement such AI-based policies due to limited budget constraints [46].

3. AI in the farming industry also requires a high-speed Internet connection all the time and in developing countries this is not always possible in rural areas because of lower connectivity [47]. Apart from this, professionals with a high skill set in AI are required to manage the data and resources.

4. The most important problem associated with AI in smart farming is that it is making many jobs related to farming outdated as machines are able to do more work as compared to humans thus leading to the unemployment [48].

## 11.7   CHALLENGES OF AI AND IoT APPROACH IN AGRICULTURE

As AI provides a lot of opportunities in the field of agriculture, still there are many problems associated with using ML and IoT-based solutions in farming [49].

1. Increasing risk of Cybersecurity: All of the technology depends upon the Internet and when data has been collected, the chances of cyberattacks increase. Eavesdroppers can access important and sensitive data; attackers can control IoT devices by attacking them. So, in order to use IoT and AI-based application in farming, the farmers must carry out the cybersecurity in order to protect the data and the devices [50].

2. Higher cost of Implementation: Usage of IoT devices, sensors, drones, etc., can potentially increase the overall budget of the farmers which sometimes is not feasible. Also, for the analysis, organization, and storage of data, some cloud-based system or other system is also required. Hence the overall cost of farming increases multiple times [51–53].

3. Harm to the environment: Smart farming requires a lot of devices, sensors [54], etc., which needs huge amounts of energy to create [55]. In addition, a large amount of fuel is required by these devices. Data storage devices also produce a large volume of heat and energy, and all these things are not at all suitable for a safe environment. Thus, increases in technology has consequences for the environment [56–58].

4. Durability: Apart from constant connectivity to the agricultural IoT [59] system, durability and robustness of the associated devices is a major issue which needs to be focused on before the installation of the system otherwise it will not be fruitful at all [60, 61].

## 11.8   CONCLUSION

The emphasis on intelligent, finer, and more proficient farming techniques is needed in order to fulfill the increasing food demand of the world. The growth of novel techniques for the improvisation of crop management and control is increasing with each passing day. This research paper considers all of the features of technology-enabled farming, especially AI and IoT, to make farming increasingly smart, well organized, and well planned in order to fulfill the upcoming future expectancy. For this reason, drones, sensors, cloud-based technology, etc., are defined properly. Further, an extensive understanding of previous research attainments is also elaborated. Additionally, IoT-based structures and AI techniques are defined that support farming applications. An overview of the challenges that the farming industry is facing due to the increasing usage of technology is also listed. Hence, it can be concluded that the main objective of farming is to increase the yield production, and in order to achieve this target, the usage of IoT and AI techniques is mandatory.

## REFERENCES

[1] S. Yahata et al., "A hybrid machine learning approach to automatic plant phenotyping for smart agriculture," *2017 International Joint Conference on Neural Networks (IJCNN)*, pp. 1787–1793, 2017, doi: 10.1109/IJCNN.2017.7966067.

[2] F. Balducci, D. Impedovo, and G. Pirlo, "Machine learning applications on agricultural datasets for smart farm enhancement," *Machines*, vol. 6, no. 3, 2018, doi: 10.3390/machines6030038.

[3] S. R. Rajeswari, P. Khunteta, S. Kumar, A. Raj Singh, and V. Pandey, "Smart farming prediction using machine learning," *Int. J. Innov. Technol. Explor. Eng.*, vol. 8, no. 7, pp. 190–194, 2019.

[4] P. Khandelwal, R. T. Maharaj, P. M. Khandelwal, and H. Chavhan, "Artificial intelligence in agriculture: An emerging era of research article," *Researchgate*, vol. 1, pp. 1–8, 2019, [Online]. Available: https://www.researchgate.net/publication/335582861.

[5] M. Chetan Dwarkani, R. Ganesh Ram, S. Jagannathan, and R. Priyatharshini, "Smart farming system using sensors for agricultural task automation," *2015 IEEE Technological Innovation in ICT for Agriculture and Rural Development (TIAR)*, pp. 49–53, 2015, doi: 10.1109/TIAR.2015.7358530.

[6] T. Satish, T. Bhavani, and S. Begum, "Agriculture productivity enhancement system using IOT," *Int. J. Theor. Appl. Mech.*, vol. 12, no. 3, pp. 543–554, 2017, [Online]. Available: http://www.ripublication.com.

[7] A. Dengel, "Special Issue on Artificial Intelligence in Agriculture," *KI - Künstliche Intelligenz*, vol. 27, no. 4, pp. 309–311, 2013, doi: 10.1007/s13218-013-0275-y.

[8] H. Murase, "Artificial intelligence in agriculture," *Comput. Electron. Agric.*, vol. 29, no. 1–2, pp. 1–2, 2000, doi: 10.1016/S0168-1699(00)00132-0.

[9] Y. Ampatzidis, "Applications of artificial intelligence for precision agriculture," pp. 1–5, 2019, [Online]. Available: https://edis.ifas.ufl.edu.

[10] P. P. Ray, "Internet of things for smart agriculture: Technologies, practices and future direction," *J. Ambient Intell. Smart Environ.*, vol. 9, no. 4, pp. 395–420, 2017, doi: 10.3233/AIS-170440.

[11] L. Nóbrega, P. Gonçalves, P. Pedreiras, and J. Pereira, "An IoT-based solution for intelligent farming," *Sensors (Switzerland)*, vol. 19, no. 3, pp. 1–24, 2019, doi: 10.3390/s19030603.

[12] J. Roux, C. Escriba, J.-Y. Fourniols, and G. Soto-Romero, "A new Bi-frequency soil smart sensing moisture and salinity for connected sustainable agriculture," *J. Sens. Technol.*, vol. 09, no. 03, pp. 35–43, 2019, doi: 10.4236/jst.2019.93004.

[13] S. Rajeswari, K. Suthendran, and K. Rajakumar, "A smart agricultural model by integrating IoT, mobile and cloud-based big data analytics," *2017 International Conference on Intelligent Computing and Control (I2C2)*, pp. 1–5, 2018, doi: 10.1109/I2C2.2017.8321902.

[14] A. Kamilaris, A. Kartakoullis, and F. X. Prenafeta-Boldú, "A review on the practice of big data analysis in agriculture," *Comput. Electron. Agric.*, vol. 143, pp. 23–37, 2017, doi: 10.1016/j.compag.2017.09.037.

[15] M. S. Mekala and P. Viswanathan, "A survey: Smart agriculture IoT with cloud computing," *2017 International Conference on Microelectronic Devices, Circuits and Systems (ICMDCS)*, pp. 1–7, 2017, doi: 10.1109/ICMDCS.2017.8211551.

[16] L. M. Bhar, V. Ramasubramanian, A. Arora, S. Marwaha, and R. Parsad, "Era of artificial intelligence: Prospects for Indian agriculture," *Indian Farming*, vol. 3, no. 69, pp. 10–13, 2019.

[17] I. A. Lakhiar, G. Jianmin, T. N. Syed, F. A. Chandio, N. A. Buttar, and W. A. Qureshi, "Monitoring and control systems in agriculture using intelligent sensor techniques: A review of the aeroponic system," *J. Sensors*, vol. 2018, 2018, doi: 10.1155/2018/8672769.

[18] V. Mokaya, "Future of precision agriculture in india using machine learning and artificial intelligence," *Int. J. Comput. Sci. Eng.*, vol. 7, no. 3, pp. 422–425, 2019, doi: 10.26438/ijcse/v7i3.422425.

[19] S. Wolfert, L. Ge, C. Verdouw, and M. J. Bogaardt, "Big data in smart farming – A review," *Agric. Syst.*, vol. 153, pp. 69–80, 2017, doi: 10.1016/j.agsy.2017.01.023.

[20] R. Shukla, G. Dubey, P. Malik, N. Sindhwani, R. Anand, A. Dahiya, and V. Yadav, "Detecting crop health using machine learning techniques in smart agriculture system," *J. Sci. Ind. Res.*, vol. 80, no. 8, pp. 699–706, 2021.

[21] T. Feng, Y. Chai, Y. Huang, and Y. Liu, "A real-time monitoring and control system for crop," *Proceedings of the 2019 2nd International Conference on Algorithms, Computing and Artificial Intelligence*, pp. 183–188, 2019, doi: 10.1145/3377713.3377742.

[22] N. Zhu et al., "Deep learning for smart agriculture: Concepts, tools, applications, and opportunities," *Int. J. Agric. Biol. Eng.*, vol. 11, no. 4, pp. 21–28, 2018, doi: 10.25165/j.ijabe.20181104.4475.

[23] K. G. Liakos, P. Busato, D. Moshou, S. Pearson, and D. Bochtis, "Machine learning in agriculture: A review," *Sensors (Switzerland)*, vol. 18, no. 8, pp. 1–29, 2018, doi: 10.3390/s18082674.

[24] A. Kamilaris and F. X. Prenafeta-Boldú, "Deep learning in agriculture: A survey," *Comput. Electron. Agric.*, vol. 147, pp. 70–90, 2018, doi: 10.1016/j.compag.2018.02.016.

[25] N. Sindhwani, R. Anand, R. Vashisth, S. Chauhan, V. Talukdar, and D. Dhabliya, "Thingspeak-Based Environmental Monitoring System Using IoT," in *2022 Seventh International Conference on Parallel, Distributed and Grid Computing (PDGC)*, 2022, pp. 675–680, doi: 10.1109/PDGC56933.2022.10053167.

[26] T. Duckett et al., "Agricultural robotics: The future of robotic agriculture," 2018, [Online]. Available: http://arxiv.org/abs/1806.06762.

[27] V. Dharmaraj and C. Vijayanand, "Artificial intelligence (AI) in agriculture," *Int. J. Curr. Microbiol. Appl. Sci.*, vol. 7, no. 12, pp. 2122–2128, 2018, doi: 10.20546/ijcmas.2018.712.241.

[28] M. Ayaz, M. Ammad-Uddin, Z. Sharif, A. Mansour, and E. H. M. Aggoune, "Internet-of-Things (IoT)-based smart agriculture: Toward making the fields talk," *IEEE Access*, vol. 7, pp. 129551–129583, 2019, doi: 10.1109/ACCESS.2019.2932609.

[29] A. Nayyar and V. Puri, "Smart farming: IoT based smart sensors agriculture stick for live temperature and moisture monitoring using arduino, cloud computing & solar technology," *Proceeding of the International Conference on Communication and Computing Systems (ICCCS-2016)*, pp. 673–680, 2017, doi: 10.1201/9781315364094-121.

[30] R. Dagar, S. Som, and S. K. Khatri, "Smart farming - IoT in agriculture," *2018 International Conference on Inventive Research in Computing Applications (ICIRCA)*, pp. 1052–1056, 2018, doi: 10.1109/ICIRCA.2018.8597264.

[31] S. Dimitriadis and C. Goumopoulos, "Applying machine learning to extract new knowledge in precision agriculture applications," *2008 Panhellenic Conference on Informatics*, pp. 100–104, 2008, doi: 10.1109/PCI.2008.30.

[32] D. R. Vincent, N. Deepa, D. Elavarasan, K. Srinivasan, S. H. Chauhdary, and C. Iwendi, "Sensors driven ai-based agriculture recommendation model for assessing land suitability," *Sensors (Switzerland)*, vol. 19, no. 17, 2019, doi: 10.3390/s19173667.

[33] T. Gundu and V. Maronga, "IoT security and privacy: Turning on the human firewall in smart farming," vol. 12, pp. 95–104, 2019, doi: 10.29007/j2z7.

[34] M. De Clercq, A. Vats, and A. Biel, "Agriculture 4.0: The future of farming technology," *Proceedings of the World Government Summit, Dubai, UAE*, p. 30, 2018, [Online]. Available: https://www.worldgovernmentsummit.org/api/publications/document?id=95df8ac4-e97c-6578-b2f8-ff0000a7ddb6.

[35] D. Vasisht et al., "Farmbeats: An IoT platform for data-driven agriculture," *Proceeding 14th USENIX Symposium on Networked Systems Design and Implementation, NSDI 2017*, pp. 515–529, 2017.

[36] C. Brewster, I. Roussaki, N. Kalatzis, K. Doolin, and K. Ellis, "IoT in agriculture: Designing a Europe-wide large-scale pilot," *IEEE Commun. Mag.*, vol. 55, no. 9, pp. 26–33, 2017, doi: 10.1109/MCOM.2017.1600528.

[37] Annu Dhankhar, et al., "Kernel parameter tuning to tweak the performance of classifiers for identification of heart diseases," *Int. J. E-Health Med. Commun. (IJEHMC)*, vol. 12, no. 4, pp. 1–16, 2021.

[38] A. Juneja, S. Juneja, S. Kaur, and V. Kumar, "Predicting diabetes mellitus with machine learning techniques using multi-criteria decision making," *Int. J. Inf. Retr. Res.*, vol. 11, no. 2, pp. 38–52, 2021, doi: 10.4018/ijirr.2021040103.

[39] Sandhya Sharma, et al., "Recognition of gurmukhi handwritten city names using deep learning and cloud computing," *Sci. Program.*, vol. 2022, pp. 1–16, 2022.

[40] T. Thomas, A. P. Vijayaraghavan, and S. Emmanuel, *Machine Learning Approaches in Cyber Security Analytics*, 2020, doi: 10.1007/978-981-15-1706-8.

[41] U. Cortés, M. Sànchez-Marrè, L. Ceccaroni, I. R-Roda, and M. Poch, "Artificial intelligence and environmental decision support systems," *Appl. Intell.*, vol. 13, no. 1, pp. 77–91, 2000, doi: 10.1023/A:1008331413864.

[42] Sapna Juneja, et al., "computer Vision-Enabled character recognition of hand Gestures for patients with hearing and speaking disability," *Mobile Inf. Syst.*, vol. 2021, pp. 1–10, 2021.

[43] R. Vinuesa et al., "The role of artificial intelligence in achieving the sustainable development goals," *Nat. Commun.*, vol. 11, no. 1, pp. 1–10, 2020, doi: 10.1038/s41467-019-14108-y.

[44] A. Abdullayeva, "Impact of artificial intelligence on agricultural, healthcare and logistics industries," *Ann. Spiru Haret Univ. Econ. Ser.*, vol. 19, no. 2, pp. 167–175, 2019, doi: 10.26458/1929.

[45] V. Saiz-Rubio and F. Rovira-Más, "From smart farming towards agriculture 5.0: A review on crop data management," *Agronomy*, vol. 10, no. 2, 2020, doi: 10.3390/agronomy10020207.

[46] Gaurav Dhiman, et al., "A novel machine-learning-based hybrid CNN model for tumor identification in medical image processing," *Sustainability*, vol. 14, no. 3, p. 1447, 2022.

[47] V. R. Palanivelu and B. Vasanthi, "Role of artificial intelligence in business transformation," *Int. J. Adv. Sci. Technol.*, vol. 29, no. 4, pp. 392–400, 2020.

[48] J. I. Z. Chen and P. Hengjinda, "Applying AI technology to the operation of smart farm robot," *Sensors Mater.*, vol. 31, no. 5, pp. 1777–1788, 2019, doi: 10.18494/SAM.2019.2389.

[49] M. Ryan, "Ethics of using AI and big data in agriculture: The case of a large agriculture multinational," *ORBIT J.*, vol. 2, no. 2, 2019, doi: 10.29297/orbit.v2i2.109.

[50] Rohit Anand, Nidhi Sindhwani, and Sapna Juneja, "Cognitive Internet of Things, Its Applications, and Its Challenges: A Survey," in *Harnessing the Internet of Things (IoT) for a Hyper-Connected Smart World.* Apple Academic Press, 2022, pp. 91–113.

[51] J. Kaur, N. Sindhwani, and R. Anand, "Implementation of IoT in Various Domains." In *IoT Based Smart Applications*, Springer International Publishing, USA, 2022, pp. 165–178.

[52] O. Elijah, T. A. Rahman, I. Orikumhi, C. Y. Leow, and M. N. Hindia, "An overview of internet of things (IoT) and data analytics in agriculture: Benefits and challenges," *IEEE Internet Things J.*, vol. 5, no. 5, pp. 3758–3773, 2018, doi: 10.1109/JIOT.2018.2844296.

[53] Junaid Rashid, et al., "An augmented artificial intelligence approach for chronic diseases prediction," *Front. Pub. Health*, vol. 10, 2022, doi: 10.3389/fpubh.2022.860396.

[54] Neha Gupta, et al., "Enhanced virtualization-based dynamic bin-packing optimized energy management solution for heterogeneous clouds," *Math. Probl. Eng.*, vol. 2022, 2022, doi: 10.1155/2022/8734198.

[55] Sonam Aggarwal, et al., "A convolutional neural network-based framework for classification of protein localization using confocal microscopy images," *IEEE Access*, vol. 10, pp. 83591–83611, 2022.

[56] Sapna Juneja, et al., "COVID-19 and Machine Learning Approaches to Deal With the Pandemic," in *Enabling Healthcare 4.0 for Pandemics: A Roadmap Using AI, Machine Learning, IoT and Cognitive Technologies*, Wiley, USA, 2021, pp. 1–19.

[57] Himani Chugh, et al., "Image retrieval using different distance methods and color difference histogram descriptor for human healthcare," *J. Healthcare Eng.*, vol. 2022, 2022, doi: 10.1155/2022/9523009.

[58] S. Sharma, S. Gupta, D. Gupta, J. Rashid, S. Juneja, J. Kim, and M. M. Elarabawy, "Performance evaluation of the deep learning based convolutional neural network approach for the recognition of chest X-ray images," *Front. Oncol.*, vol. 12, p. 932496, 2022, doi: 10.3389/fonc.2022.932496.

[59] K. Kour, D. Gupta, K. Gupta, G. Dhiman, S. Juneja, W. Viriyasitavat, … M. A. Islam, "Smart-hydroponic-based framework for saffron cultivation: A precision smart agriculture perspective," *Sustainability (Switzerland)*, vol. 14, no. 3, 2022, doi:10.3390/su14031120.

[60] G. Dhiman, et al., "A novel machine-learning-based hybrid CNN model for tumor identification in medical image processing," *Sustainability*, vol. 14, p. 1447, 2022.

[61] Ashima Arya, et al., "Heart Disease Prediction with Machine Learning and Virtual Reality: From Future Perspective," in *Extended Reality for Healthcare Systems*, Academic Press (Elsevier), UK, 2023, pp. 209–228.

# 12 Parivem–Parivahan Emulator

*Dimple Chawla and Yukta Malhotra*
Vivekananda Institute of Professional Studies, New Delhi, India

## CONTENTS

12.1 Introduction ................................................................................................191
12.2 Material and Methods ..................................................................................192
12.3 State of Art Methods ...................................................................................196
12.4 Proposed System Architecture ....................................................................198
12.5 Conclusion ..................................................................................................201
References ............................................................................................................201

## 12.1 INTRODUCTION

The total production of vehicles in India in the fiscal year 2020–21 surged to 22.7 million units making India a leading country in vehicle production [1]. The sales of passenger vehicles grew to 27.11 lakhs units, two-wheelers grew to 151.19 lakhs units, the sales of commercial vehicles surged to 5.69 lakhs units, and for three-wheelers it was 2.16 lakhs units [2]. A proper vehicle management system is inevitably needed with such massive numbers.

This fast-growing vehicle fleet is a major contributor to pollution and climate change. Polluted gas emissions from vehicles beyond the defined standard values contribute to air pollution, leading to various environmental and health hazards. Hence, it is necessary to check the level of gases emitted. This can be done using sensors. A sensor is a physical device that records a physical parameter. It can be a simple button or a chip capable of encoding and decoding digital and analog signals. Some sensors can detect light, noise, smoke, alcohol, and other toxic gases. A few of them are metal oxide semiconductors such as MQ-2 and MQ-7 sensors that can measure the content of flammable gases and carbon monoxide emitted from a vehicle. These sensors are widely used for air quality monitoring.

Another growing concern is the accidents due to drunk driving in India rising to 12,256, as reported by Mr. Nitin Gadkari, Road Transport and Highways Minister, in Rajya Sabha [3]. The alcohol content in a driver's breath can be measured using an MQ-3 sensor. The MQ-3 sensor operates at +5V and can detect alcohol and benzene concentrations anywhere from 25 to 500 ppm.

These sensors are connected to a microcontroller that controls and monitors the entire connection in the circuit. The microcontroller, Arduino Uno, is an open-source board that is built on ATmega328P that controls and monitors the system [4]. The Interaction Design Institute Ivrea (IDII) in Ivrea, Italy first initiated the Arduino Project. The Arduino board holds several digital and analog input/output pins interfaced with other circuits and expansion boards. To automate the entire vehicle management system, the Radio Frequency Identification (RFID) tag is installed in the vehicle [6]. The RFID reader recognizes and scans the tag through electromagnetic fields and saves the collected data in the cloud. It is based on automatic detection and data capture. It has a processor that makes it capable of storing extra information. It was first developed in the U.S. in the 1920s during World War II to identify airplanes. The airplanes radiated signals to read a unique identification number that differentiated the aircraft of allies or enemies [5]. Unlike the older barcode technology, it is not required to be in the line of sight of an RFID reader [6].

The Global Positioning System (GPS) is a network that uses satellites and radio frequencies for navigation and supplies information about time and location to a GPS receiver anywhere on Earth.

DOI: 10.1201/9781003319238-12

It was first started by the U.S. Department of Defense in 1973 [7]. European Telecommunication Standards Institute developed the Global System for Mobile Communications (GSM) to describe the protocols for the cellular connection of mobile devices [8]. It was first developed in Finland in December 1991.

Since these sensors can be used on multiple different devices, they include voluminous information and generate massive amounts of semi-structured or unstructured data. The cloud is one of the most cost-effective and suitable solutions when working with the huge amount of data collected by the sensors [9]. It is a system that enables the delivery of data over the Internet to the targeted data centers. Computing services such as databases, computer networking, virtual hosts, servers, and intelligence when provided over the cloud are called cloud computing. The agile platform it supplies to build IoT devices and applications has gained much importance in the last few years. The pay-as-you-go service helps lower operating costs and drive infrastructure more efficiently. There are two concepts of clouds – "public cloud" and "private cloud." Public clouds are publicly available with a price to pay as you use them while private clouds are exclusively built for businesses' private use. Examples of public cloud include Azure, Amazon Web Services (AWS), Google App Engine, etc. Any data center run by a large enterprise is categorized as a private cloud [10].

This paper aims to introduce an RFID-enabled vehicle management system that can track pollution levels using MQ-7 and MQ-2 sensors and keep the customer and Regional Transport Office (RTO) informed if it reaches beyond the permissible levels. It also enables live GPS tracking of the vehicle, automatic toll deduction, drunken driving prevention, and supports storing various vehicle-related documents like registration certificates (RCs). All the information collected is stored in the cloud. This paper is divided into five sections. The recent and traditional approaches to the vehicle management system, i.e., the Literature Review, are discussed in Section 12.2. Section 12.3, the State of Art Methods, elaborates on different problems prevailing in this system, followed by their solution in Section 12.4, Proposed Architecture. Section 12.5 concludes the paper.

## 12.2  MATERIAL AND METHODS

RFID technology has been appearing in different fields; the RFID reader and the RFID tag are the main components of RFID technology. The reader comprises a transceiver and an antenna. Weak radio signals that can range from a few feet to a few yards are generated by the transceiver [11]. The antennas then send these signals to wake or activate the tag as shown in Figure 12.1; the antenna propagates the waves in horizontal and vertical directions. The RFID tag consists of a transponder that transforms these radio frequencies into electric power and uses it for two-way communication of messages. RFID is capable of finding objects without any human intervention and fetching useful information from radio frequency signals.

In the views of Wei Wang and Shidong Fan, RFID implementation includes information systems that play a vital role in the success of RFID projects. Since such systems manage information, it is important to protect them from theft, alteration, or destruction. Security is a pivotal issue that must be taken care of while installing RFID systems [12]. These security issues are because these systems often require a connection to the Internet. However, this problem was resolved by centralizing all

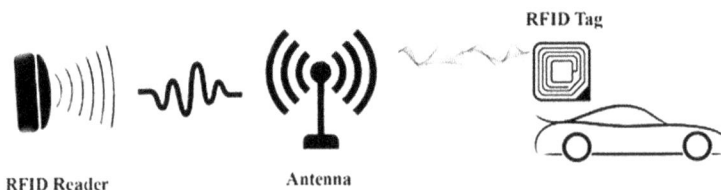

RFID Reader                          Antenna                          RFID Tag

**FIGURE 12.1**  Working of RFID.

the information on the web-based database. Now, the records could be retrieved only with a valid user ID and password, along with the tag number on the vehicle, hence increasing data security [13].

S. Dharanya, A. Umamakeswari et al. have provided a unique way to protect vehicles and automate the toll collection process. GPS gives an automated warning in case of over speeding and deduces the challan if the driver takes no notice of the warning. It also traces the car's location and alerts the concerned traffic police department using GSM [14].

All the researchers in [15] have given an approach for air pollution monitoring and control systems using MQ-7 and MQ-2 sensors. All these gas sensors are analog. Analog sensors are the sensors that output a continuous voltage or current proportional to the gas level, while the digital sensors send and receive signals that are simply strings of characters and can be read by a microprocessor or a computer. Here, an RFID smart card is provided to every license holder which acts as a key for the vehicle. If the gas emissions exceed the permissible level, a warning is sent and, if ignored by a driver, a direct message is sent to the RTO through GSM technology for license cancellation. They also incorporated automatic toll collection that regulates the entire process in a much more efficient way.

Based on the ATmega328, the microcontroller Arduino Uno has features such as a processing speed of 20 MHz, Power supply of 1.8–5.5, Operational range of –100°C to 850°C, 32 KB Flash, 1 KB EEPROM, and 2 KB RAM [16]. Arduino has the following components that are suitable for supporting any microcontroller:

- Microcontroller – the brain of Arduino where the programs are loaded.
- USB port – used to establish a connection between the Arduino board and the computer.
- USB to Serial Chip – enables translation of data that comes from the computer to the microcontroller. This part makes programming the Arduino board possible from the computer.
- Digital pins – these are the pins that use digital logic to perform operations such as turn on/off. There are 14 digital input/output pins, of which six can be used as PWM outputs.
- Analog pins – are needed when reading analog values. It can read in a 10-bit resolution. There are six analog inputs.
- 5V/3.3V pins – are used to power external components connected to Arduino.
- GND – referred to as ground or negative is used to complete the circuit.
- VIN – referred to as Voltage In, it connects external power supplies.

In 2018, the authors created a gas detection unit to detect the emission of toxic gases such as carbon monoxide using an MQ-7 sensor and volatile gases such as methane, hydrogen, LPG, and i-butane using an MQ-2 sensor, in a brief period. The alarm is raised if the gas concentrations exceed the threshold limit [17]. The MQ-7 gas sensor is a semiconductor sensor that comprises of Aluminum Oxide ($Al_2O_3$) ceramic tube and a Tin Dioxide ($SnO_2$) sensitive layer, measuring electrode, and heater mounted on a plastic and stainless-steel net. The working environment for the sensitive components is provided by the heater. The sensor is based on the principle of conductivity change. As the conductivity of the semiconductor material changes, the concentration of the target gas at a particular temperature also changes, as shown in Figure 12.2. The semiconductor material is heated to 1.5 V. The material's conductivity increases with the rising concentration of carbon monoxide. This linearity is suitable for up to 5 V of supply [17]. It is known for its stability and long life.

The MQ-2 is a semiconductor-type gas sensor with bakelite standard encapsulation [18]. This sensor operates from 10°C to 50°C and has a consumption below 150 mA at 5 v. As the number of flammable gases in the atmosphere rises, the conductivity of this sensor also increases. MQ-2 has a built-in potentiometer that has a variable resistance that varies concerning the flammable gases' concentration [17]. Its operating voltage is +5V and its output voltage ranges from 0 V to 5 V. It has an operational amplifier comparator and a digital output pin and comes with 4-pins viz: VCC, GND (Ground), DO (Digital Out), and AO (Analog Output). Both MQ-7 and MQ-2 sensors can detect different gases at different temperatures [18].

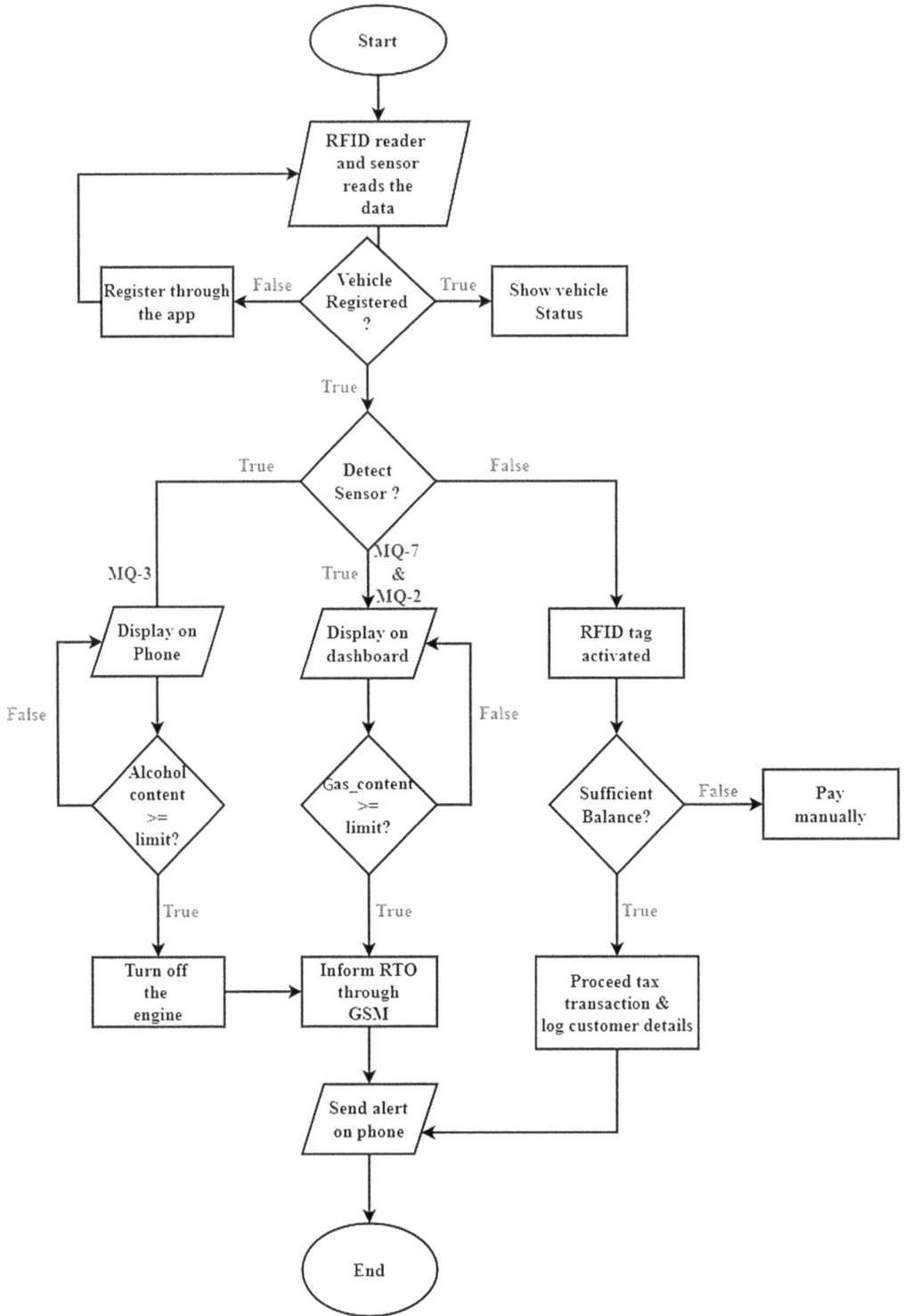

**FIGURE 12.2** Flowchart for the Solution to the Problems.

The electronic toll lanes or car-parking areas are set up with special antennas that identify the vehicle's signals emitted from the RFID tag fitted on the vehicle. The driver needs to set up an account with all his details like the transponder's number, driver's name, balance, etc. A radiofrequency (microwave) pulse is continuously sent out by the antenna, which then hits a transponder and returns. These pulses are returned from the transponder and are received by the antenna [19].

MQ-3 is an optimum sensor to be used as a breath analyzer that detects the presence of alcohol content in the driver's breath. The sensor's conductivity increases if it finds any trace of alcohol content in the air. It operates in the temperature range of $-10°C$ to $50°C$ and has a power requirement of 5 VDC at 60 mA and 165 mA. Like the MQ-2 sensor, it also has four pins viz: GND, AO, VCC, and DO. It has extremely high sensitivity towards alcohol, while gasoline, benzene, smoke, and vapor are low. The potentiometer can be used to adjust the sensitivity in MQ-3 ($SnO_2$), which shows lower conductivity in clean air [20].

The execution of the system that manages accident-location detection, anti-collision, and accident prevention, is done by GSM and GPS technologies. The Global Positioning System (GPS) comprises 24 satellites packed in a network. These satellites are present in orbit and can work extremely well irrespective of weather circumstances or location. To estimate the 2D position that comprises latitude and longitude and its respective location, the GPS receiver must be connected to the signal of at least three satellites [21]. When multiple satellites provide the data they collected, it becomes easier for the receiver to determine the exact 3D position of the user (latitude, longitude, and altitude). After the calculation of the vehicle's position, other information such as distance, speed, and time can be determined using the GPS unit [21]. The GPS takes the coordinates of the accident location, and the GSM sends the coordinates to the reference contact. This input location from GPS in latitudes and longitudes is provided to Arduino, which performs necessary actions on the information obtained while ensuring the key points [22].

The operable frequency band of GSM lies in the range of 900 MHz or 1800 MHz. It consists of five separate parts that combine to function as a whole:

1) The mobile device connects to the network through the system hardware.
2) The Base Station Subsystem (BSS) is responsible for controlling the traffic between the Network Switching Subsystem (NSS) and the mobile phone. It can be further divided into two sub-components: Base Transceiver Station (BTS) and Base Station Controller (BSC). The equipment that enables communication with radio transceivers, mobile phones, and antennas is BSC. The intelligence that makes this possible is BSC.
3) The NSS is a system owned by mobile users. It has further components such as Mobile Switching Center (MSC) and the Home Registry (HLR). Functions such as Short Text Message Service (SMS), call routing, and authentication and storage of calls and associated account information are performed by these components.
4) The Operations and Support Subsystem (OSS).
5) A Subscriber Identity Module (SIM) card provides identification information about a mobile user to the network.

The sensors embedded use the cloud to store the data that they have collected. The data collected is often referred to as big data because of its size and category. This big data has three characteristics: variety (e.g., data types), velocity (e.g., data generation frequency), and volume (e.g., data size) [9]. This massive data must be stored somewhere for easy access, data transfer and integration, monitoring of response time and data processing, and – lastly – communication between multiple hardware devices [23]. An obvious question of data security and privacy arises when we are dealing with such large data. Cloud computing is one of the efficient and secure ways of storing data and delivering computing services over the Internet through Internet standards and protocols by using a pay-as-you-go pricing model.

Cloud computing services are offered at three distinct levels: the Software as a Service (SaaS) model, the Platform as a Service (PaaS) model, and the Infrastructure as a Service (IaaS) model [24]. IoT services for an application must be wisely chosen keeping in mind these models. SaaS is a business-centric software delivery model where applications are hosted by service providers and made available to customers on a subscription basis. PaaS provides hosted application servers to build custom applications as services. These servers have near-infinite scalability. It is not only a service provider but also offers necessary services like storage, security, integration, and development tools. IaaS is akin to traditional hosting wherein a business uses the hosted server as a logical extension of the remote data center. It supplies more flexibility in hardware configuration making it more expensive than the other two services. The software includes operating systems, application platforms, database servers, frameworks, and monitoring software [25].

Microsoft Azure offers all the benefits of PaaS and promises to be as flexible as IaaS. Azure is a cloud computing platform offered by Microsoft. It provides several services that enable customers to build products and solutions for their businesses. As a part of service computing, Azure supports the platform, infrastructure, and software. It offers services such as virtual machines running in the cloud, website and database hosting, and advanced computing services like artificial intelligence, machine learning, and IoT. Azure IoT services enable storage and manipulation of data along with updating devices to fix issues by sending software updates. There are multiple services under Azure IoT viz: Azure IoT Hub, Azure IoT Central, and Azure Sphere [25].

## 12.3   STATE OF ART METHODS

With this State of Art Methods section, we would like to bring knowledge to the problems in the vehicle management system.

- It is necessary to carry all vehicle-related documents, such as driving license or RC, while driving but at times, we forget to carry them. This causes heavy challans. In allusion to amendments made by the Road Transport and Highway Ministry in Vehicle Act enforced in 2019, the fine imposed on the driver for not carrying a driving license is Rs. 5,000; one can be charged up to Rs. 10,000 if they are not carrying the vehicle permit, and driving without insurance is fined at Rs. 2,000 [26].
- Rapid urbanization in India has increased the emission of several pollutants from vehicles depending on the quality of fuel and efficiency of the engine. The major pollutants released are carbon monoxide (CO), nitrogen oxides (NOx), photochemical oxidants, air toxins, hydrocarbons (HCs), particulate matter (PM), oxides of sulfur (SO2), polycyclic aromatic hydrocarbons (PAHs) [27]. As per the surveys conducted by various organizations, transportation is a major source of air pollution and the causes are due to numerous factors such as the use of non-eco-friendly vehicles, lack of vehicle maintenance, unburnt hydrocarbons, evaporation of fuels, and release of burnt fuel by-products. This leads to multiple environmental hazards such as climate change, global warming, acidic rain, and poor air quality affecting health adversely. Some of the health-related risks are mentioned in Table 12.1.

If the emission from the vehicle reaches beyond these permissible limits, it becomes a matter of concern and needs monitoring and controlling. To resolve this issue, the Indian Government proposed a plan where citizens must get the Pollution Under Certificate (PUC) for his/her vehicle every 3–6 months. According to the amended vehicle act, the vehicles without the valid PUC will have to pay a challan of Rs. 10,000 [26]. But there are some problems associated with this system as well. There are long queues at the station due to which people avoid going there and, in some cases, they even forget to get the pollution checked or they consider it a waste of time (Table 12.2).

**TABLE 12.1**
**Health Hazards Caused by Pollutants [28]**

| Pollutant | Organ Malfunction | Health Risks |
|---|---|---|
| Carbon Monoxide | Heart and brain | • Angina<br>• Infertility<br>• Anemia<br>• Vision Impairment |
| Nitrogen Oxides | Lungs and eyes | • Pulmonary disease<br>• Nose and throat irritations<br>• Dyspnea<br>• Respiratory disease |
| Sulfur Dioxide | Lungs and mucous membranes of eyes, throat, and nose | • Bronchitis<br>• Asthma<br>• Cardiovascular disease<br>• Respiratory disease |
| Particulate Matter and Respirable Particulate Matter (SPM and RPM) | Heart and lungs | • Heart attacks<br>• Pre-mature deaths<br>• Cardiac arrhythmias<br>• Asthma attacks |
| Lead | Liver, kidney, and brain | • Brain damage<br>• Hyperactivity<br>• Less concentration<br>• Stomach disorders |
| Benzene | Liver, kidney, lungs, heart, and brain | • Leukemia<br>• Anemia<br>• Weak immune system<br>• Respiratory tract irritation |
| Hydrocarbons | Heart, lungs, stomach, kidney, liver, and skin | • Cancer<br>• Pneumonia<br>• Seizures<br>• Irregular heart rates |

- Drunk driving is another serious concern in India. It was reported in 2020 that over 38,000 road accidents over the last three years were due to drunk driving [29]. Anyone with a blood alcohol level above 30 mg per 100 mL of blood, detected by a breathalyzer test, is driving under the influence of alcohol. The same applies to any person under the influence of drugs to such an extent that they are unable to exercise proper control over the vehicle [30]. Under the new amendment to the Motor Vehicles Act, the penalty for drunken driving has been increased from Rs 2,000 to Rs 10,000 for the first offense, and Rs 15,000 for the second offense. The amended law also established a prison sentence of up to two years for repeat offenders [30].
- The traditional toll fee collection method results in long queues, longer waiting times, and more traffic jams, increasing fuel emissions and leading to air pollution. Air Pollution Program Manager, Vivek Chattopadhyay, explained that the engine at the optimal speed (40 km/hour) releases less fuel than the vehicle moving at an idle speed. The engine emits twice as many pollutants and burns more fuel which gets wasted [31].
- In the last ten years (between 2011 and 2020) 307,000 vehicles were stolen in the capital of India, Delhi, as reported by the Delhi Police. The most common items stolen across the city after mobile phones are cars and two-wheelers. Vehicle theft cases are still the least-solved crime in India, with cases rising every year [32].

**TABLE 12.2**
**Permissible Limit of Pollutants from the Vehicle's Exhaust [28]**

| Vehicle Type | CO (gm/Km) | HC (gm/Km) | NOx (gm/Km) | SO$_2$ (gm/Km) | PM (gm/Km) |
|---|---|---|---|---|---|
| Two-Wheelers | 1.50 | 0.12 | 0.25 | 0.02 | 0.10 |
| Four-Wheelers (Petrol engine) | 2.25 | 0.28 | 0.40 | 0.04 | 0.10 |
| Four-Wheelers (Diesel engine) | 1.00 | 0.28 | 0.85 | 0.04 | 0.18 |
| Six-wheelers (Petrol engine) | 3.20 | 0.40 | 0.60 | 0.10 | - |
| Six-wheelers (Diesel engine) | 2.20 | 0.40 | 1.20 | 0.10 | - |

After discussing various issues in vehicle management, we tried to provide a solution with the help of a flowchart for the above problems in Figure 12.2.

The working mechanism for providing a solution to the problems by referring to the RFID reader and sensors present in the vehicle management system keeps reading the data as long as the vehicle's engine is on. Before interpreting any reading, it is important to check whether the vehicle is registered in the RFID device database. If it is not registered, the driver must be prompted to register himself.

On verifying the vehicle record in the RFID device database, the readings by the sensor and RFID reader are checked. There will be three sensors: MQ-2 and MQ-7 measure the air quality of the vehicle's exhaust, and MQ-3 detects the alcohol content in the driver's breath.

Condition one, where the reading is offered by the gas-detecting sensors MQ-2 and MQ-7, the content is displayed on the vehicle's dashboard meter and is compared with the permissible limits. If the content surpasses the limit, the driver is alerted through the phone, and in case the driver ignores the warning for up to three days, an SMS is sent to the RTO using GSM for the license cancellation. Otherwise, if the gas content is within the permissible limit, the content is displayed on the dashboard, and the process continues.

Condition two, where the alcohol detecting sensor MQ-3 offers the reading, the alcohol content in the driver's breath is displayed on the phone. If the content surpasses the predefined limit, the engine is turned off and a warning is sent to the driver's phone, and an SMS is sent to the RTO using GSM to take necessary action.

We also tried to cover a few more additional features, like automatic toll deduction and vehicle tracking in case of theft. The infrared sensors at the toll plaza detect the vehicle's presence. The RFID tag is activated, and the RFID reader scans the tag. It verifies the RFID number in the RFID device database; if the vehicle is registered and there is a sufficient balance in their e-wallet, the tax is automatically deducted, and the customer information along with the time and date of toll deduction is logged in the customer database. On successful payment, the user receives an alert on the registered mobile number. If there is insufficient balance, the driver is directed to another lane to pay the tax manually.

In the case of vehicle theft, the driver registers the complaint through the mobile application. The vehicle's location is immediately traced using GPS. The vehicle's engine is turned off, and its exact location along with the time and date is sent to the police station for strict and prompt action.

## 12.4 PROPOSED SYSTEM ARCHITECTURE

In pursuing the addressing solution to the problem, the proposed architecture is divided into certain phases. The first phase includes the user and cloud, followed by RFID and Arduino in the second phase. The third phase deals with all the sensors, and the last – fourth – phase establishes the vehicle satellite connection mentioned in Figure 12.3.

Initially, it is required that the user first registers the vehicle as soon as it is purchased. The RFID tag fitted on the vehicle is scanned using the RFID reader [33] to authenticate the purchase by

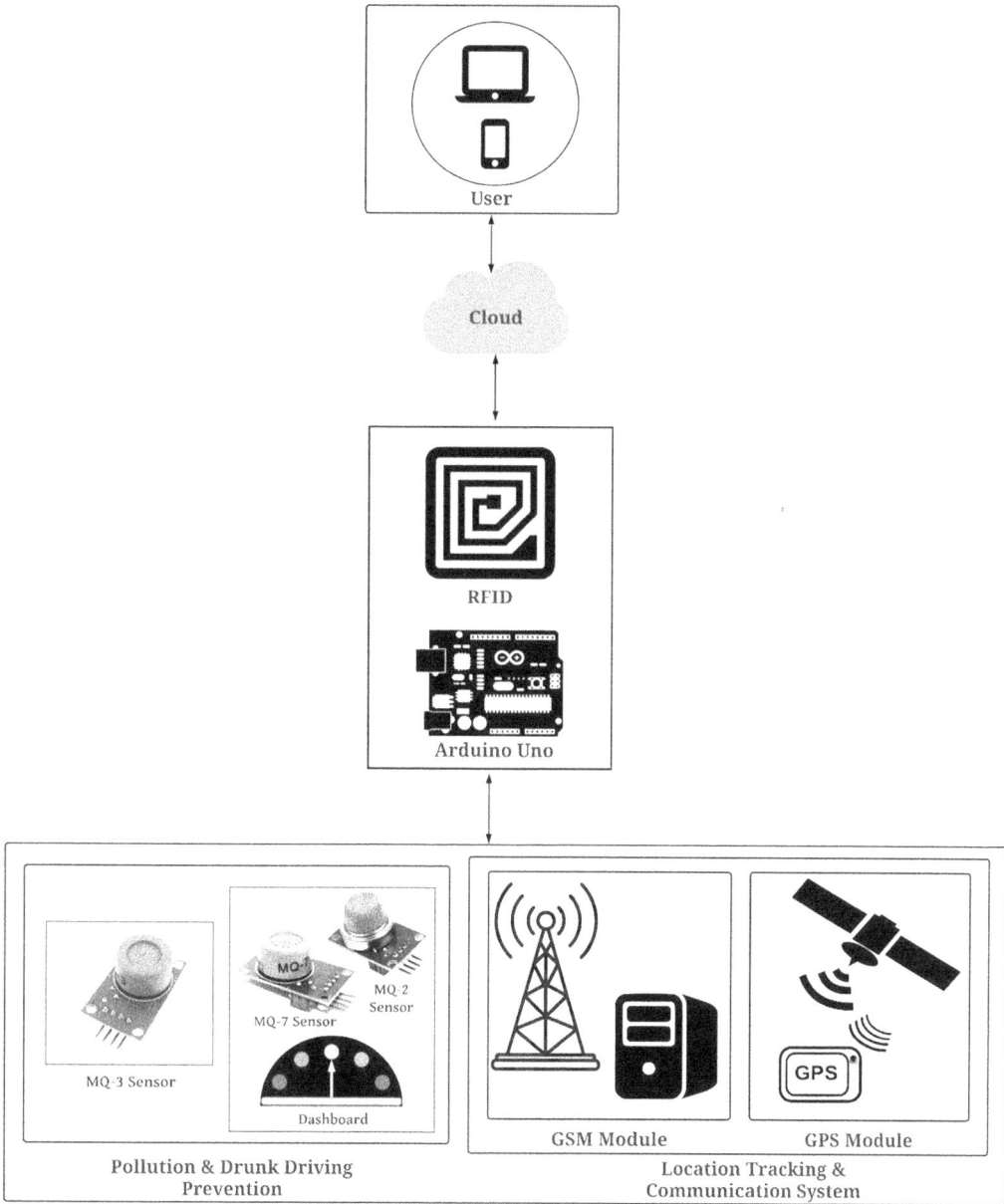

**FIGURE 12.3**   RFID-enabled Vehicle Management System Architecture.

generating a unique ID used by that vehicle only. As a first step, an account needs to be maintained with that unique ID where the user must fill all their details in the (initial/commence) registration form like their Name, Address, Age, Phone Number, E-Mail ID, Aadhar Card Number, and Bank Details which will be stored on the cloud. The user will be asked to install the mobile application and log in using the registered mobile number authenticated by a one-time password (OTP) that they must enter for successful login. The RFID reader stores all the vehicle-related information decoded in the RFID tag on the cloud. All the documents such as RC and insurance papers can be accessed through the same application that has access to the cloud. Users will also receive a notification alert of all the pending services from time to time. Another important feature is that it keeps a record of all the vehicle owners in case of second-hand or third-hand purchases with their complete details to avoid any fraudulent activity.

The proposed architecture works in the following phases.

*Phase 1 – User and Cloud*: The user will access the whole management system using a 'Parivem' mobile application. All the vehicle-related documents such as RC, PUC, and insurance papers, readings by the sensors, toll tax deductions, alerts, service reminders, and the vehicle's location will be stored on the cloud, which is accessible through this mobile application. All the information read by the sensors will also be stored on the cloud to be reviewed later and shared with the concerned authorities. By using IoT services offered by Azure, the sensors can connect to the Internet cloud and send their reading to a specific endpoint in Azure through a message. This message is then collected, aggregated, and converted into reports and alarms.

*Phase 2 – RFID and Arduino*: This system comprises the RFID reader that scans the RFID tag fitted on the vehicle and stores all the scanned information on the cloud.

There are three ways of embedding RFID tags in vehicles:

- Windshield tags adhere to the vehicle's windshield with a permanent or semi-permanent adhesive. They have a read range of 20 feet.
- Hangtags/rear-view mirror tags are hung from the vehicle's rear-view mirror. They have a read range of 20 feet in ideal conditions.
- RFID license plate tags are placed on the license plate of the vehicle. They have a read range of 50 feet in ideal conditions.

In our architecture, we are using RFID for license plate tags that will be suitable for all vehicles. Another functionality of the RFID reader is at the toll plaza. It scans the RFID tag fixed on the vehicle that contains all the information about the vehicle and its owner, such as the transponder's number, patron's account, bank details, balance, etc. After scanning, these details are checked for verification in the RFID database. If the vehicle is registered and there is a sufficient balance, the tax is deducted; otherwise, the user is proceeded to pay the tax manually. The notification of successful payment and the receipt are automatically sent on the phone and the e-mail ID of the owner. Arduino Uno, the microcontroller, controls, and monitors the entire connection. The readings by the sensor are offered to Arduino, which checks the values against the permissible limits and performs the necessary action.

*Phase 3 – Sensors*: This unit comprises MQ-7, MQ-2, and MQ-3 sensors [34], and the dashboard that will be installed on the vehicle. These sensors are used to detect the amount of carbon monoxide (CO) and other gases such as Nitrogen and Sulfur Dioxide, respectively. These two sensors are fitted in the gas outlet of the vehicle to constantly monitor the pollutant levels and report the readings to the microcontroller, Arduino Uno.

Then Arduino checks these readings against the predefined limits to determine the dangerous content in the smoke discharged. A pollution-level meter is fixed on the vehicle's dashboard to display the pollution status, which can also be accessed through the application. The constant monitoring of the pollution level and their timely controlling help curb the pollution levels and cut down the time and cost spent on getting the PUC every three to six months. This phase also consists of an MQ-3 sensor with high sensitivity [35] and a fast response toward alcohol. It analyses the amount of alcohol in the driver's breath and sends the reading to Arduino Uno's microcontroller, which then checks it against the permissible limits.

*Phase 4 – Vehicle Satellite Connection*: The last phase comprises GPS and GSM modules that are used to track a vehicle's location using satellites and establish a cellular connection of the entire system. In case of theft, the user files a complaint by pressing a button on the application. The Arduino gets activated and the GPS starts recording the exact longitude and latitude of the position of the vehicle and sends the information to the GSM module. The SMS is then sent to the police station via the GSM module so that immediate action can be taken.

## 12.5  CONCLUSION

The proposed system describes an effective vehicle management system that makes customer's life much easier and automates most vehicle-related tasks that save time and cost. It enables an automatic pollution monitor using gas-detecting sensors and the prevention of drunken driving using an MQ-3 sensor. This system also takes care of the security aspect of the vehicle. In case of any thefts, the vehicle's location can be immediately traced using GPS and sent an alert to the police station for immediate action using GSM. Apart from this, it uses RFID technology for vehicle attestation and enables automatic toll tax deductions without any contact. Documents such as RC, PUC, insurance papers, information regarding services-due alerts, and petrol and pollution levels are stored on the cloud which increases scalability, flexibility, and ease in accessibility. Sensors transport all the data to the cloud for data analysis. All this information can be tracked through the mobile application. Thus, it provides full access control by monitoring surveillance.

## REFERENCES

[1] https://www.statista.com/statistics/607818/vehicle-production-volume-by-segment-india/
[2] https://www.ibef.org/industry/india-automobiles.aspx
[3] https://auto.economictimes.indiatimes.com/news/aftermarket/3564-accidents-in-india-due-to-potholes-in2020/
[4] https://en.wikipedia.org/wiki/ArduinoUno
[5] Kay Li Ng, Choo W.R. Chiong, and Regina Reine, Vehicle Recognition System using RFID Technology for Parking Management System. In *IOP Conference Series: Materials Science and Engineering*, Volume 495, April 2019, pp. 012022, ISSN: 1757-8981, DOI: 10.1088/1757-899X/495/1/012022
[6] https://en.wikipedia.org/wiki/Radiofrequencyidentification
[7] https://en.wikipedia.org/wiki/Global_Positioning_System
[8] https://en.wikipedia.org/wiki/GSM
[9] Hany F. Atlam, Ahmed Alenezi, Abdulrahman Alharthi, Robert J. Walters, and Gary B. Wills, Integration of Cloud Computing with the Internet of Things: Challenges and Open Issues. *2017 IEEE International Conference on Internet of Things (iThings) and IEEE Green Computing and Communications (GreenCom) and IEEE Cyber, Physical and Social Computing (CPSCom) and IEEE Smart Data (SmartData)*, 2017.
[10] https://docs.microsoft.com/en-us/archive/msdn-magazine/2010/february/cloud-computing-microsoft-azurefor-enterprises
[11] R. Litty, G. Alpana, P.R. Divya, and R. Surya, A Survey on RFID Based Vehicle Authentication Using a Smart Card. *International Journal of Computer Engineering In Research Trends*, Volume 4, Issue 3, March 2017, pp. 106–110 ISSN (O): 2349-7084.
[12] Wei Wang, and Shidong Fan, *Shanghai Maritime Academy, RFID Technology Application in Container Transportation*. The Institute of Electrical and Electronics Engineers (IEEE), February 2010, DOI: 10.1109/JCPC.2009.5420106
[13] Fawzi M. Al-Naima, and Haider S. Hatem, Design of an RFID Vehicle Authentication System: A Case Study for Al-Nahrain University Campus. ("Design of an RFID Vehicle Authentication System: A Case"). *International Journal of Scientific and Technological Research*, Volume 1, Number 7, October 2015, ISSN: 2422-8702.
[14] S. Dharanya, and A. Umamakeswari, Embedded Based Conveyance Authentication and Notification System. ("Embedded Based Conveyance Authentication and Notification System"). *International Journal of Engineering and Technology (IJET)*, Volume 5, Number 1, March 2013, ISSN: 0975-4024.

[15] R. Sharmila Gowri, M. Gayathri, K. Bhuvaneswari, and Devi B. Kowsalya, Automated License Controlled Vehicle with Air Pollution Monitor and Control. *International Journal of Engineering Research and Technology (IJERT)*, Volume 5, Issue 7, 2017, ISSN: 2278-0181.

[16] B. Patil, H. Amrite, K. Gaikwad, J. Dighe, and S. Hirlekar, Smart Car Monitoring System Using Arduino. *International Research Journal of Engineering and Technology (IRJET)*, Volume: 05, Issue: 03, March 2018, e-ISSN: 2395-0056, p-ISSN: 2395-0072.

[17] N. Harathi, and V. Meenakshi, Smart Setup for Gas Detection Using Ardunio. *International Journal of Latest Trends in Engineering and Technology*, Volume 10, Issue 1, March 2018, pp. 238–243, DOI: 10.21172/1.101.42, e-ISSN: 2278-621X.

[18] MQ2 Arduino Gas Sensor: Datasheet, Working and its Applications, watelectronics.com

[19] Parmar Veenita, RFID Technology for vehicle identification. *International Journal of Advanced Engineering and Research Development*, Volume 4, Issue 1, January 2017, e-ISSN (O): 2348-4470, p-ISSN (P): 23486406.

[20] D.G. Jha, and Buva Swapnil, Alcohol Detection in Real-Time to Prevent Drunken Driving. *IOSR Journal of Computer Engineering (IOSR-JCE)*, 2018, e-ISSN: 2278-0661, p-ISSN: 2278-8727, pp. 66–71.

[21] G. Pavan Acharya, Bhat S. Shishira, S. Sahana, and A. Upadhyaya, Smart Vehicle Protecting System. *International Journal of Scientific Development and Research (IJSDR)*, Volume 2, Issue 6, June 2017, ISSN: 2455-2631.

[22] P. Kaur, A. Das, M.P. Borah, and S. Dey, Smart Vehicle System using Arduino. *ADBU Journal of Electrical and Electronics Engineering (AJEEE)*, Volume 3, Issue 1, May 2019, ISSN: 2582-0257.

[23] https://www.rapyder.com/blogs/role-of-cloud-computing-in-iot/

[24] Integration of Cloud Computing with the Internet of Things: Challenges and Open Issues, June 2017, DOI: 10.1109/iThings-GreenCom-CPSCom-SmartData.2017.105

[25] https://docs.microsoft.com/en-us/archive/msdn-magazine/2010/february/cloud-computing-microsoft-azurefor-enterprises

[26] https://www.livemint.com/news/india/new-traffic-rules-how-to-avoid-paying-fine-if-you-forget-drivinglicence

[27] S. Dey, and N.S. Mehta, Automobile Pollution Control Using Catalysis ("Dey, S. and Mehta, N.S. (2020) Automobile Pollution Control Using …") Resources, Environment and Sustainability, Volume 2, December 2020, 100006, DOI: 10.1016/j.resenv.2020.100006

[28] Status of the Vehicular Pollution Control Programme in India (March 2010) Central Pollution Control Board (Ministry of Environment and Forests, Govt. of India) Programme Objective Series PROBES/136 /2010, http://www.indiaenvironmentportal.org.in/files/status%20of%20the%20vehicular%20pollution.pdf

[29] https://www.sundayguardianlive.com/news/drunk-driving-led-38000-road-mishaps-three-years

[30] https://comtransport.assam.gov.in/frontimpotentdata/rules-against-drunk-driving-cases

[31] https://theprint.in/theprint-essential/how-fastag-has-become-the-easier-faster-and-environment-friendly-wayto-collect-toll-fees/

[32] https://www.hindustantimes.com/cities/delhi-news/city-of-cars-delhi-reports-maximum-vehicle-thefts-95stolen-everyday-ncrb-report-101631730601005.html

[33] A. Srivastava, A. Gupta, and R. Anand, Optimized Smart System for Transportation Using RFID Technology. *Mathematics in Engineering, Science and Aerospace (MESA)* Volume 12, Issue 4, 2021, 953–965.

[34] J. Kaur, N. Sindhwani, and R. Anand. "Implementation of IoT in Various Domains." In *IoT Based Smart Applications* (pp. 165–178). Springer International Publishing, USA, 2022.

[35] N. Sindhwani, R. Anand, R. Vashisth, S. Chauhan, V. Talukdar, and D. Dhabliya, "Thingspeak-Based Environmental Monitoring System Using IoT," in *2022 Seventh International Conference on Parallel, Distributed and Grid Computing (PDGC)*, 2022, pp. 675–680, DOI: 10.1109/PDGC56933.2022.10053167

# 13 Building Integrated Systems for Healthcare Considering Mobile Computing and IoT

*Rohit Anand*
G. B. Pant DSEU Okhla-1 Campus (formerly GBPEC), New Delhi India

*Ashy V. Daniel and A. Lenin Fred*
Mar Ephraem College of Engineering & Technology, Marthandam, India

*Tarun Jaiswal*
National Institute of Technology, Raipur, India

*Sapna Juneja and Abhinav Juneja*
KIET Group of Institutions, Ghaziabad, India

*Ankur Gupta*
Vaish College of Engineering, Rohtak, India

## CONTENTS

13.1 Introduction ..................................................................................................204
    13.1.1 Healthcare .........................................................................................204
        13.1.1.1 Applications of Technology in Healthcare ........................205
    13.1.2 IoT ......................................................................................................206
        13.1.2.1 Characteristics of IoT ........................................................206
        13.1.2.2 IoT: Things, Internet, and Human ....................................207
        13.1.2.3 Benefits of IoT ...................................................................207
    13.1.3 Mobile Computing .............................................................................209
        13.1.3.1 Elements of Mobile Computing .........................................209
        13.1.3.2 Advantages of Mobile Computing .....................................209
        13.1.3.3 Limitations of Mobile Computing are Discussed Ahead ......210
    13.1.4 Impact of Mobile Computing and IoT on Healthcare ........................210
        13.1.4.1 End to End Connectivity and Affordability .......................210
        13.1.4.2 Improving the Health of Patients .......................................211
        13.1.4.3 Simultaneous Monitoring and Support ..............................211
        13.1.4.4 Tracking and Alerts ...........................................................211
        13.1.4.5 Research .............................................................................211
        13.1.4.6 Data Analysis ....................................................................211

DOI: 10.1201/9781003319238-13

13.2  Literature Review ...................................................................................................211
13.3  Role of Deep Learning in Healthcare ............................................................213
    13.3.1  Deep Learning and Medical Imaging ..............................................214
    13.3.2  Diabetic Retinopathy (DR) ..............................................................215
    13.3.3  Gastrointestinal (GI) Disease ..........................................................215
    13.3.4  Tumor Detection ...............................................................................216
13.4  Problem Statement ..........................................................................................216
13.5  Proposed Methodology ...................................................................................216
    13.5.1  Flowchart of Proposed Work ...........................................................216
13.6  Results and Discussion ...................................................................................216
    13.6.1  Time Consumption ...........................................................................216
    13.6.2  Error Rate .........................................................................................217
    13.6.3  Accuracy Rate ..................................................................................217
13.7  Conclusion ......................................................................................................217
13.8  Future Scope ...................................................................................................218
References ....................................................................................................................221

## 13.1  INTRODUCTION

The only ways for doctors to communicate with their patients were through in-person visits, phone calls, and text messages prior to the advent of mobile computing and the Internet of Things (IoT). However, things are now beginning to shift. By remotely monitoring their patients with the help of IoT-connected gadgets, mobile computing and IoT have enabled doctors to provide exceptional treatment [1–3].

### 13.1.1  HEALTHCARE

Health care, or healthcare, is aimed at preventing, diagnosing, treating, alleviating, and curing illness, sickness, accident, and other physical and mental problems. Examples of medical healthcare are shown in Figure 13.1. Patients can get help from a wide range of medical and associated professionals. Healthcare encompasses the entire range of medical, dental, pharmacy, midwifery, nursing, optometry, audiology, psychology, occupational therapy, physical therapy, and athletic training. Included are efforts made in both private practice and the public sector, at all levels of care delivery [4].

A person's ability to receive medical attention may be influenced by their social and economic standing. One definition of caregiving is "the efficient application of individual health services in pursuit of an optimum state of health." Lack of financial resources, physical distance, and the constraints imposed by the individual all operate as roadblocks to receiving healthcare. Limiting access to healthcare has negative effects on people's health, treatment outcomes, and quality of life as a whole. The needs of specific populations are met by establishing healthcare systems [5].

**FIGURE 13.1**  Examples of Medical Healthcare.

According to the World Health Organization (WHO), effective healthcare systems require resources (such as money and trained personnel), information (in the form of reliable data on which to base policy), and infrastructure (healthcare facilities that are both well-maintained and stocked with high-quality pharmaceuticals and technological aids). There may be a positive correlation between a country's healthcare system and its GDP, level of development, and level of industrialization. It is commonly held that healthcare is just one of many aspects that contribute to people's entire physical and emotional health and well-being [6–8]. The WHO declared smallpox eliminated in 1980, making it the first disease in human history to be eradicated through deliberate healthcare measures [9].

### 13.1.1.1 Applications of Technology in Healthcare

There are a variety of ways technology may be used to improve healthcare.

- Electronic medical records storage and retrieval.
- Fears regarding the safety of medical records have been allayed.
- Health information handling.
- Genome management in the clinic.
- Monitoring the information contained within electronic health records (EHRs).

A general healthcare ecosystem is shown in Figure 13.2.

#### 13.1.1.1.1 Research Purposes

At present, EHRs are locked down to a single organization or network, preventing their owners from updating or sharing them with anybody else. The data might be organized in this way, with non-protected health information (non-PHI) or Personally Identifiable Information (PII) kept in the

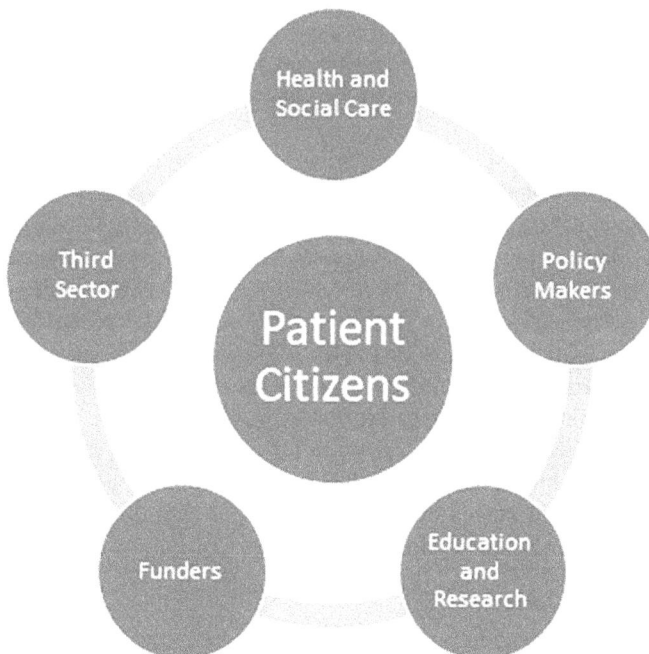

**FIGURE 13.2** Healthcare Ecosystems.

first few blocks of the blockchain (PII). Researchers and businesses of the future may have access to populations numbering in the millions [10].

### 13.1.1.1.2  Seamless Switching of Patients between Providers

Health records stored on the blockchain may theoretically be unlocked and transferred between individuals and institutions via a shared private key. This may lead to an increase in the capacity of Health Information Technology (HIT) for interoperability and collaboration [11].

### 13.1.1.1.3  Faster, Cheaper, Better Patient Care

A centralized database that is routinely updated by authorized individuals could house patients' medical records. In many cases, errors could be avoided if different doctors working on the same patient shared information. This could lead to more individualized care for patients [12].

### 13.1.1.1.4  Interoperable Electronic Health Records

As long as the blockchain continues to use a consistent set of data and secure encrypted links to supplementary data, a unified layer of transactions may be made available. Using smart contracts and standardized authorization processes could help keep gadgets networked [13].

## 13.1.2  IoT

Kevin Ashton originally coined the term "Internet of Things" (IoT) in a presentation to Procter & Gamble about the usage of radio-frequency identification (RFID) in the company's supply chain. Innovations in the IoT require the widespread use of new technologies that automatically link together all smart objects in a network. On the other hand, everything that can be used in a new way after being connected to the internet, it is possible for an IoT device to gather and send information. In recent years, the IoT has become a hot topic for academics to study. As part of the IoT revolution, technological and social components are being merged into existing healthcare systems. The future of the economy and society, it's a radical departure from the current healthcare system and is transforming it from the ground up. Healthcare that is less complicated, has better outcomes, and is more closely monitored. The importance of IoT in healthcare systems is growing as it enables better care at lower costs and more satisfying experiences for patients. There are many potential applications for this technology [14–16].

The term "Internet of Things" is used to describe an approach to doing business that allows for interoperability across many types of smart devices. Many different technologies from many different fields can be used to create hybrid experiences that integrate digital and physical interactions. Smart homes are a common IoT application because they combine sensors, hardware, software, and computer platforms. The term "smart" is currently used to describe a wide variety of high-tech devices, not just those typically seen in the home, that incorporate artificial intelligence (AI) of some form. System integration, analysis, and automation are all made easier with the help of the IoT. By doing so, they broaden and refine the scope of these fields. IoT technologies include sensors, networks, and automated systems. The expansion of the IoT is due to several factors, including the development of better software, the decrease in price of hardware, and the acceptance of new technologies. They have a significant impact on the distribution of goods and services. Energy, water, and other vital natural resources are being depleted at an alarming rate, making environmental monitoring through the IoT essential to their preservation [17].

### 13.1.2.1  Characteristics of IoT

1. Connection: Little explanation is needed here. IoT devices that rely on sensors have been linked to the web. A combination of hardware and software is required for proper use of IoT devices [18].

2. Things: Connectable objects, or "things," are those that were made to be joined together. It's not uncommon to utilize sensors and other commonplace electronics to keep tabs on pets. These can detect and react to shifting environmental conditions [19].
3. Data: The data is what makes the IoT work. After data is collected, the nodes in the sensing network share it for use in making decisions. Smarter devices acquire their functionality from pre-programmed actions based on analyzed data [20].
4. Communication: The interconnected nature of IoT devices allows for data transmission and analysis. Information could be transmitted across extremely short distances. It's possible that information could travel a great distance. Like Low Power Wide Area (LPWA) networks like LoRa or NB-IoT, or Wi-Fi [21].
5. Intelligence: IoT devices with sensors may detect and respond to changes. Now, IoT devices can learn and adapt. This clever technique allows the IoT device to perform huge data analytics [22].
6. Action: A smart system has been acknowledged as having completed a step or task. Taking action could be done by hand or automatically. Taking the next step involves trusting the system's shrewd assessments [23].
7. Ecosystem: Location of the IoT. The IoT ecosystem analyses its underlying technologies, goals, and underlying concept [24–26].

### 13.1.2.2 IoT: Things, Internet, and Human

IoT (as shown in Figure 13.3) is the result of interaction among three variables, i.e., things or objects, the Internet, and human beings.

Adoption of the IoT is a gradual process. As a matter of fact, it was a rather gradual procedure. It is also possible that in the not-too-distant future, AI may be integrated into IoT-based products and apps in order to boost their functionality [27].

### 13.1.2.3 Benefits of IoT

The benefits of IoT (as shown in Figure 13.4) are discussed here.

a. **Protection:** Financial management organizations need procedures and arrangements to proactively address these risks, as well as a convention for handling exceptional circumstances. We need to go back to our information consumption and upkeep strategies. The goal is to guarantee consumer safety and abide by appropriate guidelines without sacrificing the value of IoT data [28].

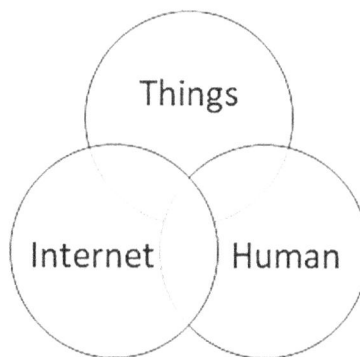

**FIGURE 13.3** Depicts Tri-sectional Intersection among Things, Internet and Human.

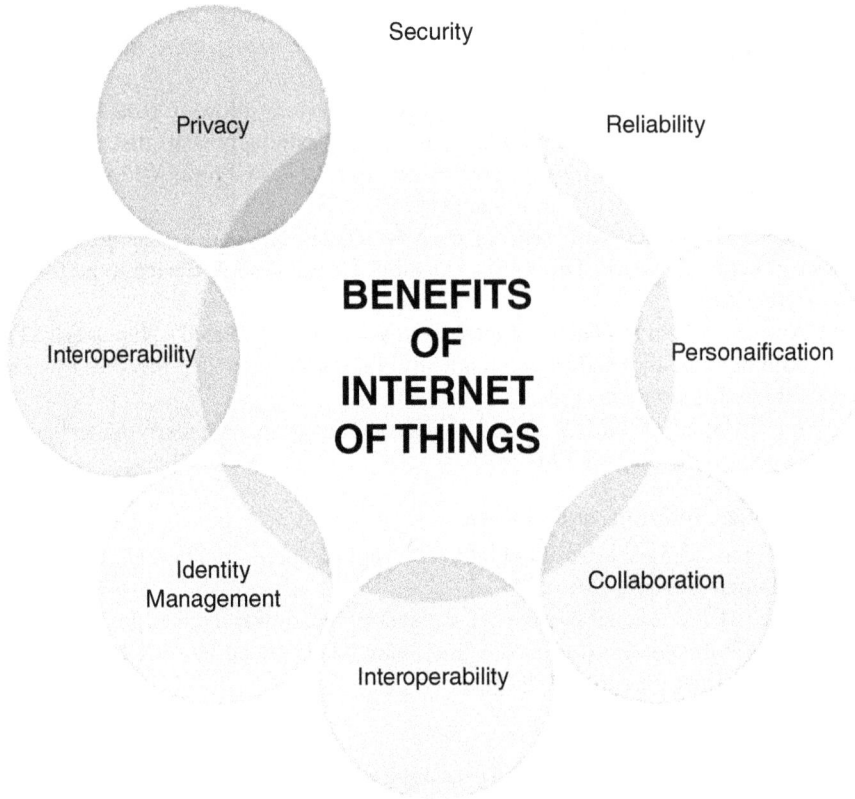

**FIGURE 13.4**  Benefits of IoT.

b. **Security:** The confidentiality of all data collected, distributed, and tracked is of the utmost importance. Most technology will be novel and partially untested in real life, making users and their actions somewhat more exposed [29].

c. **Identity Management:** As the number of electronic devices in the average home rises, so does the complexity of the web of relationships between those devices and the people who live there. Autos, health tools, and home appliances are just a few examples of gadgets that could have many users. Conventional, personality-based board configurations are inadequate for handling the intricacies inherent in these interdependencies [30].

d. **Interoperability:** Companies are rushing to market with new ways to transmit and coordinate these devices in cars, machinery, systems, computers, mobile phones, wearables, and other devices. The IoT experience will fall flat without compromises on interoperability principles. Even if there are many steps taken toward institutionalization, the process is still anticipated to take some time [31].

e. **Joint effort:** IoT devices, especially enabled devices, need to share their internal settings with one another and their users in order to realize their full potential. Once compatibility difficulties are resolved, these devices will be able to efficiently share information and perform the duties of a value-added appliance [32].

f. **Personalization:** As these tools can aid in the medical industry's ability to predict and prevent problems, their impact will be significant [33].

g. **Unwavering quality:** Basic instances in which electronics fail or become disconnected must be addressed and resolved. Goals and reinforcement plans should be in place for every-day emergencies such as automobile accidents, water holes, fires, and health checks [34].

### 13.1.3 MOBILE COMPUTING

Data, voice, and video can all be transmitted from a mobile computer to another computer or wire-lessly equipped device. This doesn't need to rely on any kind of hardwired connection. It allows people to go from one place to another while talking to each other. Only because of advancements in mobile computing have we been able to view and transmit data from any faraway areas without really being there. The range of modern mobile computing devices is extremely large. It's a super-efficient and trustworthy part of the IT industry [35].

#### 13.1.3.1 Elements of Mobile Computing

The major elements of mobile computing (as shown in Figure 13.5) are mobile communication, mobile hardware, and mobile software.

A. **Mobile Communication:** Connectivity and communication can only function properly if the proper infrastructure is in place, and this includes things like wireless network architec-ture, protocols, data formats, bandwidths, and portals [36].
B. **Mobile Hardware:** The hardware consists of the various mobile computing devices and their ancillary devices, which can run the necessary software and establish network con-nections [37].
C. **Mobile Software:** Operating systems, the "brain" of any computer system, are the most crucial piece of software. Operating systems include Windows, Linux, and macOS on lap-tops, and Android and iOS on mobile devices. Mobile software also includes the various apps installed on a smartphone [38].

#### 13.1.3.2 Advantages of Mobile Computing

a) **Portability:** The freedom of movement is undoubtedly the most attractive feature of mobile computers. Thanks to the development of mobile computing, advanced processing power can now be carried by the average person. Now, you may use any piece of software or browse the web without ever leaving the comfort of your own home [39].
b) **Affordability:** Each year brings cheaper and more feature-rich mobile computing devices. People with little financial resources can nonetheless afford to purchase a smartphone.

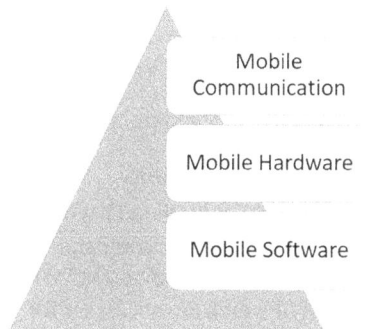

**FIGURE 13.5**  Elements of Mobile Computing.

When prices drop, more people can afford to use them, which finally helps bridge the digital gap [40].

c) **Data Access:** With the advent of the Internet and the proliferation of mobile computing devices, people now have access to an unprecedented wealth of information. The days of scouring thick tomes in a library for information in order to write a research paper, of searching through phone books, and of stopping at every street corner to ask for directions, are done [41].

d) **Increased Productivity:** The aforementioned advantages allow for a significant boost in output. The ability to conduct business from one's own residence is available 24/7. Due to the abundance of resources at their disposal, they are capable of learning and mastering any skill set [42].

e) **Entertainment:** Without Netflix and Amazon Prime, many of us would have snapped during the lockdown times. It is now very easy to get your hands on a large range of movies, as well as instructional and informational content. Nowadays, anywhere with access to a reasonably priced high-speed data connection may host a live broadcast of virtually any content imaginable [43].

f) **Cloud Computing:** The potential of mobile computing has been vastly expanded by developments in cloud computing. Mobile computing device constraints can be alleviated by storing data and using apps in the cloud, which is accessible from any Internet-connected device [44].

### 13.1.3.3   Limitations of Mobile Computing are Discussed Ahead

1) **Security:** The proliferation of mobile devices has raised a number of security problems. Leaving a device online all the time leaves it open to attacks. It's become more and harder to rein in cybersecurity issues like data breaches and immoral actions like hacking, piracy, etc. [45].

2) **Issues with Connectivity:** The vast majority of a mobile device's features require either Wi-Fi or mobile network connectivity. Most apps now require access to the Internet in order to work, rendering gadgets useless without it [46].

3) **Device Size Limitations:** Computers designed to be carried about need to be compact in order to meet this requirement. Features like processing speed, storage space, and display quality are all capped by the device's small form factor [47].

4) **Power Consumption:** The lifespan of rechargeable batteries is always bounded by the time before they must be recharged. Because of the risk of insufficient or no access to electricity, mobile computing devices should be treated with caution [48].

5) **Dependency:** As was discussed above, there are limitations to mobile computing devices that make our rising reliance on them risky [49].

### 13.1.4   IMPACT OF MOBILE COMPUTING AND IoT ON HEALTHCARE

This section focuses on the major impact of IoT and mobile computing on the healthcare industry.

### 13.1.4.1   End to End Connectivity and Affordability

Patient care workflows can be easily automated by the IoT with the use of healthcare mobility solutions, next-generation healthcare facilities, and other comparable technologies [50].

It has the potential to facilitate machine-to-machine (M2M) connection, interoperability, data mobility, and information exchange, all of which are necessary for the healthcare industry to provide an efficient service. In addition, by reducing duplicated efforts, prioritizing high-quality assets, and optimizing allocation, technology-driven solutions can significantly reduce overheads [51, 52].

### 13.1.4.2 Improving the Health of Patients

If something could alert doctors to when a patient's heart rate is abnormal or if they aren't taking care of their health, it would be a huge aid to them while they are at work [53].

To this end, IoT can ensure that every last detail is taken into account when making treatment decisions for patients by automatically updating their personal data in the cloud and removing the need to add the information into EMRs. In addition, it can be utilized as a home monitoring and medical adherence aid. Here we see how the IoT is influencing people's daily routines [54].

### 13.1.4.3 Simultaneous Monitoring and Support

During medical situations such as heart attacks, asthma attacks, diabetes, etc., real-time monitoring with the aid of linked devices can save many lives. The data connection allows the IoT devices to gather all the health-related data and send it to the doctor in real time using a smart medical device connected to a smartphone application. The IoT gadget can be used to record and transmit information about the user's health, including vitals, glucose levels, oxygen saturation, body mass index, and electrocardiograms [55].

Cloud storage makes it simple to back up this information and share it with your doctor, insurance company, or outside experts as needed. And this will facilitate a more rapid and thorough comprehension of the problem [56].

### 13.1.4.4 Tracking and Alerts

IoT devices have the potential to drastically improve the outcome of several potentially fatal circumstances. Medical IoT devices can collect crucial data and transmit it to clinicians for real-time tracking by sending push notifications to mobile apps or other associated devices [57].

These alerts provide a definitive assessment of the patient's condition at any time and from any location. Better decisions and timely care are made possible by this advancement. So, the Combination of IoT & mobility is a future trend of the healthcare niche [58].

### 13.1.4.5 Research

The IoT can play a crucial role in scientific inquiry. It's mostly due to the fact that IoT lets us compile a massive amount of data about the patient's disease and health that would take years to gather by hand. These records can then be used for further statistical analysis and research. In this way, IoT not only saves lives, but it also prevents waste of resources that may be used toward medical and scientific discovery [59].

### 13.1.4.6 Data Analysis

It is difficult to store and manage the massive volumes of data sent by a healthcare device in a short amount of time if cloud services are not available. Acquiring data in real time from multiple devices and then manually interpreting it presents a challenge even for medical professionals. The data may be collected, stored, and analyzed in real time, all of which are areas where IoT devices can shine. Because of this, IoT app development companies have become extremely popular. All of this will occur on the cloud, with only the results being shared with the doctors [60].

## 13.2 LITERATURE REVIEW

The authors in [1] focused on diabetic retinopathy in fluorescein angiography photographs, identified automatically. Diabetic retinopathy (DR) was detected, and its severity assessed, using cutting-edge convolutional neural networks (CNNs) and denoising methods applied to images obtained by fluorescein angiography. EyePACS supplied data, which consisted of fudus photos annotated by physicians to indicate varied degrees of DR severity. The 5-class severity classification task was best handled by a CNN classifier built from GoogLeNet, which achieved an area under the curve (AUC) of 0.79 and an accuracy of 0.45. To the best of our knowledge, no other published work on

DR screening works with a dataset of our size, making our article an improvement above previous work in both scale and heterogeneity.

The authors in [4] introduced multiple-source evidence-based computational gene prediction. This article details a computational technique for building gene models using information gathered from many sources, including those often used in a genome annotation workflow. The software, dubbed Combiner, accepts as input a genomic sequence and the sites of gene predictions from ab initio gene finders, protein sequence alignments, expressed sequence tag and cDNA alignments, splice site predictions, and other data. In order to determine the best way to combine evidence in the Combiner, three distinct algorithms were constructed and put to the test on 1783 Arabidopsis thaliana verified genes. Consistently outperforming the best individual gene finder, and in some instances producing huge gains in sensitivity and specificity, is what they see when we combine gene prediction evidence.

The authors in [5] reviewed perspectives on developments in EEG single-trial detection and discrimination during the 2003 BCI (Brain–Computer Interface) competition. Therefore, BCI technology may provide a new means of communication and control to those who are unable to do so via conventional means. The improvement of BCI technology requires the use of signal processing and categorization techniques. For the purpose of gauging how far down the technology curve BCIs have come, we held the BCI Competition 2003. Six datasets were submitted in a well-documented manner by four labs with experience in EEG-based BCI research. We published online both the data sets themselves and their explanations, including both labeled training sets and unlabeled test sets. To win the competition, participants needed to get the highest possible performance on the test labels' performance metric. Researchers from all across the globe competed to see who could get the best categorization results from their algorithms. The report details the six datasets used, along with the findings and functionality of the top methods.

The authors in [9] presented epilepsy seizure prediction using multivariate time series analysis: a statistical test, Chaos Interdiscip. The dynamics of electroencephalography have been reported to alter before epileptic episodes, and nonlinear time series analysis approaches have been proposed to identify these shifts. Because there aren't any currently acknowledged criteria by which to judge their effectiveness, they aren't practical for use in predicting the beginning of seizures in clinical settings. To evaluate the efficacy of multivariate seizure-prediction systems, we present an analytical strategy. Given that prediction techniques are employed over several time series and across numerous seizures, statistical tests are added to evaluate patient outcomes at the individual level. An example of their efficacy was provided by a bivariate seizure-prediction system based on synchronization theory.

The authors in [10] provided inaccurate predictions of seizures on the basis of vigilance. In order to be very sensitive, the current seizure-prediction algorithms produce a large number of false positives. The degree to which shifts in EEG dynamics contribute to erroneous forecasts is poorly understood. The sleep-wake cycle and its association with erroneous predictions are investigated here. The specificity of both seizure-prediction systems studied is limited in part by variations in EEG dynamics associated with the sleep-wake cycle. Possible improvement to generic prediction algorithms may be gleaned from this. Combining reference states has shown encouraging outcomes and may provide avenues for improving prediction technique performance.

The authors in [11] reviewed this book by Pierrebaldi and Sorenbrunak, titled *The Machine Learning Approach*, presents the fundamental machine learning (ML) methodologies and applies them to the computational issues seen in the processing of biological data. The book was written for readers with a main background in physics, mathematics, statistics, or computer science who were interested in learning more about the applicability of these fields to molecular biology.

The authors in [12] introduced probabilistic models of proteins and nucleic acids for biological sequence analysis. The beauty of hidden Markov models (HMMs) and their stochastic grammar counterparts convinced us that they are ideal for extracting the hidden meaning from biological sequences. Both the Santa Cruz and Cambridge groups separately expanded HMM approaches to stochastic context-free grammar analysis of RNA secondary structures and created two publicly accessible HMM software packages for sequence analysis. At around the same time, a team at JPL/Caltech headed by Pierre Baldi was similarly motivated to work on HMM-based techniques after being exposed to the research presented at the Snowbird conference.

In [13], the authors focused on automatic handwriting recognition using semi-supervised learning. They provide an adaptive (self-training) semi-supervised-learning system for solving the handwriting recognition issue. A generic model, trained on a collection of labeled data, was transformed into a problem-specific model by treating each issue occurrence as a set of unlabeled "training" data. By continuing to learn until convergence was obtained, they may improve upon the outcomes of the generic model. An implementation of the framework was tested on English and Arabic handwritten papers. Word recognition performance in English was 81% and in Arabic 67% using the original supervised-learning model. Improvements in word recognition performance of 86% and 77% were obtained for English and Arabic, respectively, after using semi-supervised learning.

In [14], the authors utilized structural and evolutionary data, they may anticipate the phenotypic implications of non-synonymous single nucleotide mutations. There is high hope that knowing someone's genotype will allow for more accurate risk assessments and more tailored treatment plans. Nearly half of the known genetic variants connected to human hereditary disorders are non-synonymous single nucleotide polymorphisms (nsSNPs), and they are of special relevance because they produce changes in amino acid sequence in the protein product. It was crucial to create computational techniques to anticipate the phenotypic consequences of nsSNPs in order to enable the discovery of disease-associated nsSNPs from a vast number of neutral nsSNPs.

In [17], the authors predicted which genes will be expressed using a Bayesian network of genes and their products. Motivation: Identifying operons, the basic units of transcription in a specific bacterial species, was essential for understanding transcription control in that genome. The number of species for which we have sequence and gene coordinates was expanding, yet we still don't know much about their operons. To predict operons, they introduce a Bayesian network-based probabilistic method. Our method makes use of several lines of evidence, including sequencing and expression data. Our method was tested on the K-12 genome of Escherichia coli, and the findings show that they can correctly identify approximately 78% of its operons with a false positive rate of fewer than 10%.

The authors in [18] applied the use of statistical data compression models to the problem of spam filtering. They look at an alternative method of spam detection based on dynamic statistical models of data compression. These models are well-suited for use as probabilistic text classifiers that analyze sequences of characters or binary digits. Messages were modeled as sequences, which eliminates the need for time-consuming and prone-to-error pre-processing processes like tokenization. They can be built quickly and updated in small chunks. They compare the filtering abilities of dynamic Markov compression and prediction by partial matching, two popular compression methods. Based on our empirical assessment, compression models perform better than both state-of-the-art spam filters and many other approaches suggested in the literature.

## 13.3 ROLE OF DEEP LEARNING IN HEALTHCARE

Acquiring knowledge at deep level unsupervised learning from unstructured data is the focus of deep learning, also known as deep neural learning, and the corresponding network is referred to as a deep neural network (DNN). AI does this task by modeling the way the human brain processes information in order to perform tasks like object detection, voice recognition, language translation, and decision making. It is able to learn independently from both organized and unstructured input. Among its many uses, it aids in the uncovering of fraudulent transactions and illegal monetary transactions. Conventionally referred to as "big data," the rise of digital information in almost all its forms throughout the globe is where deep learning got its start. Search engines, online stores, social media, and mobile applications are just few of the numerous online resources that contribute to big data.

As a result of deep learning, massive amounts of unstructured data that would take humans decades to understand and extract useful information may be deciphered in a matter of hours. The data analysis methods used by conventional ML methods are linear. Artificial neural networks used in deep learning are organized in a hierarchical structure that is theoretically and architecturally based on the human biological nervous system. As a result, the data is processed in a nonlinear fashion by deep learning algorithms. The perceptron was one of the first types of neural network, and it was inspired by the structure of the human brain. This method was developed at the Cornell Aeronautical laboratory with support from the United States Office of Naval Research. In the context of ML, it is a supervised-learning approach for binary classifiers. Patterns that are easily separated by a linear gradient may be classified using its input layer, which is directly linked to the output layer. Neural networks, which have a layered architecture consisting of an input layer, an output layer, and one or more hidden layers, were developed as the complexity of data grew. Since the data includes non-linear interactions, these concealed layers are equipped to handle this complexity. The neurons in this network receive data as input, process it, and send on the results as an output to the next layer. Each neuron does a summation of the input data, activates the summated data with a function, and sends that information on to the next layer, where it may be used as-is or further processed. This means that deep learning networks are composed of many interconnected layers of neurons, with the number of these layers having grown to over a thousand at this point. With such a high modeling capability, deep learning can lock down a database of all potential mappings in memory. In the beginning, though, it must be able to train well with a large database and make sound judgments about what actions to take. CNNs, recurrent neural networks (RNNs), DNNs, multilayer perceptrons (MLPs), a deep belief network (DBN), an auto encoder, a deep Boltzmann machine (DBM), a deep conventional extreme ML (DCELM), and many more are all used in healthcare research today.

Therefore, deep learning can not only recognize and extract useful traits, but also create new ones. In the medical field, it is used for both illness diagnosis and prediction of a therapy model with a defined goal. Some examples of deep learning algorithms include CNNs, RNNs, radial basis function networks (RBFNs), long short-term memories (LSTMs), auto encoders, extreme learning models (ELMs), self-organizing maps (SOMs), and generative adversarial networks (GANs). Virtual assistants, entertainment, medicine, robotics, picture coloring, etc. are just a few of the many areas where deep learning algorithms have proven effective. Disease identification and individualized therapy are only two of the many areas where deep learning has been put to use. Ophthalmology, pathology, cancer detection, and radiology are just a few of the specialties where deep learning algorithms have had a profound impact. Although deep learning's revolutionary effects are initially shown in ophthalmology, the fields of pathology and cancer detection are now receiving more research funding and developing more accurate applications. Using the NinaPro DB2 and DB3 datasets, we trained a CNN to regulate neuroproteins with an 83% success rate. With the use of the kinematic and EMG data, were able to train a CNN for movement intention decoding, concluding with an accuracy rate of above 90%. Using data from eight healthy participants, research estimated limb motions using an RNN suggested that the RNN outperformed other approaches for predicting a 3D trajectory. Through the use of 18 people and seven hand gestures, research demonstrates the effectiveness of CNN in robotic arm guiding, achieving an estimated accuracy of 97.9%. With auto encoder, they examine sleep state recognition and find an accuracy of 80.4%. Based on a comparison of CNN and DNN using a single person and 180 trials, they found that DNN performed better than CNN. Emotion recognition based on a deep learning network has been studied, who used separate datasets to arrive at a result of a valence accuracy of 49.52% and an arousal accuracy of 46.03%.

### 13.3.1 Deep Learning and Medical Imaging

When analyzing images for diagnostic purposes, it is essential to be able to spot abnormalities and quantify their severity. The quality of automated image analysis systems that use ML algorithms may, in turn, enhance the accuracy of their interpretations. Numerous resources in this field make available a mountain of data for doctors to go through. Radiological imaging (such as X-rays, CT scans, and MRIs), pathological imaging, and genetic sequencing all contribute to this body of knowledge. Even while deep learning approaches can handle a lot of data, there isn't enough software to transform it all [61–65].

### 13.3.2 Diabetic Retinopathy (DR)

The metabolic condition diabetes mellitus (DM) may cause elevated blood sugar levels. Both a pancreas that doesn't make enough insulin (Type I diabetes) and tissues that don't respond properly to that insulin (Type-II diabetes) are to blame. Diabetic retinopathy (DR) is an eye condition induced by diabetes, and severe DR may lead to permanent blindness if left untreated. A cure is possible with early detection by retinal screening. Automatic DR detection using a deep learning model outperforms the human technique of identifying DR by a wide margin. Using deep convolutional neural networks (DCNNs), we analyzed a database of 10,000 retinal pictures from the EyePACS-I system and found a sensitivity of 97.5% and a specificity of 93.4%, on the Messidor-2 dataset, which includes 1,700 pictures from 874 patients; we also employed a DCNN for classification and detection of intermediate and worse, claiming a sensitivity of 96% and a specificity of 93.4%. To classify fundus using a DCNN with a dropout layer relied on publicly accessible datasets, such as Kaggle fundus, DRIVE, and STARE, reaching an accuracy of 94%–96%. By training a 5-layered connection mechanism on the Messidor dataset, we were able to diagnose DR in its earliest stages on the Retinopathy Online Challenge (ROC), achieving up to 97% sensitivity, 96% specificity, 96% accuracy, and 0.988 AUC. We identified de-noised angiography pictures from EyePACS and made a diagnosis across five severity classes. Following the CNN's implementation, the results were 79% AUC and 45% accuracy. Research extracted characteristics from the indicated area, classified them using a DCNN, and tested their model on the SiDRP and DIARETBD1 datasets. To analyze more than 80,000 digital fundus photos from Kaggle, we used the NVIDIA CUDA DCNN package. Also, 5,000 photos were used to verify the network's accuracy. Images were downscaled to 512x512 and sharpened for display. After that, Cu-DCNN was given the features vector. Using characteristics including exudates, hemorrhages, and micro-aneurysms, they were able to classify the pictures into 5 categories with an accuracy of up to 75%, a sensitivity of up to 30%, and a specificity of up to 95% [66–70].

### 13.3.3 Gastrointestinal (GI) Disease

Detection those illnesses that manifest in the digestive system are called gastrointestinal ailments (either upper GI tract or lower GI tract). The esophagus, stomach, large intestine, small intestine, and rectum are all examples of organs that make up the gastrointestinal (GI) tract. Inflammation, hemorrhage, infection, and cancer are just some of the illnesses and conditions that may interfere with digestion. Other possible complications include diverticulitis from colon bleeding and a disorder of the small intestine called arteriovenous malformation (angiodysplasias or angio-ectasias). With the development of computer-assisted diagnostic systems, image processing and ML play a crucial role in the diagnosis of various diseases. Recently, many imaging tests that rely on ML algorithms have become commonplace. These include procedures like wireless capsule endoscopy and interoscope, colonoscopy or sigmoidoscopy with radiopaque dyes and X-ray studies, deep small bowel enteroscopy, intra operative enteroscopy, computed tomography, and magnetic resonance imaging (MRI). Research used DCNNs on a dataset of 10,000 wireless capsule endoscopy pictures. Automatic feature extraction from endoscopic pictures is a prominent approach used to identify GI lesions using CNNs. Following this, a support vector machine is used to categories these attributes, with an accuracy of 80%. In a study compared 599 non-inflammatory pictures of the KID digestive system with 337 photos that were labeled as having inflammation. In 2016, they used CNN architecture to extract features from WCE movies of GI disorders and then fed those characteristics into a support vector machine [71–77].

### 13.3.4  Tumor Detection

Tumors and neoplasms refer to abnormal, often destructive, cell development in any region of the body. Cancer may or may not start with a tumor. This leads us to classify tumors as either malignant (cancerous) or benign (noncancerous). Benign tumors are significantly safer than malignant ones since they don't metastasize (spread to other regions of the body). Malignant tumors, on the other hand, may metastasize to other organs, making a cure much more unlikely. In a 2016 study, Wang et al. analyzed 482 pictures from patients aged 32–70; they found tumors in 246 women. The photos were de-noised, and then the breast tumor was segmented utilizing morphological operations and region growth in a modified wavelet transform analysis. The morphological and textural characteristics were then sent to an extreme learning machine and support vector machine for breast cancer classification and diagnosis. The overall error rate was 84 when using ELM and 96 when using SVM as the output. In their 2015 study "automatic coronary calcium scoring in cardiac angiography using CNN," Jeimer et al. reported an area under the curve of 87% utilizing just data on malignant mass and benign single cyst. The AUC reported, all of which employed CNN, was between 80% and 85% [78–82].

## 13.4  PROBLEM STATEMENT

There has been much research in the area of healthcare that has made significant differences in real life. But it has been observed that it is challenging task to implement integrated system for healthcare that should be supported by IoT. There is a need to introduce scalable, efficient model for IoT-based healthcare system. In other words, several healthcare-related studies have produced useful, practical results. However, it has been noted that it is a difficult job to develop an integrated system for healthcare that should be supported by IoT. It is essential to implement a model that is both scalable and effective for the IoT-based healthcare system.

## 13.5  PROPOSED METHODOLOGY

IoT methods have been used in a number of recent healthcare-related research. Studies in mobile computing have often focused on finding solutions to practical issues. The research methodology is shown in Figure 13.6 and the flow of the proposed work is shown in Figure 13.7.

However, there are challenges associated with implementing an IoT strategy in healthcare, such as the need to include an optimization mechanism to guarantee the integrity of a health app in a mobile and IoT setting. In addition, there has to be more scalability in conventional research methods. There is a pressing need to suggest a new method for the healthcare system in order to boost precision and address performance difficulties [83].

### 13.5.1  Flowchart of Proposed Work

## 13.6  RESULTS AND DISCUSSION

Integration of healthcare systems for mobile computing and IoT has been a challenging and complex operation. Simulation results present the comparative study of time consumption and accuracy in case of an integrated system [84].

### 13.6.1  Time Consumption

Simulation of time consumption has been shown in Table 13.1 and Figure 13.8 where conventional and proposed work has been compared [85].

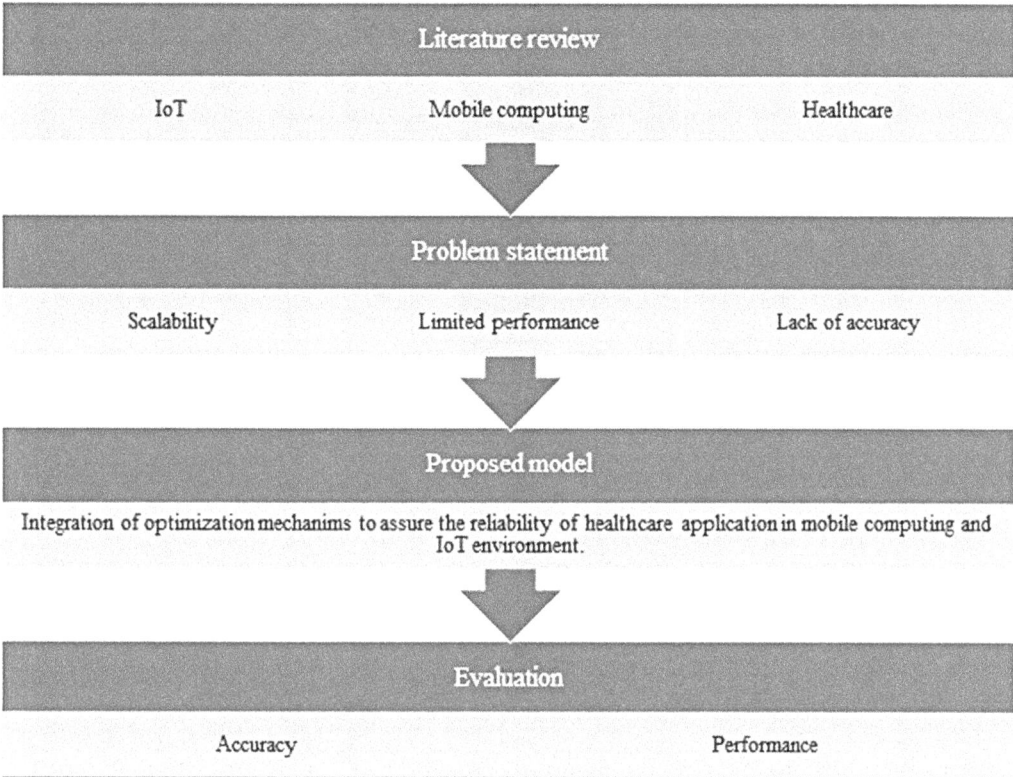

**FIGURE 13.6**    Research Methodology.

### 13.6.2  ERROR RATE

Simulation of the error rate has been shown in Table 13.2 and Figure 13.9 where conventional and proposed work has been compared.

### 13.6.3  ACCURACY RATE

Simulation of the accuracy comparison has been shown in Table 13.3 and Figure 13.10 where conventional and proposed work has been compared.

## 13.7  CONCLUSION

Machine learning in a mobile IoT setting has been the subject of much healthcare-related research. Recent research suggests that traditional AI and ML studies have focused on solving practical challenges [86–88]. There are a number of obstacles to using deep learning in healthcare, including the aforementioned lack of accuracy and subpar performance. In order to provide better results, conventional research must be more adaptable and scalable. The suggested method is shown to be more efficient. In addition, the suggested work provides a more accurate model since its error rate is lower and its accuracy rate is higher [89–92].

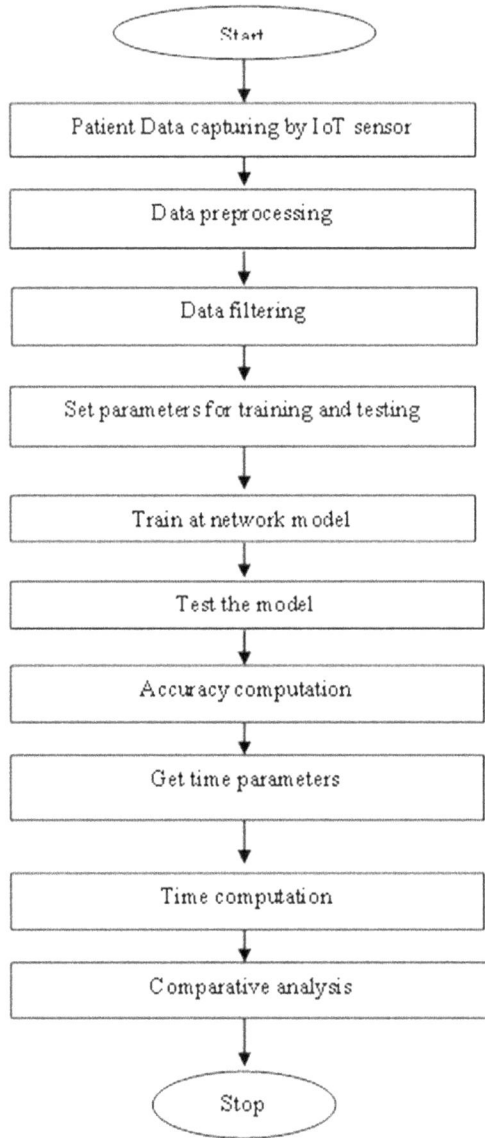

**FIGURE 13.7**    Flowchart of Proposed Work.

## 13.8   FUTURE SCOPE

Although AI is only getting off the ground in India, it is already being used in a wide variety of fields, from farming and medicine to education and transportation, from banking and manufacturing to hospitality. ML algorithms and other cognitive technologies have several uses in the healthcare industry. AI refers to the ability of machines to learn and think like humans and carry out tasks that would previously have required human intellect. AI is finding applications throughout the healthcare industry, including in diagnostics, imaging, virtual assistance, hospital administration, and risk assessment and management. These types of systems are expected to play a crucial part in the future of mobile computing and the Internet of Things.

## TABLE 13.1
## Comparison of Time Taken

| Case | Conventional Work | Proposed Work |
|------|-------------------|---------------|
| 1 | 81.01233011 | 79.71532697 |
| 2 | 183.9042034 | 183.6643128 |
| 3 | 90.33714352 | 89.66976169 |
| 4 | 48.45784579 | 47.5459412 |
| 5 | 357.6121517 | 357.2400042 |
| 6 | 66.90465775 | 66.50224862 |
| 7 | 561.7787612 | 561.2489864 |
| 8 | 463.6349531 | 463.0067552 |
| 9 | 644.4990407 | 644.0749001 |
| 10 | 714.7083657 | 713.7248563 |

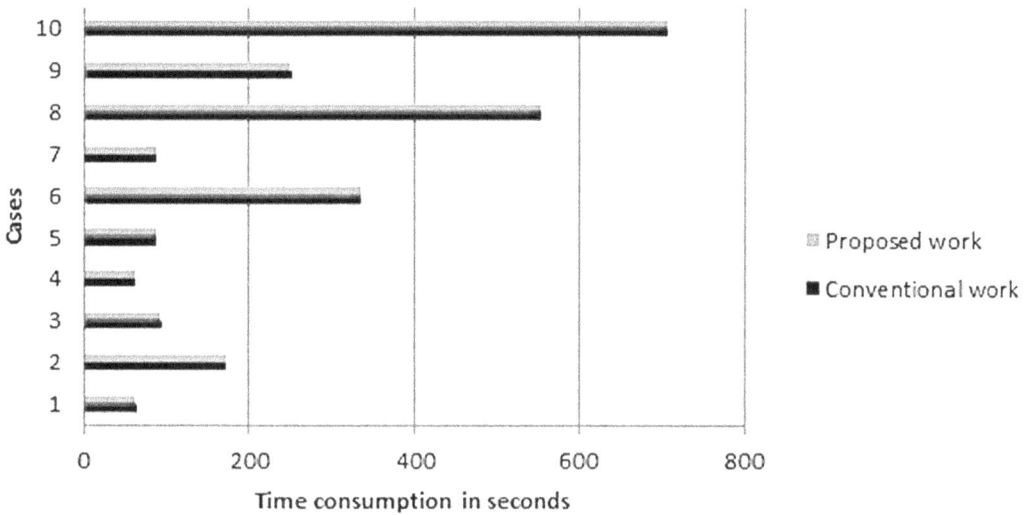

FIGURE 13.8   Comparison of Time Taken.

## TABLE 13.2
## Comparison of Error Rates

| Cases | Conventional Work | Proposed Work |
|-------|-------------------|---------------|
| 1 | 5.041214988 | 4.367477136 |
| 2 | 5.743879489 | 5.410381139 |
| 3 | 5.798970901 | 5.087869036 |
| 4 | 5.596501147 | 4.999681904 |
| 5 | 5.838422034 | 5.304782693 |
| 6 | 5.216455656 | 4.251046781 |
| 7 | 5.279164488 | 5.110319216 |
| 8 | 5.094251625 | 5.084468296 |
| 9 | 5.258979073 | 4.575220731 |
| 10 | 5.282439736 | 4.983472955 |

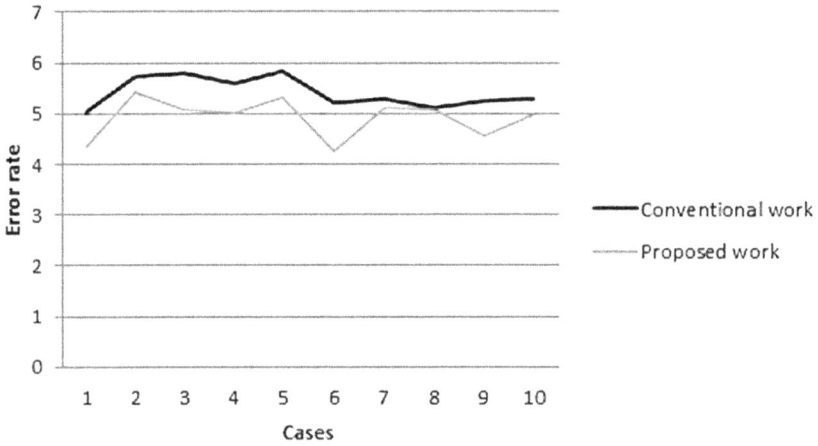

**FIGURE 13.9**    Comparison of Error Rates.

**TABLE 13.3**

**Comparison of Accuracy Rate**

| Cases | Conventional Work | Proposed Work |
| --- | --- | --- |
| 1 | 94.95878501 | 95.63252286 |
| 2 | 94.25612051 | 94.58961886 |
| 3 | 94.2010291 | 94.91213096 |
| 4 | 94.40349885 | 95.0003181 |
| 5 | 94.16157797 | 94.69521731 |
| 6 | 94.78354434 | 95.74895322 |
| 7 | 94.72083551 | 94.88968078 |
| 8 | 94.90574837 | 94.9155317 |
| 9 | 94.74102093 | 95.42477927 |
| 10 | 94.71756026 | 95.01652705 |

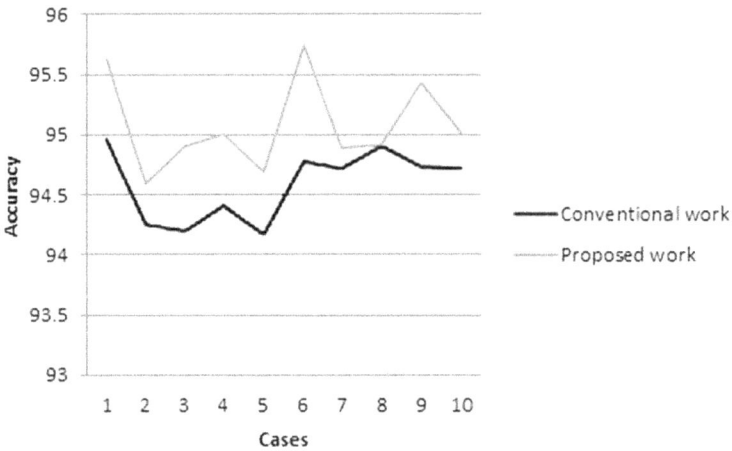

**FIGURE 13.10**    Comparison of Accuracy Rate.

# REFERENCES

1. Alban, M., & Gilligan, T. (2016). Automated detection of diabetic retinopathy using fluorescein angiography photographs. Report of Stanford Education.
2. Kaur, J., Sindhwani, N., Anand, R., & Pandey, D. (2023). Implementation of IoT in Various Domains. In *IoT Based Smart Applications* (pp. 165–178). Springer, Cham.
3. Tripathi, A., Sindhwani, N., Anand, R., & Dahiya, A. (2023). Role of IoT in Smart Homes and Smart Cities: Challenges, Benefits, and Applications. In *IoT Based Smart Applications* (pp. 199–217). Springer, Cham.
4. Allen, J. E., Pertea, M., & Salzberg, S. L. (2004). Computational gene prediction using multiple sources of evidence. *Genome Research*, 14:142–148.
5. Blankertz, B., Müller, K. R., Curio, G., Vaughan, T. M., Schalk, G., Wolpaw, J. R., Schlögl, A., Neuper, C., Pfurtscheller, G., Hinterberger, T., Schröder, M., & Birbaumer, N. (2004). The BCI Competition 2003: progress and perspectives in detection and discrimination of EEG single trials. *IEEE Transactions on Biomedical Engineering* 51(6):1044–1051. https://doi.org/10.1109/TBME.2004.826692. PMID: 15188876.
6. Sharma, S., Rattan, R., Goyal, B., Dogra, A., & Anand, R. (2023). Microscopic and Ultrasonic Super-Resolution for Accurate Diagnosis and Treatment Planning. In *Communication, Software and Networks* (pp. 601–611). Springer, Singapore.
7. Bommareddy, S., Khan, J. A., & Anand, R. (2022). A Review on Healthcare Data Privacy and Security. In *Networking Technologies in Smart Healthcare* (pp. 165–187). CRC Press, UK.
8. Juneja, S., Juneja, A., & Anand, R. (2020). Healthcare 4.0-digitizing healthcare using big data for performance improvisation. *Journal of Computational and Theoretical Nanoscience*, 17(9–10), 4408–4410.
9. Schelter, B., Winterhalder, M., Maiwald, T., Brandt, A., Schad, A., Schulze-Bonhage, A., & Timmer, J. (2006). Testing statistical significance of multivariate time series analysis techniques for epileptic seizure prediction, *Chaos: An Interdisciplinary Journal of Nonlinear Science* 16(1), 013108.
10. Schelter, B., Winterhalder, M., Maiwald, T., Brandt, A., Schad, A., Timmer, J., & Schulze-Bonhage, A. (2006). Do false predictions of seizures depend on the state of vigilance? A report from two seizure-prediction methods and proposed remedies. *Epilepsia* 47(12), 2058–2070.
11. Baldi, P., & Brunak, S. (2001). *Bioinformatics: The Machine Learning Approach*. MIT Press, USA.
12. Durbin, R., Eddy, S., Krogh, A., & Mitchison, G. (1998). *Biological Sequence Analysis: Probabilistic Models of Proteins and Nucleic Acids*. Cambridge University Press, Cambridge. https://doi.org/10.1017/CBO9780511790492.
13. Ball, G. R., and Srihari, S. N. (2009). Semisupervised Learning for Handwriting Recognition. In *10th International Conference on Document Analysis and Recognition, ICDAR'09*. IEEE.
14. Bao, L., & Cui, Y. (2005). Prediction of the phenotypic effects of nonsynonymous single nucleotide polymorphisms using structural and evolutionary information. *Bioinformatics* 21(5), 2185–2190.
15. Gupta, A., Asad, A., Meena, L., & Anand, R. (2023). IoT and RFID-Based Smart Card System Integrated with Health Care, Electricity, QR and Banking Sectors. In *Artificial Intelligence on Medical Data* (pp. 253–265). Springer, Singapore.
16. Jain, S., Sindhwani, N., Anand, R., & Kannan, R. (2022). COVID Detection Using Chest X-Ray and Transfer Learning. In *International Conference on Intelligent Systems Design and Applications*, pp. 933–943. Springer, Cham.
17. Bockhorst, J., Craven, M., Page, D., Shavlik, J., & Glasner, J. (2003). A Bayesian network approach to operon prediction. *Bioinformatics* 19(10), 1227–1235. doi: 10.1093/bioinformatics/btg147. PMID: 12835266.
18. Bratko, Andrej, Cormack, Gordon, Filipic, Bogdan, Lynam, Thomas, & Zupan, Blaz (2006). Spam filtering using statistical data compression models. *Journal of Machine Learning Research* 6, 2673–2698.
19. Kathirvel, T. R. (2016). Classifying diabetic retinopathy using deep learning architecture. *International Journal of Engineering Research Technology*, 5(6), 19–24.
20. Caragea, Cornelia, & Honavar, Vasant (2008). *Machine Learning in Computational Biology*. doi: 10.1007/978-0-387-39940-9_636.
21. Carter, R. J., Dubchak, I., & Holbrook, S. R. (2001). A computational approach to identify genes for functional RNAs in genomic sequences. *Nucleic Acids Research* 29(19), 3928–3938.
22. Cho, Sung-Bae & Won, Hong-Hee. (2003). Machine Learning in DNA Microarray Analysis for Cancer Classification. In *Proceedings of the First Asia-Pacific bioinformatics Conference*, vol. 34, pp. 189–198.

23. Degroeve, S., De Baets, B., Van de Peer, Y., & Rouzé, P. (2002). Feature subset selection for splice site prediction. *Bioinformatics* 18(Suppl 2), S75–83. doi: 10.1093/bioinformatics/18.suppl_2.s75. PMID: 12385987.

24. Nurse, E., Mashford, B. S., Yepes, A. J., Kiral-Kornek, I., Harrer, S., & Freestone, D. R. (2016). Decoding EEG and LFP Signals Using Deep Learning: Heading True North. In *Proceedings of the ACM International Conference on Computing Frontiers*. ACM, pp. 259–266.

25. Anand, R., Sinha, A., Bhardwaj, A., & Sreeraj, A. (2018). Flawed Security of Social Network of Things. In *Handbook of Research on Network Forensics and Analysis Techniques* (pp. 65–86). IGI Global, USA.

26. Srivastava, A., Gupta, A., & Anand, R. (2021). Optimized smart system for transportation using RFID technology. *Mathematics in Engineering, Science & Aerospace (MESA)*, 12(4), 953–965.

27. Shukla, R., Dubey, G., Malik, P., Sindhwani, N., Anand, R., Dahiya, A., & Yadav, V. (2021). Detecting crop health using machine learning techniques in smart agriculture system. *Journal of Scientific & Industrial Research*, 80(08), 699–706.

28. Singh, S.K., Thakur, R.K., Kumar, S., & Anand, R. (2022). "Deep Learning and Machine Learning based Facial Emotion Detection using CNN," in *2022 9th International Conference on Computing for Sustainable Global Development (INDIACom)*, pp. 530–535, https://doi.org/10.23919/INDIACom54597.2022.9763165

29. Huve, G., Takahashi, K., & Hashimoto, M. (2017). Brain Activity Recognition with a Wear- Able FNIRS Using Neural Networks. In *2017 IEEE International Conference on Mechatronics and Automation (ICMA)*. IEEE, pp. 1573–1578.

30. Fogel, Gary B., & Corne, David W. (2002). *Evolutionary Computation in Bioinformatics*. Morgan Kaufmann, USA.

31. San, Gilbert Lim Yong, Lee, Mong Li, & Hsu, Wynne (2012). Constrained-MSER Detection of Retinal Pathology. In *2012 21st International Conference on Pattern Recognition (ICPR)*, 2059–2062. IEEE.

32. Lim, Gilbert, Lee, Mong Li, Hsu, Wynne, & Wong, Tien Yin (2014). Trans-formed representations for convolutional neural networks in diabetic retinopathy screening. *Modern Artif Intell Health Anal* 55, 21–25.

33. Guyon, Isabelle, & Elisseeff, André (2003). An introduction to variable and feature selection. *The Journal of Machine Learning Research* 3, 1157–1182.

34. Haider, Peter, Chiarandini, Luca, & Brefeld, Ulf (2012). Discriminative Clustering for Market Segmentation. In *Proceedings of the 18th ACM SIGKDD International Conference on Knowledge Discovery and Data Mining*. ACM.

35. Haykin, Simon, & Chen, Zhe (2005). The cocktail party problem. *Neural Computation* 17(9), 1875–1902.

36. Haloi, M. (2015). Improved microaneurysm detection using deep neural networks. arXiv preprint arXiv:1505.04424.

37. Higgins, D., & Taylor, W. (eds) (2000). *Bioinformatics. Sequence, Structure, and Databanks*. Oxford University Press, UK.

38. Hirosawa, M., Totoki, Y., Hoshida, M., & Ishikawa, M. (1995). Comprehensive study on iterative algorithms of multiple sequence alignment. *Computer Applications in the Biosciences* 11(1), 13–18.

39. Hong, Z., Zeng, X., Wei, L., et al. (2020). Identifying enhancer–promoter interactions with neural network based on pre-trained DNA vectors and attention mechanism. *Bioinformatics* 36, 1037–1043.

40. Shujie, Hou, Qiu, Robert C., Chen, Zhe, & Hu, Zhen (2011). SVM and Dimensionality Reduction in Cognitive Radio with Experimental Validation. arXiv preprint arXiv:1106.2325.

41. Nguyen, Hung Dinh, Yoshihara, I., Yamamori, K., & Yasunaga, M. (2002). A Parallel Hybrid Genetic Algorithm for Multiple Protein Sequence Alignment. In *Proceedings of the 2002 Congress on Evolutionary Computation CEC'02 (Cat. No.02TH8600)*, Honolulu, HI, vol. 1, pp. 309–314. doi: 10.1109/CEC.2002.1006252.

42. Husmeier, D., Dybowski, R., & Roberts, S. (eds) (2005). *Probabilistic Modelling in Bioinformatics and Medical Informatics*. Springer Verlag, Germany.

43. Kyu-Baek, Hwang, Cho, Dong-Yeon, Park, Sang-Wook, Kim, Sung-Dong, & Zhang, Byoung-Tak (2002). Applying Machine Learning Techniques to Analysis of Gene Expression Data: Cancer Diagnosis. In *Methods of Microarray Data Analysis* (pp. 167–182). Springer US, USA.

44. Kononenko, Igor (2001). Machine learning for medical diagnosis: history, state of art and perspective. *Artificial Intelligence in Medicine* 23, 89–109.

45. Ishikawa, M., Toya, T., Hoshida, M., Nitta, K., Ogiwara, A., & Kanehisa, M. (1993). Multiple sequence alignment by parallel simulated annealing. *Computer Applications in the Biosciences: CABIOS*, 9(3), 267–273.

46. Jagota, A. (2000). *Data Analysis and Classification for Bioinformatics*. Bioinformatics by the Bay Press, Canada.

47. Wolterink, Jelmer M., Leiner, Tim, Viergever, Max A., & Išgum, Ivana (2015). Automatic Coronary Calcium Scoring in Cardiac CT Angiography Using Convolutional Neural Networks. In *International Conference on Medical Image Computing and Computer-Assisted Intervention*, pp. 589–596. Springer.

48. Wolterink, Jelmer M., Leiner, Tim, de Vos, Bob D., van Hamersvelt, Robbert W., Viergever, Max A., & Išgum, Ivana (2016). Automatic coronary artery calcium scoring in cardiac ct angiography using paired convolutional neural networks. *Medical Image Analysis*, 34, 123–136.

49. Jia, X., & Meng, M. Q. H. (2016). A Deep Convolutional Neural Network for Bleeding Detection in Wireless Capsule Endoscopy Images. In *2016 IEEE 38th Annual International Conference of the Engineering in Medicine and Biology Society (EMBC)*, pp 639–642. IEEE.

50. Jiang, T., Xu, X., & Zhang, M. Q. (eds) (2002). *Current Topics in Computational Molecular Biology*. The MIT Press, UK.

51. Jiang, F., Jiang, Y, Zhi, H. et al. (2017). Artificial intelligence in healthcare: past, present and future. *Stroke and Vascular Neurology* 2, e000101. doi: 101136/sv-2017-000101.

52. Gupta, A., Srivastava, A., Anand, R., & Tomažič, T. (2020). Business Application Analytics and the Internet of Things: The Connecting Link. In *New Age Analytics* (pp. 249–273). Apple Academic Press, Canada.

53. Quinn, John A., Nakasi, Rose, Mugagga, Pius K. B., Byanyima, Patrick, Lubega, William, & Andama, Alfred (2016). Deep convolutional neural net-works for microscopybased point of care diagnostics. arXiv preprint arXiv:1608.02989.

54. Park, K.-H., & Lee, S.-W. (2016). Movement Intention Decoding Based on Deep Learning for Multiuser Myoelectric Interfaces. In *2016 4th International Winter Conference on Brain-Computer Interface BCI*. IEEE, pp. 1–2.

55. Kim, J., Cole, J. R., & Pramanik, S. (1996). Alignment of possible secondary structures in multiple RNA sequences using simulated annealing. *Computer applications in the Biosciences* 12(8), 259–267.

56. Kiral-Kornek, I., Mendis, D., Nurse, E. S., Mashford, B. S., Freestone, D. R., Grayden, D. B., & Harrer, S. (2017). Truenorth-Enabled Real-Time Classification of EEG Data for Brain-Computer Interfacing. In *Engineering in Medicine and Biology Society (EMBC), 2017 39th Annual International Conference of the IEEE*. IEEE, pp. 1648–1651.

57. Kononenko, Igor (2001). Machine learning for medical diagnosis: history, state of the art and perspective. *Artificial Intelligence in Medicine* 23(1), 89–109.

58. Kamnitsas, Konstantinos, Ledig, Christian, Newcombe, Virginia F. J., Simpson, Joanna P., Kane, Andrew D., Menon, David K., Rueckert, Daniel, and Glocker, Ben (2017). Efficient multi-scale 3d CNN with fully connected CRF for accurate brain lesion segmentation. *Medical Image Analysis* 36, 61–78.

59. Fraiwan, L., & Lweesy, K. (2017). Neonatal Sleep State Identification Using Deep Learning Autoencoders. In *2017 IEEE 13th International Colloquium on Signal Processing & its Applications (CSPA)*. IEEE, pp. 228–231.

60. Larranaga, P., Menasalvas, E., Pena, J. M., et al. (2003). Special issue in data mining in genomics and proteomics. *Artificial Intelligence in Medicine* 31, III–IV.

61. Babu, S. Z. D. et al. (2023). Analysation of Big Data in Smart Healthcare. In Gupta, M., Ghatak, S., Gupta, A., Mukherjee, A. L. (eds) *Artificial Intelligence on Medical Data*. Lecture Notes in Computational Vision and Biomechanics, vol 37. Springer, Singapore. https://doi.org/10.1007/978-981-19-0151-5_21

62. Pandey, B. K. et al. (2023). Effective and Secure Transmission of Health Information Using Advanced Morphological Component Analysis and Image Hiding. In Gupta, M., Ghatak, S., Gupta, A., Mukherjee, A. L. (eds) *Artificial Intelligence on Medical Data*. Lecture Notes in Computational Vision and Biomechanics, vol 37. Springer, Singapore. https://doi.org/10.1007/978-981-19-0151-5_19

63. Sreekanth, N., Rama Devi, J., Shukla, A. et al. (2022). Evaluation of estimation in software development using deep learning-modified neural network. *Applied Nanoscience*, 1–13. https://doi.org/10.1007/s13204-021-02204-9

64. Veeraiah, V., Khan, H., Kumar, A., Ahamad, S., Mahajan, A., & Gupta, A. (2022). Integration of PSO and Deep Learning for Trend Analysis of Meta-Verse. In *2022 2nd International Conference on Advance Computing and Innovative Technologies in Engineering (ICACITE), 2022*, pp. 713–718. doi: 10.1109/ ICACITE53722.2022.9823883.

65. Sarier, N. D. (2021). Comments on biometric-based non-transferable credentials and their application in blockchain-based identity management. *Computers and Security* 105, 102243. https://doi.org/10.1016/j. cose.2021.102243

66. Veeraiah, V., Rajaboina, N. B., Rao, G. N., Ahamad, S., Gupta, A., & Suri, C. S., (2022). Securing Online Web Application for IoT Management. In *2022 2nd International Conference on Advance Computing and Innovative Technologies in Engineering (ICACITE), 2022*, pp. 1499–1504. doi: 10.1109/ ICACITE53722.2022.9823733.

67. Gupta, A., Singh, R., Nassa, V. K., Bansal, R., Sharma, P., & Koti, K. (2021). Investigating Application and Challenges of Big Data Analytics with Clustering. In *2021 International Conference on Advancements in Electrical, Electronics, Communication, Computing and Automation (ICAECA), 2021*, pp. 1–6. doi: 10.1109/ICAECA52838.2021.9675483.

68. Namasudra, S., Deka, G. C., Johri, P., Hosseinpour, M., & Gandomi, A. H. (2021). The revolution of blockchain: State-of-the-art and research challenges. *Archives of Computational Methods in Engineering* 28(3), 1497–1515. https://doi.org/10.1007/s11831-020-09426-0

69. Veeraiah, V., Kumar, K. R., Lalitha Kumari, P., Ahamad, S., Bansal, R., & Gupta, A. (2022). Application of Biometric System to Enhance the Security in Virtual World. In *2022 2nd International Conference on Advance Computing and Innovative Technologies in Engineering (ICACITE), 2022*, pp. 719–723. doi: 10.1109/ICACITE53722.2022.9823850.

70. Shukla, A., Ahamad, S., Rao, G. N., Al-Asadi, A. J., Gupta, A., & Kumbhkar, M. (2021). Artificial Intelligence Assisted IoT Data Intrusion Detection. In *2021 4th International Conference on Computing and Communications Technologies (ICCCT), 2021*, pp. 330–335. doi: 10.1109/ICCCT53315.2021.9711795.

71. Liang, W., & Ji, N. (2022). Privacy challenges of IoT-based blockchain: a systematic review. *Cluster Computing* 25, 2203–2221. https://doi.org/10.1007/s10586-021-03260-0

72. Kaushik, D., & Gupta, A. (2021). Ultra-secure transmissions for 5G-V2X communications. *Materials Today: Proceedings*. https://doi.org/10.1016/j.matpr.2020.12.130.

73. Veeraiah, V., Gangavathi, P., Ahamad, S., Talukdar, S. B., Gupta, A., & Talukdar, V. (2022). Enhancement of Meta Verse Capabilities by IoT Integration. In *2022 2nd International Conference on Advance Computing and Innovative Technologies in Engineering (ICACITE), 2022*, pp. 1493–1498, https://doi. org/10.1109/ICACITE53722.2022.9823766.

74. Ranjith Kumar, M. V., & Bhalaji, N. (2021). Blockchain based chameleon hashing technique for privacy preservation in E-governance system. *Wireless Personal Communications* 117(2), 987–1006. https://doi. org/10.1007/s11277-020-07907-w

75. Garg, M., Gupta, A., Kaushik, D., & Verma, A. (2020). Applying machine learning in IoT to build intelligent system for packet routing system. *Materials Today: Proceedings*. doi: 10.1016/j.matpr.2020.09.539.

76. Gupta, A., Anand, R., Pandey, D., Sindhwani, N., Wairya, S., Pandey, B. K., & Sharma, M. (2021). Prediction of breast cancer using extremely randomized clustering forests (ERCF) technique: Prediction of breast cancer. *International Journal of Distributed Systems and Technologies (IJDST)* 12(4), 1–15.

77. Singh, H., Pandey, B. K., George, S., Pandey, D., Anand, R., Sindhwani, N., & Dadheech, P. (2023). Effective Overview of Different ML Models Used for Prediction of COVID-19 Patients. In *Artificial Intelligence on Medical Data* (pp. 185–192). Springer, Singapore.

78. Wang, D., Wang, H., & Fu, Y. (2021). Blockchain-based IoT device identification and management in 5G smart grid. *Eurasip Journal on Wireless Communications and Networking*, 2021(1). https://doi. org/10.1186/s13638-021-01966-8

79. Rathee, G., Balasaraswathi, M., Chandran, K. P., Gupta, S. D., & Boopathi, C. S. (2021). A secure IoT sensors communication in industry 4.0 using blockchain technology. *Journal of Ambient Intelligence and Humanized Computing* 12(1), 533–545. https://doi.org/10.1007/s12652-020-02017-8

80. Aggarwal, B., Gupta, A., Goyal, D., Gupta, P., Bansal, B., & Barak, D. (2021). A review on investigating the role of block-chain in cyber security. *Materials Today: Proceedings*. https://doi.org/10.1016/j. matpr.2021.10.124.

81. Gupta, A., Garg, M., Verma, A., & Kaushik, D. (2020). Implementing lossless compression during image processing by integrated approach. *Materials Today: Proceedings*. https://doi.org/10.1016/j.matpr.2020.10.052

82. Garba, A., Dwivedi, A. D., Kamal, M., Srivastava, G., Tariq, M., Hasan, M. A., & Chen, Z. (2021). A digital rights management system based on a scalable blockchain. *Peer-to-Peer Networking and Applications* 14(5), 2665–2680. https://doi.org/10.1007/s12083-020-01023-z

83. Verma, A., Gupta, A., Kaushik, D., & Garg, M. (2021). Performance enhancement of IOT based accident detection system by integration of edge detection. *Materials Today: Proceedings*. https://doi.org/10.1016/j.matpr.2021.01.468

84. Pramanik, S., & Suresh Raja, S. (2020). A secured image steganography using genetic algorithm. *Advances in Mathematics: Scientific Journal* 9(7), 4533–4541.

85. Pramanik, S., Singh, R. P., & Ghosh, R. (2020). Application of bi-orthogonal wavelet transform and genetic algorithm in image steganography. *Multimedia Tools and Applications*. https://doi.org/10.1007/s11042-020-08676-2020

86. Singh, H., Ramya, D., Saravanakumar, R., Sateesh, N., Anand, R., Singh, S., & Neelakandan, S. (2022). Artificial intelligence based quality of transmission predictive model for cognitive optical networks. *Optik* 257, 168789.

87. Chawla, P., Juneja, A., Juneja, S., & Anand, R. (2020). Artificial intelligent systems in smart medical healthcare: Current trends. *International Journal of Advanced Science and Technology* 29(10), 1476–1484.

88. Raghavan, R., Verma, D. C., Pandey, D., Anand, R., Pandey, B. K., & Singh, H. (2022). Optimized building extraction from high-resolution satellite imagery using deep learning. *Multimedia Tools and Applications* 81, 42309–42323.

89. Pramanik, S. (2022). An Effective Secured Privacy-Protecting Data Aggregation Method in IoT. In Odhiambo, M. O., Mwashita, W. (eds) *Achieving Full Realization and Mitigating the Challenges of the Internet of Things*. IGI Global. https://doi.org/10.4018/978-1-7998-9312-7.ch008

90. Bansal, R., Jenipher, B., Nisha, V., Jain, Makhan R., Dilip, Kumbhkar, Pramanik, S., Roy, S., & Gupta, A. (2022). Big Data Architecture for Network Security. In *Cyber Security and Network Security*. Wiley, New Jersey, USA.

91. Pradhan, D., Sahu, P. K., Goje, N. S., Myo, H., Ghonge, M. M., Tun, M., Rajeswari, R., & Pramanik, S. (2022). Security, Privacy, Risk, and Safety Toward 5G Green Network (5G-GN). In *Cyber Security and Network Security*. Wiley, New Jersey, USA.

92. Kaushik, D., Garg, M., Annu, A. Gupta, & Pramanik, S. (2021). Application of Machine Learning and Deep Learning in Cyber Security: An Innovative Approach. In Ghonge, M., Pramanik, S., Mangrulkar, R., & Le, D. N. (eds) *Cybersecurity and Digital Forensics: Challenges and Future Trends*. Wiley, New Jersey, USA.

# 14 Clustering of Big Data in Cloud Environments for Smart Applications

*Rohit Anand*
G. B. Pant DSEU Okhla-1 Campus (formerly GBPEC), New Delhi, India

*Vipin Jain, Anushi Singh, Disha Rahal, Prachi Rastogi, and Avinash Rajkumar*
Teerthanker Mahaveer University, Moradabad, India

*Ankur Gupta*
Vaish College of Engineering, Rohtak, India

## CONTENTS

14.1 Introduction ....................................................................................................228
    14.1.1 Big Data ...........................................................................................228
        14.1.1.1 Characteristics of Big Data .............................................228
        14.1.1.2 Advantages of Big Data Processing ................................229
    14.1.2 Cluster .............................................................................................229
    14.1.3 Cloud Computing ...........................................................................229
        14.1.3.1 Types of Cloud Computing .............................................230
        14.1.3.2 Opportunities for Cluster Computing in the Cloud ............230
        14.1.3.3 Cloud for Big Data ........................................................231
    14.1.4 Smart Applications of Big Data .......................................................232
    14.1.5 Issues in Using Cloud Services .......................................................233
14.2 Literature Review ............................................................................................233
14.3 Problem Statement ..........................................................................................238
14.4 Research Methodology ....................................................................................238
14.5 Proposed Work ...............................................................................................238
14.6 Result and Discussion .....................................................................................240
    14.6.1 K-means Algorithm .........................................................................240
    14.6.2 EM Algorithm .................................................................................240
    14.6.3 MATLAB Simulation for Comparative Analysis of Security ............241
    14.6.4 Man-in-the-Middle Attack ...............................................................242
    14.6.5 Brute Force Attack ..........................................................................242
    14.6.6 Denial-of-Service Attack .................................................................242
14.7 Conclusion .....................................................................................................244
14.8 Future Scope ..................................................................................................244
References ..............................................................................................................244

DOI: 10.1201/9781003319238-14

## 14.1  INTRODUCTION

Data administration and analysis has become more challenging as a result of the rise of big data in the IT sector [1]. It's important to think about factors like scale, variety, velocity, value, and complexity. Clustering greatly simplifies the handling of massive data sets. Cheaper and more convenient access to computer resources including servers, data storage, database management systems, networking, analytics, and artificial intelligence is made possible via cloud computing. In addition to improving the cluster's availability and security, the vast array of services made available with cloud-based clustering drastically cuts the time and energy needed to get up and running [2].

### 14.1.1  BIG DATA

Big data is used to describe the massive amount of data that continues to increase at an exponential rate. Due to its sheer size and complexity, big data presents a significant challenge for existing data management systems. "Big data" refers to extremely large data sets. The following are some instances: Every day the Fresh York Stock Exchange creates around a terabyte of new trade data; Facebook claims that every day, at least 500 terabytes of data are added to the site's databases [3, 4]. The vast bulk of this information comes from user actions like sharing media files.

#### 14.1.1.1  Characteristics of Big Data

Some properties of big data are shown in Figure 14.1.

a) Volume: Even the name "Big Data" suggests a massive undertaking. The quantity of available data is crucial for extracting maximum value from it. Furthermore, whether or not anything qualifies as "Big Data" is determined by the quantity of data involved. Therefore, "volume" needs to be considered while working with big data approaches [5].

b) Variety: The second component of big data that needs to be considered. The term "variety" can refer to both organized and unstructured forms of information. Once upon a time, most programs relied only on spreadsheets and databases to save and retrieve their data [5].

c) Velocity: Used as a synonym when it comes to data creation speed. True potential is determined by the rate at which data is collected and analyzed. The term "big data velocity" is used to describe the rapidity with which information is gathered from a wide range of places [5].

d) Veracity or Inconsistency: This pertains to the fact that the data is not always consistent, which makes it challenging to handle and manage effectively [5].

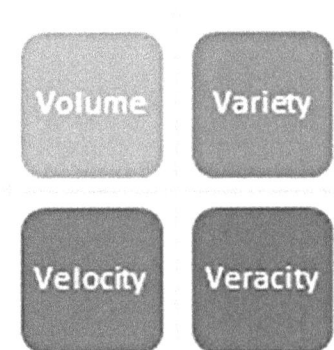

**FIGURE 14.1**   Characteristics of Big Data

### 14.1.1.2 Advantages of Big Data Processing

Big data processing capabilities are advantageous for several reasons, including:

- Businesses may now refine their strategies with the help of social data obtained from social media and search engine platforms like Facebook and Twitter [6].
- Increased quality of service to customers when big data tools supplant older forms of gathering input from those customers. In these innovative frameworks, consumer responses are studied and assessed with the use of big data and natural language processing technologies [6].
- Possible risks of the product or service are identified and mitigated before they have a chance to have a significant impact [6].
- It is possible to pre-process larger data sets in the big data environment before sending them to the warehouse, which increases operational efficiency. The combination of dig data with data warehouses allows businesses to eliminate the burden of storing data that is seldom used [6].

### 14.1.2 CLUSTER

Two or more computers – or nodes – work together to fulfill a shared objective in a computer cluster. In this way, heavy, parallelizable workloads may be distributed among the cluster's computing nodes [7]. Since each computer's combined memory and processing capabilities may help numerous workloads, performance is improved. Since the nodes in a computer cluster need to be able to talk to one another, an internode network is essential. Clustering nodes together requires special software. Each node might use its local storage device, or they could share a storage device. Typically, one of the nodes in a cluster is designated as the "leader node," which acts as the cluster's main point of entry. For example, this node may be responsible for delegating tasks to subordinates, collecting results, and then relaying the information to an external source. In addition, a cluster's communication between nodes should be optimized to decrease latency and eliminate bottlenecks [8]. There are several distinct types of cluster computing. Computer clusters and supercomputers are used to solve advanced computational problems in high-performance clusters. They are used to carrying out tasks that necessitate the utilization of nodes to communicate. Several nodes work together in a distributed fashion to increase throughput.

i. Load-balancing clusters: Demand for system resources is dispersed across computers that are all utilizing the same or comparable software or data [9, 10]. That way, no one node is overloaded with tasks. Host computer systems often employ distributed file systems.
ii. In case of failure, high-availability (HA) clusters are designed to have backup nodes ready to take over. Business operations, complicated databases, and consumer services like websites and file-sharing networks are just some of the examples of the kinds of computer services that run nonstop. They guarantee 24/7 data accessibility for their clientele.

### 14.1.3 CLOUD COMPUTING

The term "cloud computing" is used to refer to the practice of utilizing a third-party service provider to house and manage various IT infrastructure components, such as servers, data storage, development tools, and networking capabilities, and making them available over the Internet on demand. The communications service provider (CSP) charges customers every month for the use of various resources. The following are some advantages of cloud computing over on-premises IT. To save on expenses, you should consider migrating some or your entire on-premises infrastructure to a cloud service [11].

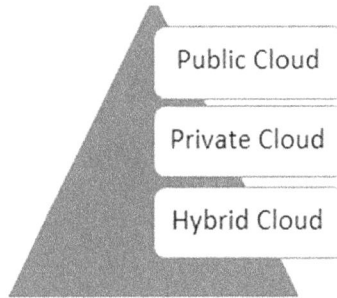

**FIGURE 14.2**   Types of Cloud Computing.

Rather than waiting weeks or months for IT to reply to a request, acquire, and set up suitable hardware and install software, your firm may start using corporate applications within minutes via the cloud. In particular, software and supporting infrastructure available in the cloud are useful for developers and data scientists [12].

- As opposed to investing in surplus infrastructure that sits idle during slow times, businesses may quickly and affordably adjust cloud-based infrastructure in response to demand spikes and valleys. To better serve users everywhere, you may use the global reach of the cloud service you choose.

Cloud computing may also be used to describe the physical and logical systems that enable cloud services. With the use of specialized software, servers, operating systems, networks, and other parts of infrastructure may be abstracted so that they can be shared and split without respect to physical constraints. For instance, a single physical server may be split up into several virtual servers. Cloud providers can get the most out of their data centers because of virtualization [13].

### 14.1.3.1   Types of Cloud Computing

We can break down cloud computing as shown in Figure 14.2 [14].

a) Public cloud: When a cloud service provider makes its computer resources accessible to consumers through the Internet, we say that they are operating in the public cloud. These tools could be provided at no cost, or you might have to pay a fee to utilize them, or both. Public cloud providers frequently offer high-bandwidth network access to their data centers and the hardware they run on to guarantee optimal performance of their client's applications [14].

b) Private cloud: In a private cloud, just one client uses the infrastructure and resources made available to them. Cloud computing's scalability, adaptability, and ease of service delivery are all present in a private cloud system. A private cloud is often hosted by the business itself.

c) Hybrid cloud: By definition, a hybrid cloud combines elements from both public and private clouds. The term "hybrid cloud" refers to the practice of merging a company's private cloud services with those of a public cloud provider to create a single, highly scalable environment for running the company's applications and workloads.

### 14.1.3.2   Opportunities for Cluster Computing in the Cloud

Cloud-based clustering can decrease the time and effort necessary to get started, while simultaneously delivering a wide range of services to increase the cluster's availability and security. In the cloud, there are several reasons to run computer clusters [15]:

- Launching a production-quality cloud cluster takes only a few minutes, regardless of whether you need a tiny cluster of just ten nodes with hundreds of active cores or a large cluster of more than 100,000 nodes. It may take weeks or months to get a new cluster up and running locally.
- The usage of preemptible virtual machines (VMs), long-term use discounts, and dynamic scalability are all features available on Google Cloud that help minimize the cost per run. Depending on how many jobs are waiting in line, more or fewer nodes may be added or withdrawn [16].
- In many situations, the computational analysis is produced by a team of professionals from various organizations. Access to data and analytic tools may be regulated at the project level with the aid of Google Cloud's identity and access management solutions. Enabling a single point of access for all authorized users ensures that no data needs to be copied, versioned, or synchronized between clusters [17].
- Running clusters in the cloud allows for each team or group to have its cluster since the cost of work is based simply on the total core hours, rather than the number of instances. It will be easier for policymakers to handle the issue of multi-group use if they utilize this approach. Any specialized cloud cluster may be adjusted to perform optimally for a certain program [17].
- To execute large-scale calculations, researchers must first devote significant time and effort to data preparation. If these scientists migrate to the cloud, they'll have access to all the big data resources available there. The results from the computers must also be analyzed [18].

### 14.1.3.3 Cloud for Big Data

**Infrastructure as a Service (IaaS) in a public cloud:** Users of big data services may get almost endless data storage and processing capacity by using the infrastructure of a cloud provider [19, 20]. By having cloud service providers handle the difficulties and costs associated with managing the underlying hardware, IaaS enables business clients to build cost-effective and easily scalable IT solutions. Instead of buying, installing, and integrating hardware on their own, company customers may simply use the cloud resource on an as-needed basis, which is especially helpful when the size of their operations varies or when they're trying to grow [21, 22].

**Platform as a Service (PaaS) in a private cloud:** Big data technologies like Hadoop and MapReduce are increasingly being included in PaaS suppliers' services. These technologies free users from the difficulties of managing software and hardware components separately. It's common practice for web developers to use isolated PaaS environments throughout the lifecycle of a project, from initial design and prototyping through testing and production hosting. PaaS may be used by organizations to construct their internal software, in particular, to set up isolated environments for testing and development [23].

**Software as a Service (SaaS) in a hybrid cloud:** Business owners increasingly recognize the need of listening to their customers, particularly through social media. The software and social media data for the analysis are both supplied by the same SaaS provider. The most common type of SaaS used by companies is office productivity software. SaaS may be used for a variety of business processes, including bookkeeping, sales, invoicing, and strategic planning. Companies can choose to utilize either a single program that can handle everything or a collection of programs that specialize in different areas. Any employee with a subscription to the service may log in from any work computer to utilize it. To find a solution that works better for them, they may just move to a different program. Any number of people, from a single individual to an entire organization with hundreds of employees, may be permitted to use a program [24–26].

| Tracking Shopping Behavior | Recommendation | Smart Traffic System | Secure Air Traffic System |
|---|---|---|---|
| Auto Driving Cars | Virtual Personal Assistant tool | IoT | Education Sector |
| | Energy Sector | Media and Entertainment Sector | |

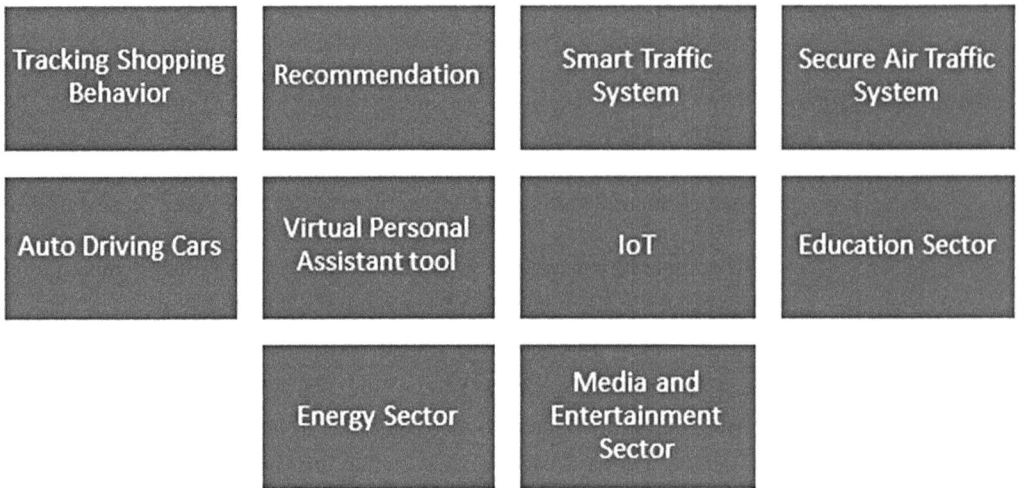

**FIGURE 14.3**   Smart Applications of Big Data.

### 14.1.4 SMART APPLICATIONS OF BIG DATA

This section focuses on smart applications of big data (as shown in Figure 14.3).

1. **Tracking Shopping Behavior:** Stores like Amazon, Walmart, Big Bazar, etc., have to keep track of their customer's purchasing habits, brand preferences, favorite goods, and other personal information [27, 28]. The banking industry makes use of client spending behavior data to make personalized offers, such as a rebate or discount when a customer uses the bank's credit or debit card to make a purchase.
2. **Recommendation:** Big retail stores can make recommendations to their customers based on their spending habits and buying patterns. Online retailers like Amazon, Walmart, and Flipkart offer suggestions for related items. Customers' product searches are monitored, and suggestions are made based on those findings.
3. **Smart Traffic System:** Information regarding the state of traffic on various roads is gathered through cameras mounted by the side of the road, at city entrances and exits, and from GPS devices installed in vehicles [29, 30]. All of this information is examined so that faster, less time-consuming alternatives can be suggested. By analyzing massive amounts of data, cities can create intelligent transportation systems. Saving money on gas is a bonus.
4. **Secure Air Traffic System:** Sensors are located in various parts of the aircraft to monitor flying conditions. Information on factors, including air velocity, humidity, temperature, and pressure are all gathered by these sensors [31, 32]. Using this information, flying parameters may be established and adjusted to optimize the experience. Using the data collected during a flight, engineers may determine how much longer certain equipment will last before it has to be serviced or replaced [33, 34].
5. **Auto Driving Cars:** A automobile can be driven with no human interpretation thanks to big data analysis. A sensor is installed at various points in the automobile camera system and collects information about the environment, including the size of vehicles and obstacles, the distance between them, and so on [35, 36].
6. **Virtual Personal Assistant Tool:** Virtual personal assistant tools (like Siri on Apple Devices, Cortana on Windows, and Google Assistant on Android) may employ big data research to better respond to consumers' questions. This app remembers the user's location, local time, season, and other variables relevant to the query asked. It analyzes all of the information and returns a result [37].

7. **IoT:** Sensors connected to the Internet of Things are embedded in manufacturing equipment to record machine performance. By analyzing this data, businesses can foresee how long their machines will run smoothly before breaking down and plan maintenance accordingly. This prevents having to replace the entire equipment and the associated costs [38–40].

8. **Education Sector:** To find people who are interested in taking their courses online, companies who offer them use big data. When someone looks for a how-to video on YouTube, providers of related courses, both online and offline, will advertise their offerings to them.

9. **Energy Sector:** Every 15 minutes, smart electric meters record the amount of energy used and transmit that information to a server. There, the data is processed to determine when in the day the city's power load is lowest. If a factory or a housekeeper wanted to save money on their electricity bill, the system might tell them to run their power-hungry machinery at night, when demand is lower [41].

10. **Media and Entertainment Sector:** Data acquired from consumers is analyzed by media and entertainment service providers like Netflix, Amazon Prime, and Spotify. Business decisions are made based on collected and evaluated data, such as the most popular types of videos and music, the average time spent on the site, etc. [42].

### 14.1.5  Issues in Using Cloud Services

The following are a few major concerns with cloud services:

a. **Data Security:** Data security is paramount; therefore, businesses should check that their cloud provider agreement has that provision. Some people feel uneasy about disclosing personal information to third parties. The inability to guarantee the security of their data may discourage business leaders from using cloud computing [42, 43].

b. **Performance:** The agreement should include a discussion of, and where feasible a quantification for, key cloud performance metrics. To be valid, an exception must have a clear explanation. The service level agreement (SLA) between a consumer and provider should spell out every detail of the relationship between the two parties [44].

c. **Compliance:** The compliance requirements of a company must be met by the cloud service provider. Concerns regarding government oversight have been voiced by several businesses. Market researchers estimate that around half of consumers worry that they will be stuck with a single cloud storage supplier [45].

d. **Legal Issues:** The location of the cloud's physical resources should not raise any legal concerns for the company. Confidentiality and privacy breaches are made more likely due to the cloud, which raises several legal concerns related to privacy issues with data kept in different cloud locations.

e. **Costs:** Since the cloud provides a pay-as-you-go form of the cost paid by the organization, businesses need to be aware of all associated expenses and use the services in a measured manner.

## 14.2  LITERATURE REVIEW

Georgios Skourletopoulos et al. (2017) focused on applications, big data services, and potential future developments in mobile cloud computing (MCC) over 5G mobile networks. The availability of more computationally powerful mobile devices and the ever-rising need for high data rates and mobility required by several mobile network services have paved the way for research on 5G mobile networks, which are scheduled to be implemented around 2020. Large and intricate location-aware datasets, however, are beyond the scope of current spatial computing methods. Cloud computing and mobile computing were combined to form MCC, which allows users to have access to cloud-based services via their mobile devices. When it comes to processing, storing, and other expensive tasks, mobile apps make use of cloud technology by being run on external resource providers rather than on the devices themselves. To help novice researchers get up to speed on MCC in the 5G era, this article gives an overview of the current technologies, techniques, and applications as well as an examination of open research problems and future difficulties in this field [1].

Kai Peng et al. (2018) introduced intrusion detection over large data in a mobile cloud setting utilizing hierarchies and principal component analysis. With the advent of the widespread use of big data and MCC, cyber security has become an increasingly pressing issue. As a result, it was crucial to make it possible to use a technique of intrusion detection applied to huge data in a mobile cloud setting. In our previous work, we proposed a technique for enormous data called Mini Batch K-means with Principal Component Analysis (PMBKM), which may quickly handle the clustering problem for intrusion detection in massive data, but it necessitates the number of clusters to be established in advance. To find the best value for clustering, it is necessary to regularly test different values and compare the results. In this study, we propose a unique clustering method, dubbed Balanced Iterative Reducing and Clustering Using Hierarchies with Principal Component Analysis (PBirch), to address the aforementioned issue. In contrast to PMBKM, the experimental results show that PBirch may provide a reasonable clustering result without presetting clustering values and that this result can be improved upon by optimization of the necessary parameters. Data clustering in PBirch takes less and less time as more and more clusters are generated. PBirch's processing time drops down exponentially when more clusters are added. Overall, our suggested approach has broad applicability for large data in mobile cloud settings [2].

Samiya Khan et al. (2018) presented big data analytics on the cloud: a review of the literature and prospects. Since the dawn of the information era, data of all shapes and sizes have mushroomed. The majority of data is expected to be stored in the cloud by 2016. The only way for businesses to learn anything from this data is if they have a system in place to collect it, clean it up, and analyze it. Because of this, cloud-based analytics has become an area of study that may be pursued. However, before this synergistic model can be widely implemented, several challenges must be overcome and dangers reduced. This article examines the work already done on the topic, as well as its obstacles, unanswered questions, and potential future research directions [3].

Wenzhun Huang et al. (2019) looked at Hadoop-based signal clustering and analytical hierarchy model: a new method for cluster computing. Countless resources are now available online thanks to the explosion of the Internet. For the source to offer accurate information, it must first be processed, evaluated, and connected. To access and secure such massive amounts of data, a reliable storage and management mechanism is essential. Both organized and unstructured information may be found online but processing it requires sophisticated machinery. Cloud computing is the answer to the problem of where and how to keep data. However, there are several powerful technologies available for data access and security, including parallel and map reduction, approaches. On the other hand, these approaches struggle when dealing with massive amounts of data. In this post, they look at how the Hadoop data model may be used to achieve high-performance computing on big data in the cloud. The investigation compared the suggested approach to earlier studies. Therefore, with a receiver throughput of 770 kbps, the suggested technique obtained a packet delivery ratio of 0.51 with an elapsed time/word transfer of 0.71, significantly outperforming earlier studies. Using Hadoop and a single cluster with an Analytic Hierarchy Process (AHP) to compute data reduces processing time and communication issues while providing fault tolerance[4].

R. Joseph Manoj et al. (2019) presented for massive datasets, an Ant Colony Optimization (ACO) based Artificial Neural Network (ANN) feature selection technique. Feature selection is a method for narrowing down a large dataset to a more manageable size by selecting individual records based on their shared characteristics. It can help reduce the size of the massive dataset. To ensure precise prediction or output in big data analytics, it filters out irrelevant information from the original data. In the work presented here, ACO and ANN algorithms are used to build a feature selection algorithm method for text classification (ANN). The simulations conducted using Reuter's data set demonstrated the efficacy of this hybrid strategy [5].

Safanaz Heidari et al. (2019) provided density-dependent clustering in large datasets using MapReduce. The Density-based Spatial Clustering of Applications with Noise (DBSCAN) algorithm was a well-known technique among density-based clustering algorithms, and its primary strength was in its capacity to identify clusters of varying forms and sizes, as well as noise data. However, this approach has certain issues, such as its inability to locate clusters of varying densities. As the information age continues to advance rapidly, however, new technologies were needed to store and extract knowledge from the massive amounts of data that are being generated every day. Big data refers to a massive amount of information that exceeds the processing power of commonly used programmers. In this work, we make an effort to provide a novel approach for clustering huge data of varying densities on a Hadoop platform using MapReduce. The central concept of this study is the use of local density to determine the density of individual points. If you use this method, you won't have to worry about joining together clusters with unequal densities. Using the MapReduce paradigm, the suggested method is built and compared to existing algorithms, demonstrating superior scalability and performance while dealing with data sets of varied densities [6].

P. Praveen et al. (2019) reviewed clustering in large datasets, or big data: adapting standard data mining procedures for use with massive datasets. The phrase "big data" has gained widespread use in the modern data environment to describe these exponentially growing datasets. Processing this massive amount of information requires very effective knowledge extraction methods. There are various methods for handling massive amounts of data, but one that stands out is cluster analysis, a data mining method that has many practical applications and is thus frequently utilized in fields like information retrieval, image processing, machine learning, etc. Data objects are clustered into classes based on their similarities or differences using this procedure. There are a lot of hidden costs involved with large data clustering, such as coming up with new methods or converting effective data mining techniques for a distributed setting. In this study, we provided a high-level review of both classic clustering methods and contemporary clustering model developments for processing big data, allowing for more effective processing and analysis of today's massive datasets. Today, the clustering of large datasets is a booming area of study with plenty of room for the development of new and better clustering methods [7].

Ahmed Ismail et al. (2019) introduced intelligent big data analytics for healthcare analysis: lessons learned future directions, and open questions. Due to huge populations and the difficulties of covering all patients by the available physicians, rising demand and expenditures for healthcare were an issue. Healthcare data processing and administration become difficult to handle due to issues with the data itself, such as its irregularity, high-dimensionality, and sparsity. Several academics have addressed these issues, developing healthcare solutions that are both effective and scalable. They introduce healthcare analytics and application techniques and systems along with some relevant solutions. Our strategy necessitates the addition of middleware between the many heterogeneous data sources and the MapReduce Hadoop cluster. The issue of efficiently processing data from several sources was addressed by the solution [8].

S. Sudhakar Ilango et al. (2019) presented big data clustering inspired by simulations of bee colonies, for optimization. As one of the main issues, the conventional algorithm takes more time to execute and has difficulty processing big amounts of data. In contrast to the high degree of similarity across clusters, the degree of similarity between different clusters is very modest. Clustering optimization algorithms are a method for making the most efficient use of limited means. When dealing with data sets with a lot of dimensions, the standard optimization procedure breaks down. Minimizing execution time and optimizing optimum clusters for different dataset sizes are primary goals of the proposed artificial bee colony (ABC) technique. To cope with this, we are transitioning to a distributed environment, which will improve both speed and accuracy. For numerical optimization problems, especially those involving clustering, the suggested ABC method emulates the foraging behavior of actual bees. The algorithm's timings are plotted against a range of dataset sizes. The ABC algorithm's observer and worker phases provide the results from which the classification error % is calibrated for a range of fitness and probability values. Hadoop's MapReduce framework was used to bring the suggested ABC algorithm to life. The suggested ABC approach decreases both the execution time and the classification error for choosing the best clusters, as shown by testing results. The results demonstrate that the suggested ABC scheme outperforms particle swarm optimization (PSO) and differential evolution (DE) in terms of efficiency of time [9].

T. Ramalingeswara Rao et al. (2019) provided a look at the parts, pieces, and tech that make up the big data system. Unstructured data and large, real-time datasets were beyond the capabilities of conventional database management systems. Big data was the result of a large number of little datasets that have not been unified or standardized in any way, making it difficult to store, manage processes, analyze, display, and extract relevant insights using conventional database techniques. While big data is trending in the right direction, numerous technological challenges must be overcome before it can benefit society. The purpose of this article was to provide a high-level overview of a comprehensive big data system, including the many steps and crucial components involved in processing large data at each step. In this paper, they focus on the characteristics involved in the data management process and make comparisons between different distributed file systems and MapReduce-supported NoSQL databases. In addition, we provide several distributed/cloud-based machine learning (ML) instruments that are important in the process of designing, developing, and deploying data models. Some of the distributed ML tools examined in this work include Mahout, Spark MLlib, and FlinkML. Furthermore, we divide analytics into categories according to data type, industry, and use case. Different visualization tools are distinguished by their functionality, analytical power, and development environment support. Moreover, we take a systematic, comparative look at the state of the art in big data tools and technologies, including distributed/cloud-based stream processing tools. Additionally, we compare and contrast the features of several SQL Query tools for Hadoop across ten criteria. Finally, we provide some key considerations for future big data research possibilities and objectives. By researching big data infrastructure tools in light of current advancements, one gains a deeper comprehension of how various tools and technologies might be used to address practical problems [10].

Somnath Mazumdar et al. (2019) looked at methods for storing and locating huge data in the cloud: a survey. These days, the amount of data that computers need to sift through and use grows exponentially. The so-called "Big Data" phenomenon has placed a strain on already-existing technology to handle such enormous datasets in a scalable, quick, and efficient manner. Recent applications and the current user support from multi-domain computing helped ease the shift from data-centric to knowledge-centric computing. However, there is still a problem with storing and locating such large data sets and migrating them across data centers. Due to the dynamic nature of applications and data centers, it is essential to analyze trends of data access or consumption. One of the main goals was to find a better place to store data that would both cut down on the overall cost of data placement and improve application performance. In this white paper, we provide a comprehensive review of the current state of the art in cloud-centric big data deployment and data storage strategies. It's an effort to draw attention to the real connection between the two so that big data management can benefit from it more effectively. They are particularly interested in management issues when seen through the lens of structural rather than functional qualities. When everything is said and done, readers will be able to appreciate the in-depth study of various big data management systems and be led in making an informed choice based on how well it meets their functional and non-functional application needs. In addition, obstacles are provided to draw attention to the gaps in big data management and indicate the direction it must change shortly [11].

Li Zhu1 et al. (2019) focused on studies of enhanced parallel collaborative filtering algorithms for massive data mining. In the age of big data, the traditional parallel collaborative filtering approach became unsuitable for the efficient and precise processing of data. This paper improves the standard parallel collaborative filtering algorithm by analyzing its execution flow, discussing its drawbacks, and finally detailing the steps of improving the algorithm, beginning with the generation of node scoring vectors and continuing through the retrieval of neighboring nodes and the formation of recommendation information. The expanded parallel collaborative filtering approach was verified in terms of its performance on three metrics: load, speedup, and accuracy. Compared to the usual parallel technique based on the co-occurrence matrix, the improved parallel collaborative filtering algorithm proposed in this study shows significant improvements in both running efficiency and suggestion accuracy [12].

Rabindra Kumar Barik et al. (2019) presented possibilities and constraints of hybrid mist-cloud systems for massively parallel geographic big data analysis and processing. Big geographic data was becoming a growing field for the cloud and fog computing paradigms to store, process, and analyze. Mist computing was the newest fad, and it improves upon fog and cloud ideas for the computing process by making use of edge devices to improve throughput and decrease latency at the client's edge. In this study, the authors highlighted the recent surge in the use of mist computing for geospatial analysis. In addition, it developed a mist computing framework, dubbed MistGIS, to extract mining-related insights from geographic large data. The MistGIS platform is used by the Tourism Information Infrastructure Management as well as the Faculty Information Retrial System. Management of the geospatial data about tourism destinations like lakes, mountains, rivers, and woods, and religious structures like churches, mosques, and monuments is what Tourism Information Infrastructure is all about. All of the involved parties or consumers may benefit from having more information to utilize in future studies. The city of Bhubaneswar, India (also known as the "Temple City") was used as a case in this research. The Faculty Information Retrial System enabled a variety of features for locating specific faculty members' biographical information, research interests, contact information, and email addresses throughout India's 31 NITs. Raspberry Pi is used for the framework's microprocessor. The MistGIS platform was shown effective in preliminary analyses that made use of cluster and overlay techniques. From what we can see, mist computing improves cloud and fog computing capacity to analyze geographically extensive data [13].

Table 14.1 shows the brief literature survey.

## TABLE 14.1
## Literature Survey

| S.No. | Author/Year | Title | Methodology | Limitation |
|---|---|---|---|---|
| [1] | Skourletopoulos/ 2017 | Applications, Big Data Services, and Future Opportunities for Mobile Cloud Computing in 5G Mobile Networks. | Big data, cloud computing | Lack of technical work |
| [2] | Peng/2018 | Intrusion detection over huge data in a mobile cloud context employing hierarchies for balanced iterative reduction and clustering using principal component analysis (PBirch). | Cloud computing, big data | There is no implication for future |
| [3] | Khan/2018 | Insights on the state of cloud-based big data analytics and where the field is headed next. | Cloud computing | There is limited scope in future |
| [4] | Huang/2019 | Hadoop-based signal clustering and analytical hierarchy model: a new approach to cluster computing. | Cluster computing | The scope of this research is very narrow |
| [5] | Joseph Manoj/2019 | The Big Data Feature Selection Algorithm Based on Aggregate Coefficients and ANN. | Big data | The performance of this research is very low |
| [6] | Heidari/2019 | MapReduce-powered, density-dependent grouping of massive datasets. | Big data | There is a lack of performance |
| [7] | Praveen/2019 | Cluster Analysis for Massive Data Sets: Traditional Data Mining Methods Applied to the Big Data World. | Big data, clustering | Lack of accuracy |
| [8] | Ismail/2019 | The Smart Big Data Analytics of Healthcare: Past, Present, and Future. Reviews, Problems, and Suggestions. | Big data | There is less technical work |
| [9] | Ilango/2019 | An Artificial Bee Colony-based Clustering Approach to Big Data Optimization. | Big data | Lack of accuracy |
| [10] | Rao/2019 | A look at the many parts, software, and hardware that make up the big data infrastructure. | Big data | There is a lack of performance |
| [11] | Mazumdar/2019 | Data storage and deployment strategies in the Cloud-Big Data ecosystem: a survey. | Big data, cloud environment | There is a lack of security |
| [12] | Zhu/2019 | Investigations into massive data mining using a refined parallel collaborative filtering method. | Big data, cloud environment | There is a lack of technical work |
| [13] | Barik/2019 | Possibilities and constraints of hybrid mist-cloud systems for processing and analyzing enormous quantities of geographic big data. | Big data, cloud environment | The performance of this research is very low |

## 14.3   PROBLEM STATEMENT

It has been observed that there are several types of research in the area of big data processing. Clustering is playing a significant role in the management of big data. But it is quite challenging to implement big data processing in a cloud environment. There is a need to propose a robust clustering approach to manage the clustering of big data in a cloud environment. Issues with previous research work are lack of scalability and flexibility. However, there are hybrid approaches in existence, but these approaches have several limitations such as lack of security. Thus, there is a need to propose an application that should make use of clustering in big data processing over the cloud for smart applications.

## 14.4   RESEARCH METHODOLOGY

In the proposed research the applicability of big data clustering in a cloud environment has been considered for smart application. Different clustering mechanisms are considered, and a comparative study of clustering mechanisms is made. A hybrid model has been proposed that is making use of clustering in big data processing in association with encryption to enhance security. Then the issues in previous research such as performance, security, scalability, and flexibility are considered. The complete methodology is shown in Figure 14.4.

## 14.5   PROPOSED WORK

The suggested study takes into account the feasibility of large data clustering in a cloud setting. Multiple clustering methods are taken into account, and each method is analyzed and compared to the others. To increase security, a hybrid strategy has been developed that uses clustering in big data processing in tandem with encryption. After that, we take into account the problems that have arisen in the prior studies, including those related to performance, security, scalability, and adaptability.

In the proposed work as shown in Figure 14.5, big data has been taken as input, and preprocessing of big data takes place before applying the clustering mechanism to make data manageable.

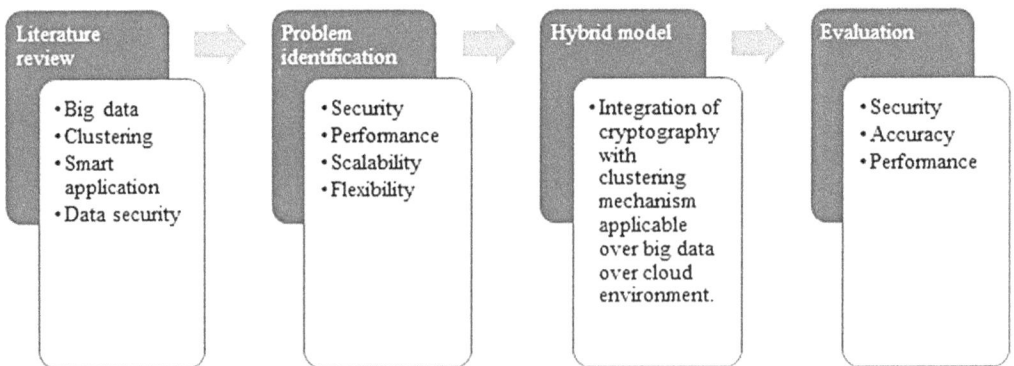

**FIGURE 14.4**   Research Methodology.

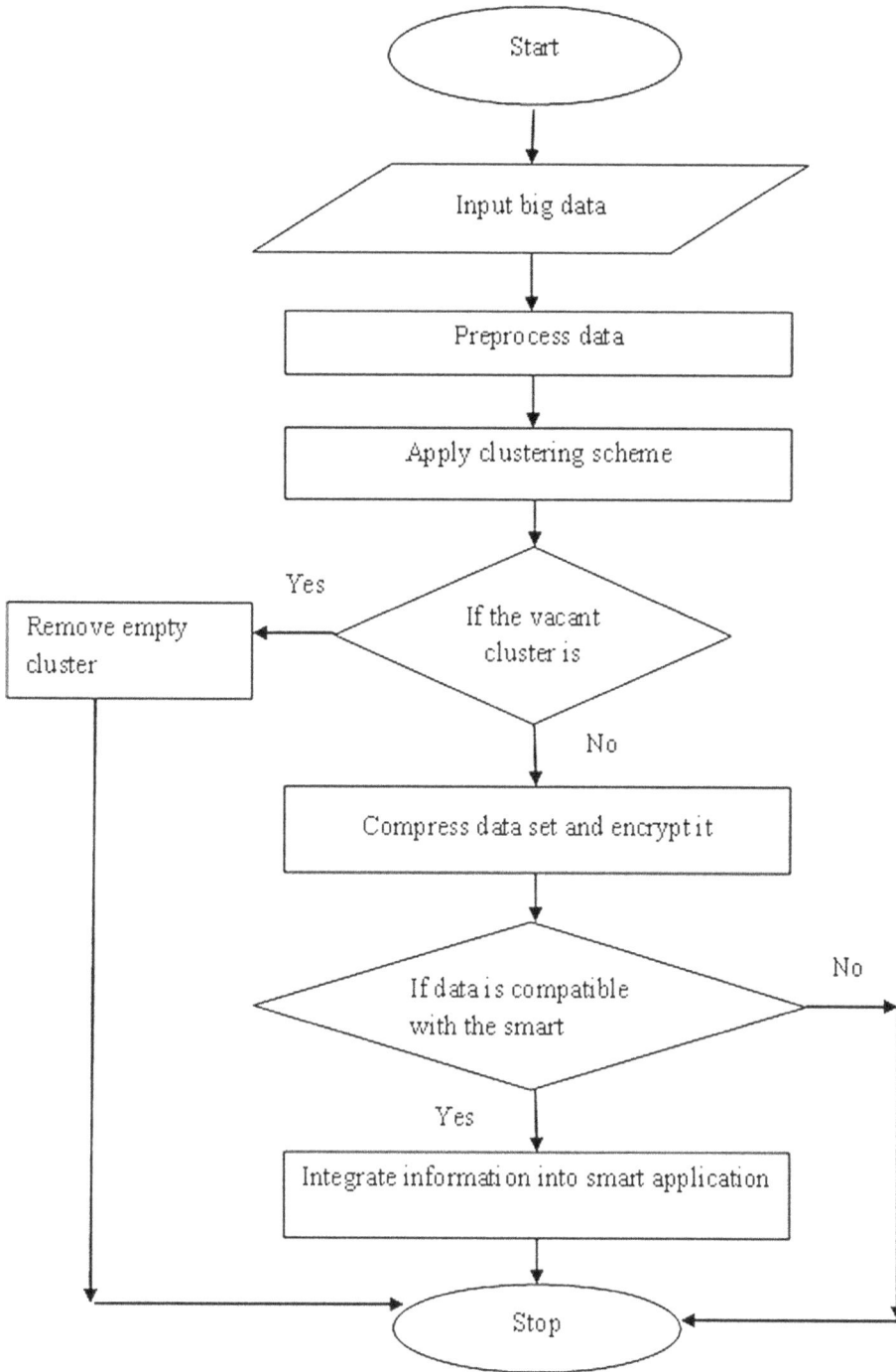

**FIGURE 14.5** Flow Chart for Proposed Work.

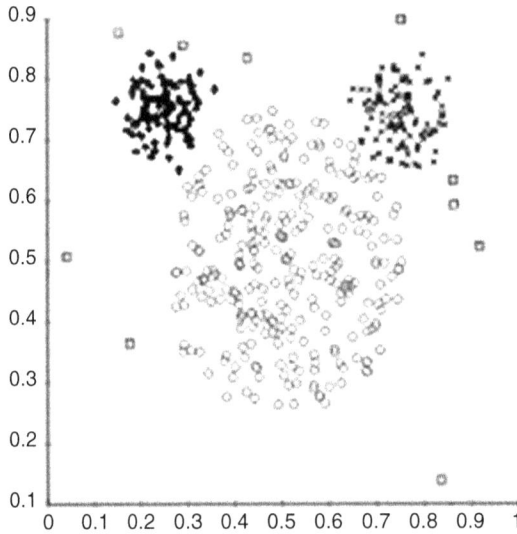

**FIGURE 14.6**   Original Data.

Preprocessed information is clustered and empty clusters are ignored. Data is considered cluster encrypted for security. Compatibility of data is considered with the smart application while data integration operation.

## 14.6   RESULT AND DISCUSSION

The present simulation is presenting a cluster of original data (as shown in Figure 14.6), k-means clustering, and expectation–maximization (EM) clustering.

### 14.6.1   K-MEANS ALGORITHM

$$J(V) = \sum_{i=1}^{c} \sum_{j=1}^{c_i} \left( \left\| x_i - v_j \right\| \right)^2$$

where,

    '$\|x_i - v_j\|$' is the Euclidean distance between $x_i$ and $v_j$,
    '$c_i$' is the number of data points in the $i^{th}$ cluster, and
    '$c$' is the number of cluster centers.

    K-means clustering is shown in Figure 14.7.

### 14.6.2   EM ALGORITHM

Combining other unsupervised ML methods, such as the k-means clustering algorithm, is what makes up the EM algorithm. It's an iterative method with two distinct phases. Estimating the unknown or latent variables is the first method. Consequently, it is known as the estimation/expectation step (E-step). In addition, the second mode is used to fine-tune the models' parameters to provide the best possible explanation for the data. "M-step" stands for "maximization step" and describes the second mode, as shown in Figure 14.8.

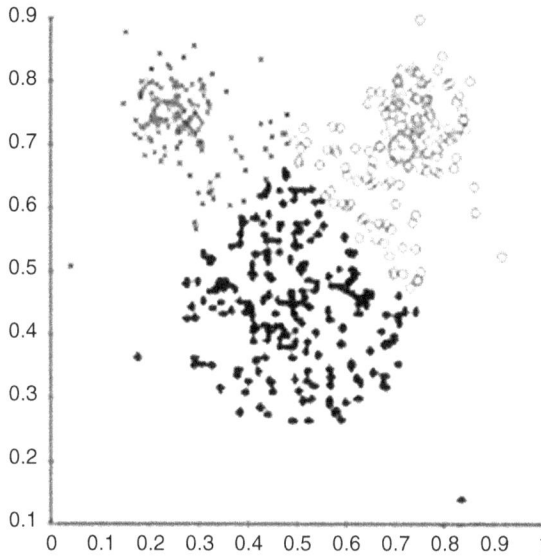

**FIGURE 14.7**    K-means Clustering.

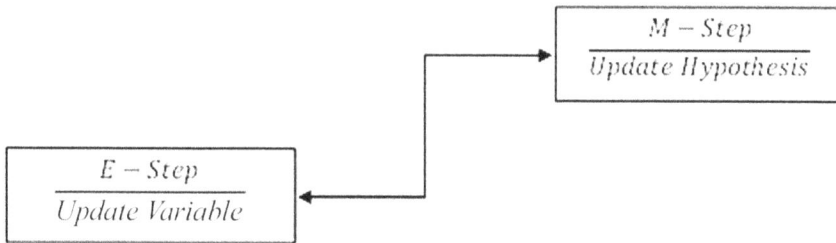

**FIGURE 14.8**    Relationship between E-step and M-step.

- Expectation step (E-step): The goal is to estimate (guess) all missing values in the dataset such that, once done, there are no more gaps in the data.
- Maximization step (M-step): Estimated information is used in the E-step, and the parameters are updated.
- When the values don't converge you'll need to keep doing E-steps and M-steps.

The main purpose of the EM method is to update the parameters in the M-step based on the estimated values of the missing data of the latent variables, which are derived from the available observed data of the dataset. The main purpose of the EM method is to update the parameters in the M-step based on the estimated values of the missing data of the latent variables, which are derived from the available observed data of the dataset. EM clustering is shown in Figure 14.9.

## 14.6.3    MATLAB SIMULATION FOR COMPARATIVE ANALYSIS OF SECURITY

In this part, we'll look at how the suggested changes will affect safety. As the number of assaults grows, fewer packets will be impacted by the suggested changes. Research has shown that DNA cryptography [12] is superior to RSA [13] and advanced RSA [14]. However, the work that is being presented is superior to DNA cryptography. Compared to RSA and DNA-based encryption methods, the number of impacted packets is lower in the case of the proposed work, as shown in the accompanying figures.

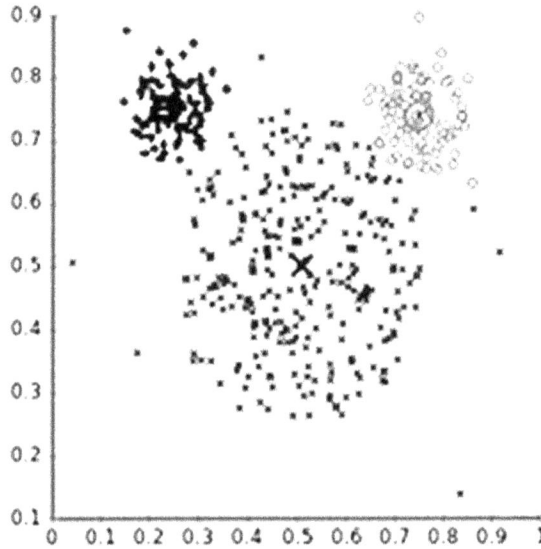

**FIGURE 14.9**    EM Clustering.

### 14.6.4  MAN-IN-THE-MIDDLE ATTACK

The effects on the packet for RSA, Advance RSA, and DNA cryptography, as well as the counter-measures recommended against these attacks, are detailed below. Comparative analysis of man-in-the-middle attack is shown in Figure 14.10 and Table 14.2.

### 14.6.5  BRUTE FORCE ATTACK

A brute force attack is an attempt to gain access to a system by repeatedly trying the login credentials. In addition, we use the usage of encryption keys and a covert website. The next section provides a comparative study of this assault. Comparative analysis of brute force attack is shown in Figure 14.11 and Table 14.3.

### 14.6.6  DENIAL-OF-SERVICE ATTACK

Denial-of-service (DoS) attacks are a specific kind of cyberattack in which the target is the availability of a service or resource on a computer or network. The likelihood of a DoS is lowered when packets are smaller and require less time to transmit across a network. As a result, the suggested work mitigates the effects of DoS. Table 14.4 and Figure 14.12 show the comparisons for the several types of DoS attacks.

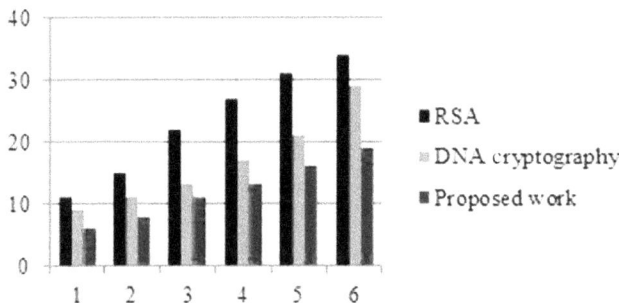

**FIGURE 14.10**    Comparative Analysis in Case of Man-in-the-Middle Attack.

**TABLE 14.2**
**Comparative Analysis of Man-in-the-Middle Attack**

| Number of Attacks | RSA | DNA Cryptography | Proposed Work |
|---|---|---|---|
| 10 | 11 | 9 | 6 |
| 20 | 15 | 11 | 8 |
| 30 | 22 | 13 | 11 |
| 40 | 27 | 17 | 13 |
| 50 | 31 | 21 | 16 |
| 60 | 34 | 29 | 19 |

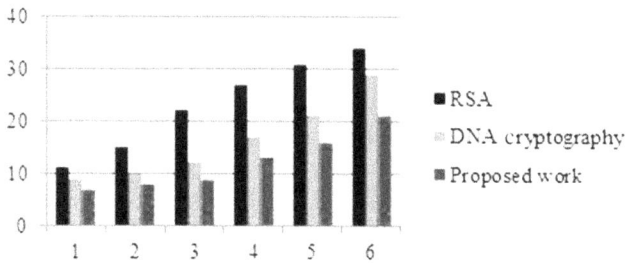

**FIGURE 14.11** Comparative Analysis in Case of Brute Force Attack.

**TABLE 14.3**
**Comparative Analysis of Brute Force Attack**

| Number of Attacks | RSA | DNA Cryptography | Proposed Work |
|---|---|---|---|
| 10 | 11 | 9 | 7 |
| 20 | 15 | 10 | 8 |
| 30 | 22 | 12 | 9 |
| 40 | 27 | 17 | 13 |
| 50 | 31 | 21 | 16 |
| 60 | 34 | 29 | 21 |

**TABLE 14.4**
**Comparative Analysis of Denial-of-Service Attack**

| Number of Attacks | RSA | DNA Cryptography | Proposed Work |
|---|---|---|---|
| 10 | 10 | 8 | 5 |
| 20 | 14 | 10 | 7 |
| 30 | 22 | 13 | 9 |
| 40 | 28 | 15 | 12 |
| 50 | 33 | 21 | 16 |
| 60 | 35 | 29 | 23 |

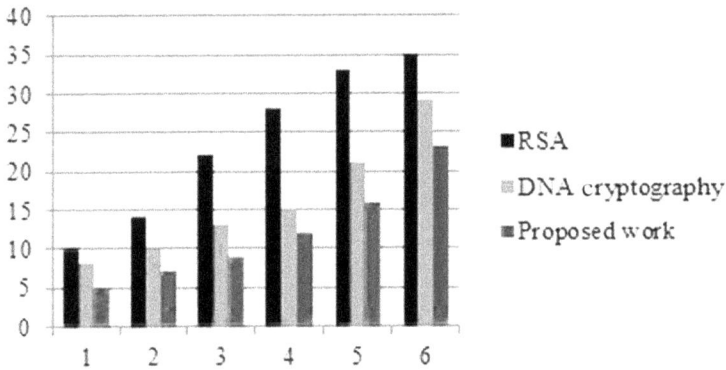

**FIGURE 14.12** Comparative Analysis in Case of Denial-of-Service Attack.

## 14.7  CONCLUSION

It has been concluded that data clustering performs a significant role in big data management [46–48]. Issues related to empty clusters have been considered in the present research. Big data security is also considered to secure contents. The significance of big data clusters and security has been considered to improve the real-time use of big data clustering for cloud environments. It is observed that the probability of the different types of attack is less in the case of the proposed work. Moreover, proposed research work is considering performance along with real-time usage.

## 14.8  FUTURE SCOPE

With the development of portable computing devices like laptops, palmtops, mobile phones, and wearable computers, it is now feasible to carry vast amounts of data with you at all times. The next logical step in the field of pervasive computing is the application of sophisticated data analysis to derive actionable information [49–52]. Problems arise when trying to access and analyze data from a device that is always in use. Classifiers, clusters, relationships, and other patterns might be hard to see on mobile devices. Data mining in an interactive setting is made more difficult by the limited-screen real estate. A further complex challenge is data management in a mobile setting. Integration between data mining technologies and our way of life has not yet been studied from a social or psychological perspective.

## REFERENCES

1. Skourletopoulos, G., Mavromoustakis, C. X., Mastorakis, G., Batalla, J. M., Dobre, C., Panagiotakis, S., & Pallis, E. (2017). Towards mobile cloud computing in 5G mobile networks: applications. *Big Data Services, and Future Opportunities*, 43–62. https://doi.org/10.1007/978-3-319-45145-9_3
2. Peng, K., Zheng, L., Xu, X., Lin, T., & Leung, V. C. M. (2018). Balanced Iterative Reducing and Clustering Using Hierarchies with Principal Component Analysis (PBirch) for Intrusion Detection over Big Data in Mobile Cloud Environment. In *Lecture Notes in Computer Science (Including Subseries Lecture Notes in Artificial Intelligence and Lecture Notes in Bioinformatics): Vol. 11342 LNCS*. Springer International Publishing. https://doi.org/10.1007/978-3-030-05345-1_14
3. Khan, S., Shakil, K. A., & Alam, M. (2018). Cloud-based big data analytics—a survey of current research and future directions. *Advances in Intelligent Systems and Computing*, 654, 595–604. https://doi.org/10.1007/978-981-10-6620-7_57
4. Sansanwal, K., Shrivastava, G., Anand, R., & Sharma, K. (2019). Big Data Analysis and Compression for Indoor Air Quality. In *Handbook of IoT and Big Data* (pp. 1–21). CRC Press, USA.

5. Bommareddy, S., Khan, J. A., & Anand, R. (2022). A Review on Healthcare Data Privacy and Security. In *Networking Technologies in Smart Healthcare* (pp. 165–187). CRC Press, USA.

6. Heidari, S., Alborzi, M., Radfar, R., Afsharkazemi, M. A., & RajabzadehGhatari, A. (2019). Big data clustering with varied density based on MapReduce. *Journal of Big Data*, 6(1). https://doi.org/10.1186/s40537-019-0236-x

7. Praveen, P., & Jayanth Babu, C. (2019). Big Data Clustering: Applying Conventional Data Mining Techniques in Big Data Environment. In *Lecture Notes in Networks and Systems* (Vol. 74). Springer, Singapore. https://doi.org/10.1007/978-981-13-7082-3_58

8. Ismail, A., Shehab, A., & El-Henawy, I. M. (2019). *Healthcare Analysis in Smart Big Data Analytics: Reviews, Challenges, and Recommendations*. Springer International Publishing, USA. https://doi.org/10.1007/978-3-030-01560-2_2

9. Ilango, S. S., Vimal, S., Kaliappan, M., & Subbulakshmi, P. (2019). Optimization using Artificial Bee Colony-based clustering approach for big data. *Cluster Computing*, 22, 12169–12177. https://doi.org/10.1007/s10586-017-1571-3

10. Rao, T. R., Mitra, P., Bhatt, R., & Goswami, A. (2019). The Big Data System, Components, Tools, and Technologies: A Survey. In *Knowledge and Information Systems* (Vol. 60, Issue 3). Springer, London. https://doi.org/10.1007/s10115-018-1248-0

11. Mazumdar, S., Seybold, D., Kritikos, K., & Verginadis, Y. (2019). A survey on data storage and placement methodologies for Cloud-Big Data ecosystem. *Journal of Big Data*, 6(1), 1–37. https://doi.org/10.1186/s40537-019-0178-3

12. Tripathi, A., Sindhwani, N., Anand, R., & Dahiya, A. (2023). Role of IoT in Smart Homes and Smart Cities: Challenges, Benefits, and Applications. In *IoT Based Smart Applications* (pp. 199–217). Springer, Cham.

13. Kaur, J., Sindhwani, N., Anand, R., & Pandey, D. (2023). Implementation of IoT in Various Domains. In *IoT Based Smart Applications* (pp. 165–178). Springer, Cham.

14. Chung, K., & Park, R. C. (2019). Chatbot-based heathcare service with a knowledge base for cloud computing. *Cluster Computing*, 22, 1925–1937. https://doi.org/10.1007/s10586-018-2334-5

15. Idwan, S., Mahmood, I., Zubairi, J. A., & Matar, I. (2020). Optimal management of solid waste in smart cities using internet of things. *Wireless Personal Communications*, 110(1), 485–501. https://doi.org/10.1007/s11277-019-06738-8

16. Babu, S. Z. D. et al. (2023). Analysation of Big Data in Smart Healthcare. In Gupta, M., Ghatak, S., Gupta, A., Mukherjee, A. L. (eds) *Artificial Intelligence on Medical Data*. Lecture Notes in Computational Vision and Biomechanics (Vol. 37). Springer, Singapore. https://doi.org/10.1007/978-981-19-0151-5_21

17. Pandey, B. K. et al. (2023). Effective and Secure Transmission of Health Information Using Advanced Morphological Component Analysis and Image Hiding. In: Gupta, M., Ghatak, S., Gupta, A., Mukherjee, A. L. (eds) *Artificial Intelligence on Medical Data*. Lecture Notes in Computational Vision and Biomechanics (Vol. 37). Springer, Singapore. https://doi.org/10.1007/978-981-19-0151-5_19

18. Sreekanth, N., Rama Devi, J., Shukla, A. et al. (2022). Evaluation of estimation in software development using deep learning-modified neural network. *Applied Nanoscience*, 2022. https://doi.org/10.1007/s13204-021-02204-9

19. Veeraiah, V., Khan, H., Kumar, A., Ahamad, S., Mahajan, A., & Gupta, A. (2022). Integration of PSO and Deep Learning for Trend Analysis of Meta-Verse. In *2022 2nd International Conference on Advance Computing and Innovative Technologies in Engineering (ICACITE), 2022*, pp. 713–718. https://doi.org/10.1109/ICACITE53722.2022.9823883

20. Sarier, N. D. (2021). Comments on biometric-based non-transferable credentials and their application in blockchain-based identity management. *Computers and Security*, 105, 102243. https://doi.org/10.1016/j.cose.2021.102243

21. Veeraiah, V., Rajaboina, N. B., Rao, G. N., Ahamad, S., Gupta, A., & Suri, C. S. (2022). Securing Online Web Application for IoT Management. In *2022 2nd International Conference on Advance Computing and Innovative Technologies in Engineering (ICACITE), 2022*, pp. 1499–1504. https://doi.org/10.1109/ICACITE53722.2022.9823733

22. Gupta, A., Singh, R., Nassa, V. K., Bansal, R., Sharma, P., & Koti, K. (2021). Investigating Application and Challenges of Big Data Analytics with Clustering. In *2021 International Conference on Advancements in Electrical, Electronics, Communication, Computing and Automation (ICAECA), 2021*, pp. 1–6. https://doi.org/10.1109/ICAECA52838.2021.9675483

23. Namasudra, S., Deka, G. C., Johri, P., Hosseinpour, M., & Gandomi, A. H. (2021). The revolution of blockchain: State-of-the-art and research challenges. *Archives of Computational Methods in Engineering*, 28(3), 1497–1515. https://doi.org/10.1007/s11831-020-09426-0

24. Veeraiah, V., Kumar, K. R., Lalitha Kumari, P., Ahamad, S., Bansal, R., & Gupta, A. (2022). Application of Biometric System to Enhance the Security in Virtual World. In *2022 2nd International Conference on Advance Computing and Innovative Technologies in Engineering (ICACITE), 2022*, pp. 719–723. https://doi.org/10.1109/ICACITE53722.2022.9823850

25. Shukla, A., Ahamad, S., Rao, G. N., Al-Asadi, A. J., Gupta, A., & Kumbhkar, M. (2021). Artificial Intelligence Assisted IoT Data Intrusion Detection. In *2021 4th International Conference on Computing and Communications Technologies (ICCCT), 2021*, pp. 330–335. https://doi.org/10.1109/ICCCT53315.2021.9711795

26. Liang, W., & Ji, N. (2022). Privacy challenges of IoT-based blockchain: A systematic review. *Cluster Computing*, 25, 2203–2221. https://doi.org/10.1007/s10586-021-03260-0

27. Kaushik, D., & Gupta, A. (2021). Ultra-secure transmissions for 5G-V2X communications. *Materials Today: Proceedings*. https://doi.org/10.1016/j.matpr.2020.12.130

28. Veeraiah, V., Gangavathi, P., Ahamad, S., Talukdar, S. B., Gupta, A., & Talukdar, V. (2022). Enhancement of Meta Verse Capabilities by IoT Integration. In *2022 2nd International Conference on Advance Computing and Innovative Technologies in Engineering (ICACITE), 2022*, pp. 1493–1498. https://doi.org/10.1109/ICACITE53722.2022.9823766

29. Ranjith Kumar, M. V., & Bhalaji, N. (2021). Blockchain based chameleon hashing technique for privacy preservation in E-governance system. *Wireless Personal Communications*, 117(2), 987–1006. https://doi.org/10.1007/s11277-020-07907-w

30. Chibber, A., Anand, R., & Singh, J. Smart Traffic Light Controller Using Edge Detection in Digital Signal Processing. In *Wireless Communication with Artificial Intelligence* (pp. 251–272). CRC Press, USA.

31. Wang, D., Wang, H., & Fu, Y. (2021). Blockchain-based IoT device identification and management in 5G smart grid. *Eurasip Journal on Wireless Communications and Networking*, 2021(1), 125. https://doi.org/10.1186/s13638-021-01966-8

32. Rathee, G., Balasaraswathi, M., Chandran, K. P., Gupta, S. D., & Boopathi, C. S. (2021). A secure IoT sensors communication in industry 4.0 using blockchain technology. *Journal of Ambient Intelligence and Humanized Computing*, 12(1), 533–545. https://doi.org/10.1007/s12652-020-02017-8

33. Aggarwal, B., Gupta, A., Goyal, D., Gupta, P., Bansal, B., & Barak, D. (2021). A review on investigating the role of block-chain in cyber security. *Materials Today: Proceedings*. https://doi.org/10.1016/j.matpr.2021.10.124

34. Gupta, A., Garg, M., Verma, A., Kaushik, D. (2020). Implementing lossless compression during image processing by integrated approach. *Materials Today: Proceedings*. https://doi.org/10.1016/j.matpr.2020.10.052

35. Garba, A., Dwivedi, A. D., Kamal, M., Srivastava, G., Tariq, M., Hasan, M. A., & Chen, Z. (2021). A digital rights management system based on a scalable blockchain. *Peer-to-Peer Networking and Applications*, 14(5), 2665–2680. https://doi.org/10.1007/s12083-020-01023-z

36. Verma, A., Gupta, A., Kaushik, D., & Garg, M. (2021). Performance enhancement of IOT based accident detection system by integration of edge detection. *Materials Today: Proceedings*. https://doi.org/10.1016/j.matpr.2021.01.468

37. Pramanik, S., & Raja, S. (2020). A Secured Image Steganography using Genetic Algorithm. *Advances in Mathematics: Scientific Journal*, 9(7), 4533–4541.

38. Pramanik, S., Singh, R. P., & Ghosh, R. (2020). Application of bi-orthogonal wavelet transform and genetic algorithm in image steganography. *Multimedia Tools and Applications*. https://doi.org/10.1007/s11042-020-08676-2020

39. Gupta, A., Srivastava, A., Anand, R., & Tomažič, T. (2020). Business Application Analytics and the Internet of Things: The Connecting Link. In *New Age Analytics* (pp. 249–273). Apple Academic Press, Canada.

40. Anand, R., Sindhwani, N., & Saini, A. (2021). Emerging Technologies for COVID-19. In *Enabling Healthcare 4.0 for Pandemics: A Roadmap Using AI, Machine Learning, IoT and Cognitive Technologies* (pp. 163–188). Wiley, New Jersey. https://doi.org/10.1002/9781119812555

41. Pramanik, S. (2022). An Effective Secured Privacy-Protecting Data Aggregation Method in IoT. In Odhiambo, M. O., Mwashita, W. (eds) *Achieving Full Realization and Mitigating the Challenges of the Internet of Things*. IGI Global, USA. https://doi.org/10.4018/978-1-7998-9312-7.ch008

42. Bansal, R., Jenipher, B., Nisha, V., Jain, M. R., Dilip, K., Pramanik, S., Roy, S., & Gupta, A. (2022). Big Data Architecture for Network Security. In *Cyber Security and Network Security*. Wiley, USA.

43. Anand, R., Shrivastava, G., Gupta, S., Peng, S. L., & Sindhwani, N. (2018). Audio Watermarking with Reduced Number of Random Samples. In *Handbook of Research on Network Forensics and Analysis Techniques* (pp. 372–394). IGI Global, USA.

44. Pradhan, D., Sahu, P. K., Goje, N. S., Myo, H., Ghonge, M. M., Tun, M., Rajeswari, R., & Pramanik, S. (2022). Security, Privacy, Risk, and Safety Toward 5G Green Network (5G-GN). In *Cyber Security and Network Security*. Wiley, USA.

45. Kaushik, D., Garg, M., Annu, A. G., & Pramanik, S. (2021). Application of Machine Learning and Deep Learning in Cyber security: An Innovative Approach. In Ghonge, M., Pramanik, S., Mangrulkar, R., Le, D. N. (eds) *Cybersecurity and Digital Forensics: Challenges and Future Trends*. Wiley, USA.

46. Chawla, P., Juneja, A., Juneja, S., & Anand, R. (2020). Artificial intelligent systems in smart medical healthcare: Current trends. *International Journal of Advanced Science and Technology*, 29(10), 1476–1484.

47. Sindhwani, N., Anand, R., Meivel, S., Shukla, R., Yadav, M. P., & Yadav, V. (2021). Performance analysis of deep neural networks using computer vision. *EAI Endorsed Transactions on Industrial Networks and Intelligent Systems*, 8(29), e3–e3.

48. Anand, R., Sindhwani, N., & Juneja, S. (2022). Cognitive Internet of Things, Its Applications, and Its Challenges: A Survey. In *Harnessing the Internet of Things (IoT) for a Hyper-Connected Smart World* (pp. 91–113). Apple Academic Press, Canada.

49. Bakshi, G., Shukla, R., Yadav, V., Dahiya, A., Anand, R., Sindhwani, N., & Singh, H. (2021). An optimized approach for feature extraction in multi-relational statistical learning. *Journal of Scientific and Industrial Research (JSIR)*, 80(06), 537–542.

50. Shukla, R., Dubey, G., Malik, P., Sindhwani, N., Anand, R., Dahiya, A., & Yadav, V. (2021). Detecting crop health using machine learning techniques in smart agriculture system. *Journal of Scientific and Industrial Research (JSIR)*, 80(8), 699–706.

51. Gupta, R., Shrivastava, G., Anand, R., & Tomažič, T. (2018). IoT-based privacy control system through android. In *Handbook of E-business Security* (pp. 341–363). Auerbach Publications.

52. Singh, S. K., Thakur, R. K., Kumar, S., & Anand, R. (2022, March). Deep Learning and Machine Learning based Facial Emotion Detection using CNN. In *2022 9th International Conference on Computing for Sustainable Global Development (INDIACom)*, pp. 530–535. IEEE.

# Index

## A

accelerometer, 115–116
Access Class Barring (ACB), 22
acoustic sensors, 7–8
Active Optical Sensors (AOSs), 6
Ad Hoc On-Demand Vector (AODV) routing
        protocol, 44–45
Agri-CLOUD, 3
agriculture, 1–4
    applications of IoT/cloud computing in, 9–16
    cattle tracking, 11
    choosing crops based on soil texture/weather, 15
    crop protection, 15–16
    crop recommendation, 10
    precision farming (PF), 11–12
    ripening of fruits, 13
    smart greenhouses, 13–15
    soil monitoring, 12–13
    cloud architecture in, 9
    future scope, 16–17
        creature detection, 16
        farm-management drones, 16
        optimized scarecrow usage, 16
        proper weather forecasting, 16
        soil heat maintenance, 17
        solar panels, 16
        weed-killing, 16–17
    needs of cloud computing in, 8–9
    role of sensors in, 4
    types of sensors in, 5–8
AI-based sensor system, 184–185
AI-ML models for detecting privacy/security issues, 73
airflow sensor, 7
Alexa, 55
algebraic manipulation detection (AMD), 68
Ali, Junade, 74
Amazon Web Services (AWS), 192
amazon webservice (AWS), 70
Analytic Hierarchy Process (AHP), 234
anonymity, 73–74
Ant Colony Optimization (ACO), 234
application layer, 57
Application Programming Interfaces (APIs), 120
Application Units (AUs), 35
applications, agriculture
    cattle tracking, 11
    choosing crops based on soil texture/weather, 15
    crop protection, 15–16
    crop recommendation, 10
    precision farming (PF), 11–12
    ripening of fruits, 13
    smart greenhouses, 13–15
    soil monitoring, 12–13
applied proxy, 26
Arduino Project, 191
area under the curve (AUC), 212

artificial bee colony (ABC), 235
artificial intelligence (AI), 54
    in agriculture, 186
    benefits of, 186
    challenges of, 187
    demerits, 187
artificial neural network (ANN), 73, 234
Ashton, Kevin, 206
Asynchronous altering Direction Method of Multipliers
        (ADMM), 104
attribute base credential (ABC), 64
audio sensor (microphone), 118
*Authlogics Password Breach Database*, 74
automated home, 166
automated transport, 166
automation, term, 83
automotive manufacturing, 101

## B

Baldi, Pierre, 213
Barik, Rabindra Kumar, 236
barometer, 115–116
Base Station (BS), 101
Base Station Controller (BSC), 195
Base Station Subsystem (BSS), 195
Base Transceiver Station (BTS), 195
Berenstein, R., 136
big data, 228
    advantages of, 229
    characteristics of, 228
    and cloud computing, 229–233
    cloud for, 231
    clustering of, 229
    issues in using cloud services, 233
    literature review, 233–237
    problem statement, 238
    proposed work, 238–240
    research methodology, 238
    results and discussion, 240–244
    smart applications of, 232–233
biometric, 72–73
blockchain, 70–71
Brain–Computer Interface (BCI), 212
bring-your-own-device (BYOD), 61

## C

Candiago, S., 136
Care-of-Address (CoA), 24
cation exchange capacity (CEC), 6
Central Processing Unit (CPU), 33–34
Chattopadhyay, 197
Chen, Z., 119
chronic obstructive pulmonary disease (COPD), 118
*cloud access security broker* (CASB), 62
cloud architecture

community cloud, 9
hybrid cloud, 9
private cloud, 9
public cloud, 9
cloud computing, 1–4, 229–230
    in agriculture
        Infrastructure as a Service (IaaS), 8
        Platform as a Service (PaaS), 8–9
        Software as a Service (SaaS), 8
    deployment models, 58
    integrating IoT with, 81–98
    privacy/security issues in infrastructure of, 59
    service delivery models, 58–59
    types of, 230
Cloud of Things (CoT), 94–97
cloud-integrated Internet of Things
    application of, 91–93
        smart city, 92
        smart environment monitoring, 92
        smart healthcare, 92–93
        smart home, 92
        smart mobility, 93
        smart surveillance, 93
        towards serverless computing, 93
    architecture of, 85
    benefits of, 85–87
        data transmission, 87
        modern capacities, 87
        processing, 87
        storage, 86
    cloud computing
        IaaS, 83–84
        PaaS, 83
        SaaS, 84
    existing solutions for, 87–91
        AWS IoT, 88–91
        CloudPlugs IoT, 91
        Nimbits, 88
        OpenIoT, 87–88
        ThingSpeak, 91
    Internet of Things (IoT), 82–83
    issues/research challenges, 94–96
        energy efficiency, 96
        interoperability, 95
        IPv6 deployment, 94–95
        resource allocation/management, 95
        scaling, 96
        security/privacy, 94
        service discovery, 95–96
        unnecessary data communication, 94
    middleware for IoT, 84–85
cluster, 35–36
cluster computing, 230–231
cluster head (CH), 35–36
communications service provider (CSP), 229
communities, 122
community cloud, 9, 58
conditional privacy-preserving authentication (CPPA), 65
confidentiality, integrity, and availability (CIA
        principles), 64
connected electric vehicles, see smart cities
content distribution networks (CDNs), 37
Content Management System (CMS), 120

convolutional neural network (CNN), 169, 211
Convolutional Neural Network (CNN), 103
Cooperative Intelligent Transportation System
        (C-ITS), 35
Cornet [PROVIDE FIRST NAME], 119
Correspondent Node (CN), 22, 25
cryptography, 67
cyber-physical systems (CPS), 166

**D**

data confidentiality, 60–62
data loss prevention (DLP), 61
data storage as a service (DaaS Cloud), 148
database management systems (DBMS), 83
Dataset Accuracy Weighted Random Forest
        (DAWRF), 104
Davis Vantage Pro2, 4
Decision Support System, 3
decision tree (DT), 10
Dedicated Short-Range Communication (DSRC), 38
deep belief network (DBN), 214
deep Boltzmann machine (DBM), 214
deep conventional extreme ML (DCELM), 214
deep convolutional neural networks (DCNNs), 215
deep neural network (DNN), 213
delay, 28
denial of service (DDoS), 55
denial of service (DoS), 55, 149
Density-based Spatial Clustering of Applications with
        Noise (DBSCAN), 234–235
Department of Homeland Security (DHS), 61
diabetic retinopathy (DR), 211, 215
dielectric soil moisture sensors, 7
distributed denial of service (DDoS), 149
Distribution Management System (DMS), 63
Do, T.-T., 118
Dynamic Source Routing (DSR), 40–44
    route cache, 41–44
    route discovery, 40–41
    route maintenance, 41

**E**

E-prescription, 120
Edan, Y., 136
edge devices, 39
electrical conductivity (EC), 5–6
electrocardiogram (ECG), 116
electrochemical sensors, 116
electromagnetic sensors, 5–6
electronic health records (EHRs)
    interoperability, 206
    patient care, 206
    research purposes, 205–206
    seamless switching, 206
Electronic Health Records (EHRs), 120
electronic medical records (EMRs), 64
Elliptic-curve cryptography (ECC), 65
elliptic-curve discrete logarithm problem (ECDLP), 65
Extensible Markup Language (XML), 58
extreme learning models (ELMs), 214

## F

Feature Driven Development (FDD), 105
Federal Communication Commission (FCC), 38
field area networks (FANs), 172
fitness wearables, 140
FIWARE, 1–14
Fog computing, 37–38
fog implementation, 155–158
force sensors, 115
Foreign Agent (FA), 23–24
frameworks, security/privacy
    biometric, 72–73
    blockchain, 70–71
    cryptography, 67
    homomorphic encryption, 67–68
    Multi-Factor Authentication (MFA), 69–70
    secret sharing, 68
    secure multi-party computation (SMPC), 68–69
    side-channel attacks, 71–72
    zero-knowledge proofs, 68
Function as a Service (FaaS), 93
Fuzzy Fault Tree Model with Bayesian Network (FFTA-BN), 101

## G

Gadkari Nitn, 191
gastrointestinal (GI) disease, 215
generative adversarial networks (GANs), 214
Geographic Information System (GIS), 135
Ghanem, S., 119
Giardini, [PROVIDE FIRST NAME], 118
Global Positioning System (GPS), 34, 61, 116, 191
Global System for Mobile Communications (GSM), 192
Goel, M., 119
Google App Engine, 3
Google Cloud Print, 60
governments, 122
Gowrishankar, V., 135
Green Normalized Difference Vegetation Index (GNDVI), 136
Group-based Fault Detection (GbFD) Algorithm, 104
gyroscope, 115–116

## H

Habibzadeh, H., 57
Hadoop Dispersed File Scheme, 99
handover management, 25–26
hard handover, 25
Hardware Trojans (HTs), 66
healthcare, building integrated systems for, 203–211
    literature review, 211–213
    problem statement, 216
    proposed methodology, 216
    results and discussion, 216–220
    role of deep learning in, 213–216
        diabetic retinopathy (DR), 215
        GI disease, 215–216
        and medical imaging, 214–215
        tumor detection, 216
heart rate (HR), 116

heart rate variability (HRV), 116
Heidari, Safanaz, 234
hidden Markov models (HMMs), 212
hierarchal state routing (HSR), 45–49
Holden, [PROVIDE FIRST NAME], 119
Home Agent (HA), 23–24
home automation systems (HANs), 172
Home Registry (HLR), 195
homomorphic encryption, 67–68
hospitals, 121
Huang, Wenzhun, 234
Hussein, Y., 119
hybrid cloud, 9
hypervisor, 58

## I

identity management system (IDM), 59
identity provider, 59
Ilango, S. Sudhakar, 235
image sensor (camera), 116–117
Industrial Control System (ICS), 103
Industrial Internet of Things (IIoT), 99–110
    analyses of work, 106–107
    applications of
        automotive manufacturing, 101
        recycling/sorting system, 102
        remote power grid, 101–102
        supply chain management, 100–101
    future scope, 107–109
    literature survey, 102–106
    previous work, 106
    proposed system, 106
information and communication technology (ICT), 4, 134, 165
information and operational technologies (IT/OT), 63
Infrastructure as a Service (IaaS), 8, 58, 62, 148
Innovative Internet of 5G Medical Robotic Things (IIo-5GMRTs), 142
integrated circuit (IC), 66
Intelligent Transport System (ITS), 4
intelligent transportation system (ITS), 65
International Mobile Station Equipment Identity (IMEI), 60
Internet of Health Things (IoHT), 139
Internet of Things (IoT), 1–4, 21, 34, 54
    and artificial intelligence (AI), 186–188
    benefits of, 131–133, 207–208
    big data clustering, 227–247
    building integrated systems for healthcare, 203–225
    characteristics of, 206–207
    in connected electric vehicles, 161–179
    integrating with cloud computing, 81–98
    and machine learning (ML), 181–186
    mobile-based health (mHealth), 111–129
    multi-variant processing model in IIoT, 99–100
    Parivem–Parivahan emulator, 191–201
    security/privacy issues, 53–80
    and smart agriculture, 133–139
    smart cities, 21–32
    and smart healthcare, 139–143
    splitter with cryptographic model, 147–160
    techniques using IoT in agriculture, 1–19
    vehicle communication, 33–51

Internet of Things Breaker Connector (IoTBC), 106
Internet of Vehicles (IoV), 34, 162
Internet Protocol (IP), 82
Internet Service Providers (ISPs), 57
Ismail, Ahmed, 235

**J**

Juneja, A., 169

**K**

Kim, S., 118
Kulkarni, N. H., 10

**L**

Larson, E. C., 119
Light Dependent Resistor, 3
Light Emitting Diodes, 3
load, 28
local area networks (LANs), 172
location sensors, 5
logistic regression (LR), 10
Londhe, S. B., 135
long short-term memories (LSTMs), 214
Long Short-term Memory Network (LSTMN), 105

**M**

Ma, [PROVIDE FIRST NAME], 119
machine learning (ML), 54, 181–182, 212
    approaches to, 182
        literature review, 182–184
    and IoT-based framework for smart farming, 184–185
*Machine Learning Approach, The* (Pierrebaldi/
        Sorenbrunak), 212
machine learning as a service (MLaaS), 73
Machine-to-Machine (M2M), 105
magnetic resonance imaging (MRI), 215
magnetometer, 115–116
Manoj, R. Joseph, 234
mart surveillance, 93
Mazumdar, Somnath, 236
mechanical sensors, 6–7
Media Access Control (MAC), 60
medical companies, 122
medical imaging, 214–215
Medical Internet of Things (M-IoT), 126
MedRealTime, 120
Message Queue Telemetry Transport (MQTT), 23
Metropolitan Area Network (MAN), 39
Micro Electro-Mechanical System (MEMS), 103
Microsoft Azure, 196
MIP version 6 (MIPv6), 22
mobile cloud computing (MCC), 233
mobile computing
    advantages of, 209–210
    elements of, 209
    impact of, 210–211
    limitations of, 210
mobile device management (MDM), 61
mobile devices, 60–62
Mobile Equipment Identifier (MEID), 60

Mobile Internet Protocol (MIP), 22
Mobile Node (MN), 22, 25
Mobile Switching Center (MSC), 195
mobile-based health (mHealth), 111–113
    benefits of, 121–124
    challenges and future trends in, 125–126
    customers in, 121–124
    future role of IoT/cloud in, 126–127
    IoT role in, 121–124
    role of cloud in, 124–125
    role of sensors in, 113–116
        different external sensors applications, 114
        sensors of smart mobile, 116–119
    software feature, 119–121
multi-factor authentication (MFA), 62
Multi-Factor Authentication (MFA), 69–70
multilayer perceptrons (MLPs), 214
Myung, D., 118

**N**

National Institute of Standards and Technology (NIST),
        59, 61, 72
neighborhood area networks (NANs), 172
network architecture, 171–172
network layer, 57
Network Switching Subsystem (NSS), 195
network topology, 170–171
neural network (NN), 73
Nitrogen-Phosphorus-Potassium (NPK), 3
non-malleable secret sharing scheme (NMSS), 68
nonsynonymous single nucleotide polymorphisms
        (nsSNPs), 213
Normalized Difference Vegetation Index (NDVI), 136

**O**

Olanrewaju, R. F., 135
On Board Units (OBUs), 35
one-time password (OTP), 69
Open Automobile Alliance (OAA), 36
Operations and Support Subsystem (OSS), 195
optical sensors, 6, 116

**P**

Parivem-Parivahan emulator, 191–192
    materials and methods, 192–196
    proposed system architecture, 198–201
    State of Art Methods, 196–198
particle swarm optimization (PSO), 235
Passive Optical Sensors (POSs), 6
Patient Data Management System (PDMS), 141
patients, 121
Peng, Kai, 234
perception layer, 57
personally distinguishable information (PDI), 59
photoplethysmograph (PPG), 118
platform as a service (PaaS Cloud), 148
Platform as a Service (PaaS), 8–9, 58, 62, 196
policymakers, 122
pollutants, 196–198
Pollution Under Certificate (PUC), 196
portable eye examination kit (PEEK), 118

*power distribution networks* (PDNs), 63
power grids, 63–64
Praveen, P., 235
Primicerio, J., 136
Principal Component Analysis (PBirch), 234
Principal Component Analysis (PMBKM), 234
*privacy-preserving authentication* (PPA), 62
privacy/security issues, 53–56
    AI-ML models for detecting, 73
    anonymity, 73–74
    countering threats, 74–76
    power grids, 63–64
    security/privacy frameworks, 67–73
    in smart city infrastructure, 59–60
    smart city overview, 56–59
    smart devices, 60–62
    smart environment, 66
    *smart healthcare systems* (SHSs), 64–65
    in smart IoT devices, 66–67
    in smart transportation, 65–66
    sustainable approaches, 76
private cloud, 9
Programmable Logic Controller (PLC), 100
public cloud, 9

**Q**

quality of service (QoS), 54

**R**

radial basis function networks (RBFNs), 214
Radio Frequency Identification (RFID), 139–140, 191, 206
random forest (RF), 10
Rao, T. Ramalingeswara, 235
recurrent neural networks (RNNs), 214
regional traffic management center (RTMC), 65
Regional Transport Office (RTO), 192
registration certificates (RCs), 192
respiration rate (RR), 118
Retinopathy Online Challenge (ROC), 215
reverse Engineering (RE), 66
Rieke, M., 136
Roadside Units (RSUs), 35, 65
ROUTE REQUEST (RREQ), 40
routing protocol
    Ad Hoc On-Demand Vector (AODV) routing
        protocol, 44–45
    Dynamic Source Routing (DSR), 40–44
routing protocols
    Ad hoc, 39
    hierarchal protocols, 45
    hierarchal state routing (HSR), 45–49
    hybrid, 40
    proactive, 39
    reactive, 40
Russo, A., 118

**S**

scientists, 121
secret sharing, 68
secure digital (SD), 61
secure multi-party computation (SMPC), 68–69

secure web gateway (SWG), 61–62
*Security-Aware Efficient Data Sharing and Transferring*
    (SA-EAST), 65
self-organizing maps (SOMs), 214
sensors
    role of, 4
    types of, 5–8
        acoustic sensors, 7–8
        airflow sensor, 7
        dielectric soil moisture sensors, 7
        electromagnetic sensors, 5–6
        location sensors, 5
        mechanical sensors, 6–7
        optical sensors, 6
serverless computing, 93
service provider, 59
service-oriented architecture (SOA), 57
Short Text Message Service (SMS), 195
side-channel attacks (SCAs), 71–72
*sidechannel attacks* (SCAs), 66
Skourletopoulos, Georgios, 233
smart agriculture
    background literature, 134–137
    benefits of, 133–134
    combined analysis, 138–139
    methodologies, 137–138
smart cities, 162–164
    communicating devices, 21–32
    components, 164–166
        agriculture, 165
        automated home, 166
        automated transport, 166
        available services, 165
        health, 165–166
        new generation industries, 166
        power systems, 165
        smart structure, 166
    countering threats, 74–76
    defining, 162–164
    different areas of, 163
    emergence of, 53–56
    power grid systems within, 63–64
    privacy/security issues, 53–80
    research gap and motivation, 164
    *smart healthcare systems* (SHSs), 64–65
    sustainable approaches, 76
    SWOT method for data analysis, 172
        coercions, 173–174
        flaws, 173
        prospects, 173
        strengths, 173
    usage of IoT in, 167–172
        architecture, 167
        challenges, 167–170
        privacy/security, 172
        technologies used for networking, 170–172
        technologies used for sensing data, 170
Smart City Architecture, 21–23
    related work, 23
    simulation, 23–27
    analysis of, 28–30
    components, 24–25
        CN, 25
        FA, 25

HA, 24
MN, 25
tunneling, 25
handover management techniques, 25–26
proposed scenario, 26–27
smart environment, 66
term, 54
smart environment monitoring, 93
smart greenhouses
advantages of, 15
applications of, 13–15
smart healthcare, 93
applications, 141
challenges, 142
current status, 141–142
future of, 142–143
how IoT helps in healthcare, 140–141
role of IoT in healthcare, 139–140
smart healthcare systems (SHSs), 64–65
smart IoT devices, 66–67
smart mobility, 93
smart system
general architecture, 55–57
general cloud-based architecture of, 57–59
smart transportation, 65–66
smartphones, 60–62
social IoT (SIoT), 57
soft handover, 25–26
Software as a Service (SaaS), 8, 58, 62, 148
Software, Platform, and Infrastructure (SPI), 58
Soil Adjusted Vegetation Index (SAVI), 136
Soil Test Crop Response (STCR), 3
SONAR, 16
splitter with cryptographic model
and cloud computing, 147–149
fog implementation, 155–158
literature review of, 150
need for proposed system, 149
objective of proposed work, 150–151
proposed model, 151–152
results, 152–155
file splitter, 153
GUI for client interface, 153
GUI for service interface, 154
secure files, 154–155
various attacks, 149
Stand-Alone Photo Voltaic System (SAPVS), 105
State of Arts Method, 196–198
Subscriber Identity Module (SIM), 195
Sujata, K., 135
*supervisory control and data acquisition* (SCADA), 63
support layer, 57
Support Vector Machine (SVM), 99
Sustainable Precision Agriculture, 3

**T**

temperature sensors, 116
Thorat, A., 135
throughput, 29–30
Tracer-based Sorting (TBS), 102
*Transport layer security* (TLS), 64
*Troy Hunt's Have I Been Pwned?*, 74
trusted authority (TA), 65

tumors, detecting, 216
tunneling, 25
two-factor authentication (2FA), 69

**U**

UN Sustainable Development Goals, 165
Unified Communications as a Service (UCaaS), 58
Unique Device Identifier (UDID), 60
Unmanned Aerial Vehicles (UAVs), 131
user experience (UX), 121
user interface (UI), 121
User-Managed Access (UMA), 23

**V**

Vegetation Indices (VI), 136
vehicle communication, 33–34
Ad Hoc On-Demand Vector (AODV) routing
protocol, 44–45
Ad hoc routing protocols, 39
cluster, 35–36
Dedicated Short-Range Communication (DSRC), 38
Dynamic Source Routing (DSR), 40–44
edge devices, 39
Fog computing, 37–38
hierarchal routing protocols, 45–49
hybrid routing protocols, 40
improvement in V2V communication using Fog
computing, 37–38
proactive routing protocols, 39
reactive routing protocols, 40
traditional VANETs, 34–35
V2I, 36
V2V, 36–37
Vehicle to Infrastructure (V2I), 166
Vehicle to Person (V2P), 166
vehicle-to-infrastructure (V2I), 33, 36
Vehicle-to-Roadside (V2R), 34
vehicle-to-vehicle (V2V), 33, 36–37, 166
vehicles, Parivem-Parivahan emulator, 191–201
Vehicular Adhoc Network (VANET), 33–34, 65
Venkatachalam, K., 135
virtual machine managers (VMMs), 64
virtual machines (VMs), 58, 64
virtual private network (VPN), 58, 61
VMware Cloud Disaster Recovery, 118

**W**

Watanabe, T., 135
Web Service (WS), 58
Wei, Z., 135
Wide Area Network (WAN), 39, 172
Wind Turbine Generator (WTG), 102
wireless body area networks (WBNs), 65
wireless sensor networks (WSNs), 57
World Health Organization (WHO), 205

**Z**

zero-knowledge proofs (ZKPs), 64, 68
Zhang, J., 136
Zhu, Li, 236

For Product Safety Concerns and Information please contact our EU
representative GPSR@taylorandfrancis.com
Taylor & Francis Verlag GmbH, Kaufingerstraße 24, 80331 München, Germany

www.ingramcontent.com/pod-product-compliance
Lightning Source LLC
Chambersburg PA
CBHW061357210326
41598CB00035B/6014

9 781032 333434